Molecular Structures and Dimensions

Vol. 2

Solid State Classes 60–86

Molecular Structures and Dimensions

Vol. 2

Complexes and Organometallic Structures

Edited by Olga Kennard and David G. Watson

Springer-Science+Business Media, B.V.

ISBN 978-94-017-2337-4   ISBN 978-94-017-2335-0 (eBook)
DOI 10.1007/978-94-017-2335-0

Copyright © 1970 by Springer Science+Business Media Dordrecht
Originally published by International Union of Crystallography in 1970.
First published 1970 by International Union of Crystallography.
Reprint of the original edition 1970

Dedicated to
PROFESSOR J. D. BERNAL, FRS
in grateful recognition of his
contributions to crystallography and
to the field of information science.

Contents

Preface

This classified bibliography of organic and organometallic crystal structures covers the period 1935–1969 and provides references to over 4,000 compounds whose structures were analysed by X-ray or neutron-diffraction methods. Volume 1 deals with general organic structures and Volume 2 with complexes, organometals and organometalloids.

The bibliography is the first of a new series of publications "Molecular Structures and Dimensions" prepared at the Crystallographic Data Centre, University Chemical Laboratory, Cambridge. The Centre has been supported since 1965 by the Office for Scientific and Technical Information as part of the British contribution to international data activities.

The main objective of the Centre is to assemble a computer-based File containing both bibliographic information and numerical data abstracted from the literature and relevant to molecular and crystal structures, as obtained by diffraction methods. The File is designed to permit the checking of published numerical values, particularly interatomic distances and bond angles and to present published results in a variety of new ways. Contents of the File will be made available through the series "Molecular Structures and Dimensions" and also by the provision of data on magnetic tapes or discs. At present the numerical data relating to molecular dimensions are being checked and will be published, at a later date, as further volumes in this series and also as magnetic tapes. The File is being extended to cover electron diffraction in the gas phase. It is hoped that by these methods the Centre will make the results of work in the specialised field of diffraction available to a wide range of scientists.

The bibliography and indexes were prepared, checked and printed by computer techniques. The basic bibliographic file was processed with typesetting programs written in Cambridge and also with special versions of the Science Abstracts (INSPEC) programs at the Institution of Electrical Engineers, London. The resulting magnetic tapes provided the input to a Photon 713 filmsetter and the output films were printed by photo-offset. Computer methods have allowed us to choose the latest possible cut-off date and to reduce the time-lag between this and the actual date of publication. The tape containing the present two volumes is available and anyone interested should contact the Centre for further details.

It is intended to issue the bibliography at yearly intervals and the 1969/1970 volume is already in preparation. Annual supplements will include cumulative formula and author indexes. It is hoped to issue a completely new edition in 1975.

The planning of the bibliography was guided by members of the Commission on Crystallographic Data of the International Union of Crystallography and by many individual crystallographers throughout the world. Its production was made possible by the generous sponsorship of the Office for Scientific and Technical Information, Department of Education and Science as part of their support, through the Crystallographic Data Centre, of international data activities.

Although every effort has been made to ensure the correctness of the entries, mistakes and omissions must inevitably have occurred. We should be most grateful if readers would inform us of these and of any suggestions for improvement of future volumes.

Cambridge July 1970   *Olga Kennard and David G. Watson*

Acknowledgements

The production of this bibliography was a collaborative effort by members of the Crystallographic Data Centre: Drs J.C. Coppola (1967–1970), J.K. Fawcett (1965–1968), K.A. Kerr (1966–1968), W.G. Town (1969–1970), W. Tundak (1967–1969), Mrs K. Watson (1969–1970), Mrs S.M. Weeds (1965–1970), Miss H. Young (1966–1970), visiting members: Drs A.C. Larson, D.L. Wampler, and part-time helpers: A. Dias, E. Doyle, C. Ingram, A. O'Brien, M. Pickett, S. Walker, M. Warne and D. Yoffe. The work of the Centre was guided by members of the OSTI Scientific Advisory Committee: Professor J.D. Bernal, Professor D.C. Phillips, Dame Kathleen Lonsdale, Professor J.W. Linnett, Mr H.W. Rooksby, Dr E. Stern, Dr M.R. Truter , Dr. M.F. Lynch, Professor A.J.C. Wilson.

Dr J.C. Coppola has checked all multiple entries in the bibliography and carried much of the burden of editing and proof-reading the final tapes and print-outs.

Dr J.K. Fawcett has contributed particularly to devising a chemical classification system and classifying the 1960–1967 material.

Dr W.G. Town was responsible for the ordered file system, wrote many of the programs for basic file handling and the preparation of the author and formula indexes.

Mrs S.M. Weeds was responsible for the literature searches and for preparing the abstracts from the original papers.

Miss M. Warne has checked most of the chemical formulae and participated in the work on classification.

Dr D.L. Wampler advised on nomenclature and formulation of metal complexes and devised the basic metal-index program.

We are grateful to the Medical Research Council for allowing a member of their External Scientific Staff (O. Kennard) to participate in this work.

We thank Professor Lord Todd for providing accommodation in

the Department of Chemistry, Cambridge and are grateful to both Lord Todd and other members of staff for their advice on chemical matters.

Our task was greatly facilitated by the excellent organisation of the Centre National de la Recherche Scientifique, which provided copies of most of the references listed in the bibliography. Dr R.S. Cahn and the Nomenclature Division of the Chemical Abstracts Service have helped with questions of nomenclature.

The present form of the bibliography owes much to the OSTI Scientific Advisory Committee and to the various scientists in some 26 countries, who examined and advised us on the pilot volume issued in 1968.

We have used the IBM 360/44 computer at the Institute of Theoretical Astronomy and we were greatly helped by both the programming staff and operators. We are grateful to INSPEC (Information Service in Physics, Electrotechnology and Computers & Control) and especially to Mr P. Simmons for the use of their computer typesetting programs, which they specially modified for our purposes. We have received valuable advice on typesetting principles from Mr H.J. Davis (OSTI Documentation Processing Centre), Mr C.J. Duncan (Computer Typesetting Research Project, University of Newcastle-upon-Tyne) and Mr A.H. Phillips (Technical Development Division, HMSO).

The bibliography was prepared in parallel with the organic section of "Crystal Data" (National Bureau of Standards, Washington D.C., USA) and both publications were strengthened through this collaboration. We are indebted to Professor J.D.H. Donnay and Dr Helen Ondik for high editorial standards which they have set in the field of crystallographic data.

Introduction

Criteria for Inclusion in the Bibliography

A crystallographic study of a single crystal of a carbon-containing substance is included if it meets one of the following requirements:—

(i) Three positional coordinates for each non-hydrogen atom in the molecule have been determined, even if not specifically recorded in the publication.

(ii) The determination of two coordinates for each non-hydrogen atom has proved sufficient to resolve an ambiguity in chemical structure.

(iii) The results were published before 1960 and were included in the critical compilation "Tables of Interatomic Distances and Conformations in Molecules and Ions" (Ed. L.E. Sutton, Chem. Soc. Special Publication Nos. 11 and 18, 1958 and 1965).

It should be noted that the bibliography refers principally to organic compounds. Purely inorganic substances such as inorganic carbides, carbonyls, carbonates, cyanides, thiocyanates are excluded even though they contain carbon. Polymers and high molecular weight compounds like proteins are excluded.

Standard Entries

The standard entry for a particular compound usually refers to an X-ray study carried out at room temperature. A qualifying phrase in parentheses appears after the name if this is not the case.

If the **absolute configuration** has been determined then this will be indicated by a parenthetical phrase. In an effort to check the completeness of our references to absolute configuration we have made use of the lists prepared by Allen and Rogers, Chem. Comm., 838, 1966; Allen, Neidle and Rogers, ibid, 308, 1968; idem, ibid, 452, 1969.

With a few exceptions the reference given is the latest publication by a particular author or group of authors which supersedes any preliminary publications. Supplementary entries, however, are often given and these fall into various categories, e.g.

(a) Independent study by different authors.
(b) Study of a different crystal form.
(c) Study at a different temperature.
(d) Study with a different radiation.

The category of the supplementary entry is indicated by a qualifying phrase after the compound name.

Source of Literature References

The main source for the post-1959 literature was the abstracting journal, *Bulletin Signalétique*, published by the Centre National de la Recherche Scientifique, Paris, France. All entries are based on original references and for pre-1960 entries a cross-reference is also given to *Structure Reports* and to the *Tables of Interatomic Distances*.

The only conference proceedings that were systematically scanned are those of the International Union of Crystallography. Abstracts of these conferences are published in *Acta Cryst.*, Vol. 13 (1960) and as special supplements to Vol. 16 (1963) and Vol. 21 (1966). They can be recognised in the bibliography by the letter A used as a prefix to the page number. American theses are covered by *Dissertation Abstracts*, as reported in *Bulletin Signalétique*.

The system of abbreviations of journal names is largely that adopted by *Bulletin Signalétique*.

Abstracts appear in *Bulletin Signalétique* some 6–10 months following the publication in the primary journal. Vols. 1 and 2, therefore, are fully comprehensive only to Dec. 31, 1968, though the the 1969 literature covers about 530 publications and is complete for two journals, viz. *Acta Cryst.* and *Chemical Communications*.

We have not been able to include at this stage recent conference proceedings of the American Crystallographic Association or the papers presented at the 8th International Congress of Crystallography, August 1969.

In future supplements we expect to improve on the cut-off date by increasing the number of journals scanned directly.

Statistical Information

Volumes 1 and 2 contain 4473 references to 4098 distinct compounds. The numbers of entries carrying 1, 2, 3, 4, 5 cross-references are 1029, 126, 89, 16, 4 respectively. The number of papers published in a given year and which have not yet been superseded are given in Table 1. References are given to 157 journals or books. 75% of the

Table 1   *Growth of structure analysis by diffraction methods*
          Number of papers per year included in the bibliography

Year	Number	Year	Number
1935	3	1953	48
1936	4	1954	38
1937	5	1955	34
1938	8	1956	48
1939	3	1957	41
1940	4	1958	69
1941	5	1959	75
1942	5	1960	118
1943	4	1961	141
1945	8	1962	157
1946	5	1963	252
1947	8	1964	259
1948	15	1965	416
1949	15	1966	569
1950	18	1967	700
1951	27	1968	817
1952	24	1969	530

Note. This table does not represent the complete growth curve since (a) certain publications which were superseded by later papers are excluded, (b) coverage for 1969 is incomplete.

Table 2   *Statistics for nine major journals*

Journal	No. of papers	% of total
Acta Cryst.	1606	35·9
Chemical Communications	496	11·1
J. Chem. Soc.	408	9·1
J. Amer. Chem. Soc.	271	6·1
Inorg. Chem.	205	4·6
Acta Chem. Scand.	189	4·2
Zh. Strukt. Khim.	103	2·3
Tetrahedron Letters	65	1·5
Kristallografija	42	0·9

publications have appeared in 9 journals and Table 2 lists the statistics for these. The distribution of compounds within major class groupings is given in Table 3.

Table 3 *Statistics for major class groupings*

Group	No. of papers	% of total
Aliphatic, Aromatic, Carbocyclic, Heterocyclic	1930	43·1
Metal Coordination Complexes	1025	22·9
Biological and Natural Products	682	15·2
Metal π-Complexes	313	7·0
Molecular Complexes	156	3·5
Others	367	. 8·3

Practical Guide to the Bibliography

In this section we provide details which should enable users to search the bibliography as efficiently as possible. There are two alternative ways of finding the bibliographic entry for a particular compound (1) through the molecular formula index, (2) through the classification system. There is an author but not a name index.

The molecular formula is expressed in terms of residues, each residue being an independent set of bonded atoms.

e.g. sodium formate has two residues: CHO_2^- and Na^+

The formula of each residue follows the common convention $C_xH_yA_aB_b$ . . . i.e. C and H come first and the other elements follow in alphabetic sequence.

Residues are listed in order of precedence as follows:—

Organic > inorganic ions > water (or organic solvent)

When more than one non-solvent organic residue is present the ordering is arbitrary though it is often suggested by the chemical name.

e.g. carbon tetrabromide: p-xylene complex:—CBr_4, C_8H_{10}

but 1,4-dithiane : iodoform complex:—CHI_3, $C_4H_8S_2$

As shown later the structure of the formula index will provide a safeguard against such anomalies as the iodoform complex.

The bibliographic entries are arranged in accordance with a classification scheme based on chemical structure and the list of classes is given on pp. xxi–xxiii. This list is intended to cover "standard" chemical classes but it is also influenced by particular crystallographic interests. Thus in view of the numerous publications relating to the structures of carboxylic acid derivatives it was decided to assign acid salts to class 2 and all other derivatives to class 1 rather than combining them in a single class. A revision of these classes will be undertaken in the next edition.

Such a classification system requires that we make adequate provision for cross-referencing. Thus a compound with only one organic residue can often be described in terms of more than one chemical class or a compound might contain two or more organic residues each of which can be classified. Within each class entries are arranged in order of increasing molecular formulae. Entries are of two types:—

(a) **main entries** which are identified by numbers of the type **m.n,** meaning the nth main entry in class m.

(b) **cross-reference entries** which are identified by numbers of the type **m.C.** m is the class number and the letter C simply denotes that this is a cross-reference entry.

The following examples illustrate the system:

1. 1,2-Dimethylallyl-dicyclopentadienyl titanium (iii) has one residue:—$C_{15}H_{19}Ti$. When first entered in the bibliography it was assigned the basic class number of 72 (metal π-complexes, open chain) and class 73 (metal π-complexes, cyclopentadiene) as the subsidiary class. In the formula index it appears as $C_{15}H_{19}Ti$ 72.44. It can also be located from the metal index where it appears under Ti.

   In class 72 entry number **72.44,** i.e. the 44th main entry appears on p. 95. The cross-reference entry in class 73 appears on p. 108 after entry **73.65** and is indicated as **73.C.**

2. Nickel (ii) bis(β-picoline) nitrite trimer benzene solvate has two residues:—$C_{36}H_{42}N_{12}Ni_3O_{12}$ and C_6H_6.
   Although the second residue is organic it will not be classified since it represents solvent.

   Thus we assign a basic class number of 83 (metal complexes, nitrogen ligand) and on p. 215 we find the main entry with number **83.215.** In the formula index it will appear as $C_{36}H_{42}N_{12}Ni_3O_{12}$, C_6H_6 83.215 and in the metal index under Ni.

3. Chloranil : bis(8-hydroxyquinolinato) palladium (ii) has two residues:—C_6Cl_4O and $C_{18}H_{12}N_2O_2Pd$.
   Residue 1 is given a basic class number of 60 (molecular

complexes) and is cross-referenced to class 18 (benzoquinones).
Residue 2 is cross-referenced to class 60 (molecular complexes), class 83 (metal complexes, nitrogen ligand) and class 84 (metal complexes, oxygen ligand).

The main entry appears on p. 11 with number **60.70** and the cross-reference entries on p. 25 with number **60.C** (last entry on page), on p. 206 with number **83.C** (after entry **83.159**) and on p. 227 with number **84.C** (2nd entry on page).

The cross-reference for residue 1 appears in Vol. 1 on p. 156 with number **18.C** (entry before **18.2**).

In the formula index this compound will appear twice as C_6Cl_4O, $C_{18}H_{12}N_2O_2Pd$ 60.70 and $C_{18}H_{12}N_2O_2Pd$, C_6Cl_4O 60.70.

In the metal index it can be located under Pd.

Use of the Indexes

As indicated in the various examples a compound whose molecular formula is known can be located through the formula index bearing in mind the conventions used in these volumes.

1 Molecular formulae are expressed in terms of residues (connected atomic groupings), e.g. CBr_4, C_8H_{10} for the complex between carbon tetrabromide and p-xylene; and CHO_2^-, Na^+ for sodium formate.

2 The arrangement of symbols is that used by Chemical Abstracts with the number of carbon atoms first, followed by the number of hydrogen atoms (if any) and the remaining elements in alphabetic order, typically as $C_xH_yA_aB_b$ . . . .

3 In the index compounds are grouped under the number of carbon atoms. Within each group, compounds are ordered by the number of hydrogen atoms (including H = zero) and then alphabetically by symbols. Thus in group C_1 the formulae CBr_4, CI_4, etc., precede the formulae of all hydrogen-containing one-carbon compounds.

4 A compound with a number of non-trivial residues is indexed under each residue with, in each case, the classified residue printed in bold type. Thus carbon tetrabromide : p-xylene complex with formula CBr_4, C_8H_{10}, can be located by searching either in group C_1 where it will appear as **CBr_4**, C_8H_{10} or in group C_8 where it will appear as **C_8H_{10}**, CBr_4.

5 In the formula index entry numbers are of the form m.n **p**

where m is the class number, n the order of main entries within the class and **p** the volume number. A + sign indicates that for the given formula there is a series of entries (c.f. multiple entries in the Introduction) starting at entry number m.n.

6 The transition metal index which appears only in Volume 2 was prepared from the entries in classes 71–86. Since metal complexes are classified in terms of the ligands this index provides an alternative search for e.g. all iridium complexes.

Searches Using the Chemical Classification System

When searching for a compound whose formula is not known exactly the likely chemical classes can be scanned and the compound identified by name. Classes may also be used for generic searches. In both these cases it is important that the user has some idea of the details of the classification conventions which have been employed Many of the class names are themselves sufficiently indicative, but the following points should be noted.

General:

(a) Inorganic residues and organic solvent residues are not classified.
(b) Heavy-atom groupings introduced to solve the crystallographic phase problem are not classified.
(c) Classes 5 and 19 (aliphatic and benzene miscellaneous) are used only when the compound cannot be assigned to a more specific class.
(d) Compounds which are clearly steroids, terpenes, alkaloids are not cross-referenced in terms of sub-structural features.
However, for class 59 (miscellaneous natural products) cross-referencing is sometimes desirable.
(e) Class 50 (antibiotics) is the only class defined with respect to function rather than structure and thus it is necessary to cross-reference all antibiotics.
(f) Classes 71–86 (metal complexes) are reserved for complexes of transition metals only. To avoid any confusion as to which metals are considered to be transition metals the user should study the transition metal index on p. M1.
For metals which do not appear in this index the user must refer to classes other than 71–86, e.g. 67, 68, 69.
It should also be noted that a metal complex is never cross-

referenced to the simple ligand class. Thus a complex MX where M is a transition metal and X is pyridine will be classified only in class 83. There will be no cross-reference to class 33.

(g) Class 83–85 (nitrogen, oxygen, sulphur or selenium ligands) are used only when the compound does not contain a more specific ligand (classes 76–82).

The words "nitrogen, oxygen, sulphur, selenium" indicate the type of chelating atom.

Specific:

Class 2: The word "ammonium" can refer to substituted ammonium. This applies also to class 14.

Class 4: Each compound contains both nitrogen and sulphur.

Class 8: The word "urea" also covers thiourea, selenourea, guanidine, etc.

Class 9: Each compound has an N–N bond.

Class 10: Each compound has an N–O bond.

Class 21: This class is not used for benzenoid compounds (classes 13–19).

Class 24: For moderately saturated naphthalenes classification may be in 27.

Class 40: A ring in the compound contains both N and O.

Class 41: A ring in the compound contains both N and S.

Class 42: This class is reserved for unusual combinations of hetero-atoms, e.g. S,O; N,Se; I,O; Se,O.

Class 46: For phosphate esters, not inorganic phosphate ions.

Class 64: For phosphorus compounds other than phosphate esters and phosphine coordination compounds.

Class 71: For compounds having a transition metal σ-bonded to carbon.

Cross-Referencing Between Volumes

It should be noted that the various volumes in this series are complementary and, in order to save space, the full entry appears only once in the particular main class where the compound has been classified. In a small number of cases a cross-reference may appear in Volume 1 and the main entry in Volume 2. We anticipate this problem to be confined mainly to class 60 (molecular complexes) and to certain entries in the metal index.

List of Classes
in Volumes 1 and 2

Vol. 1

Vol. 2

MOLECULAR COMPLEXES

60.1   **Carbon tetrabromide - p - xylene complex**
CBr_4 , C_8H_{10}
F.J.Strieter, D.H.Templeton *J. Chem. Phys.*, **37**, 161, 1962
Residue 1 also classified in 5; residue 2 classified in 60, 19

60.C   **Pyridine - cyanogen iodide complex**
CIN , C_5H_5N
For complete entry see 60.65

60.C   **Diethylether - bromodichloromethane complex (form ii)**
$CHBrCl_2$, $C_4H_{10}O$
For complete entry see 60.61

60.2   **1,3 - Iodoform - sulfur**
CHI_3 , $3S_8$
T.Bjorvatten *Acta Chem. Scand.*, **16**, 749, 1962
Residue 1 also classified in 5

60.C   **1,4 - Dithiane - iodoform**
CHI_3 , $C_4H_8S_2$
For complete entry see 60.53

60.C   **Hexamethylenetetramine - iodoform complex**
CHI_3 , $C_6H_{12}N_4$
For complete entry see 60.96

60.3   **1,3 - Iodoform - quinoline**
CHI_3 , $3C_9H_7N$
T.Bjorvatten, O.Hassel *Acta Chem. Scand.*, **16**, 249, 1962
Residue 1 also classified in 5; residue 2 classified in 60, 35

60.4   **Iodoform - 1,4 - diselenane**
$2CHI_3$, $C_4H_8Se_2$
T.Bjorvatten *Acta Chem. Scand.*, **17**, 2292, 1963
Residue 1 also classified in 5; residue 2 classified in 60, 39

1

60.5   Urea - sodium chloride monohydrate
CH_4N_2O , Na^+ , Cl^- , H_2O
J.H.Palm, C.H.MacGillavry   *Acta Cryst.*, **16**, 963, 1963
Residue 1 also classified in 8

60.6   Urea - ammonium bromide
CH_4N_2O , H_4N^+ , Br^-
C.G.C.Catesby   *Acta Cryst.*, **16**, 392, 1963
Residue 1 also classified in 8

60.7   Urea - ammonium chloride
CH_4N_2O , H_4N^+ , Cl^-
A.Rimsky   *Bull. Soc. Fr. Mineral. Cristallogr.*, **83**, 187, 1960
Residue 1 also classified in 8

60.C   2,6 - Lutidine - urea complex
CH_4N_2O , C_7H_9N
For complete entry see 60.100

60.C   L - Cysteine ethyl ester hydrochloride - urea complex
$CH_4N_2O_2$, $C_5H_{12}NO_2S^+$, Cl^-
For complete entry see 60.68

60.8   Urea - hydrogen peroxide
Hyperol
$CH_4N_2O_3$, H_2O_2
C.-S.Lu, E.W.Hughes, P.A.Gignere   *J. Amer. Chem. Soc.*, **63**, 1507, 1941
Residue 1 also classified in 8
See also *Int. Distances*, M 115; *Structure Reports*, **8**, 278, 1941

60.9   Tetra(thiourea) - cesium chloride monohydrate
$4CH_4N_2S$, Cs^+ , Cl^- , H_2O
J.C.A.Boeyens   *Acta Cryst. (B)*, **24**, 1191, 1968
Residue 1 also classified in 8

60.10   Tetra(thiourea) - cesium fluoride complex dihydrate
$4CH_4N_2S$, Cs^+ , F^- , $2H_2O$
J.C.A.Boeyens   *Acta Cryst. (B)*, **24**, 199, 1968
Residue 1 also classified in 8

60.11   Thiourea - thallium(i) dihydrogen phosphate complex
$4CH_4N_2S$, Tl^+ , $H_2O_4P^-$
L.H.W.Verhoef, J.C.A.Boeyens   *Acta Cryst. (B)*, **24**, 1262, 1968
Residue 1 also classified in 8

60.C **Thallium(i) benzoate - tetra(thiourea) complex**
$4CH_4N_2S$, $C_7H_5O_2^-$, Tl^+
For complete entry see 60.98

60.12 **Methanol - bromine complex**
$2CH_4O$, Br_2
P.Groth, O.Hassel *Acta Chem. Scand.*, **18**, 402, 1964
Residue 1 also classified in 5

60.13 **bis(Methanol) - hydrazine (high temp. form)**
$2CH_4O$, H_4N_2
R.Liminga *Ark. Kemi,* **28,** 471, 1968
Residue 1 also classified in 5

60.14 **Methanol - sodium iodide**
$3CH_4O$, Na^+, I^-
P.Piret, C.Mesureur *J. Chim. Phys. Phys.-Chim. Biol.*, **62,** 287, 1965
Residue 1 also classified in 5

60.15 **Hydrazine - tetra - methanol**
$4CH_4O$, H_4N_2
R.Liminga, A.M.Sorensen *Acta Chem. Scand.*, **21,** 2669, 1967
Residue 1 also classified in 5

60.C **bis(Diacetatobromo rhodium(iii)) - bisguanidine**
$2CH_5N_3$, $C_8H_{12}Br_2O_8Rh_2$
For complete entry see 81.56

60.C **bis(Diacetatochloro rhodium(iii)) - bisguanidine**
$2CH_5N_3$, $C_8H_{12}Cl_2O_8Rh_2$
For complete entry see 81.58

60.16 **Guanidinium chloride - N,N - dimethylacetamide complex**
$3CH_6N_3^+$, C_4H_9NO , $3Cl^-$
D.J.Haas, D.R.Harris, H.H.Mills *Acta Cryst.*, **18**, 623, 1965
Residue 1 also classified in 8; residue 2 classified in 60, 1

60.17 **Oxalyl bromide - 1,4 - dioxan complex**
$C_2Br_2O_2$, $C_4H_8O_2$
E.Damm, O.Hassel, C.Romming *Acta Chem. Scand.*, **19,** 1159, 1965
Residue 1 also classified in 1; residue 2 classified in 60, 38

60.C **Pyrazine - tetrabromoethylene complex**
C_2Br_4 , $C_4H_4N_2$
For complete entry see 60.38

3

60.18 **Oxalyl chloride - 1,4 - dioxan complex**
$C_2Cl_2O_2$, $C_4H_8O_2$
E.Damm, O.Hassel, C.Romming *Acta Chem. Scand.*, **19**, 1159, 1965
Residue 1 also classified in 1; residue 2 classified in 60, 38

60.C **1,4 - Dioxan - di - iodoacetylene complex**
C_2I_2 , $C_4H_8O_2$
For complete entry see 60.50

60.19 **Di - iodoacetylene - 1,4 - dithiane complex**
C_2I_2 , $C_4H_8S_2$
O.Holmesland, C.Romming *Acta Chem. Scand.*, **20,** 2601, 1966
Residue 1 also classified in 5; residue 2 classified in 60, 39

60.20 **Di - iodoacetylene - 1,4 - diselenane complex**
C_2I_2 , $C_4H_8Se_2$
O.Holmesland, C.Romming *Acta Chem. Scand.*, **20,** 2601, 1966
Residue 1 also classified in 5; residue 2 classified in 60, 39

60.C **Cyclohexane - 1,4 - dione - di - iodoacetylene**
C_2I_2 , $C_6H_8O_2$
For complete entry see 60.92

60.C **Pyrazine - tetraiodoethylene complex**
C_2I_4 , $C_4H_4N_2$
For complete entry see 60.39

60.C **1,4 - Diselenane - tetraiodoethylene complex**
C_2I_4 , $C_4H_8Se_2$
For complete entry see 60.58

60.21 **Acetonitrile - antimony pentachloride complex**
C_2H_3N , Cl_5Sb
H.Binas *Z. Anorg. Allg. Chem.*, **352,** 271, 1967
Residue 1 also classified in 7

60.22 **bis(Acetonitrile) - bromine complex**
$2C_2H_3N$, Br_2
K.-M.Marstokk, K.O.Stromme *Acta Cryst. (B)*, **24,** 713, 1968
Residue 1 also classified in 7

60.C **Benzenediazonium chloride - acetic acid complex (low temp.study)**
$C_2H_4O_2$, $C_6H_5N_2{}^+$, Cl^-
For complete entry see 60.82

60.C **Ethyl picarte - cesium ethylate complex**
$C_2H_5O^-$, $C_8H_7N_3O_7$, Cs^+
For complete entry see 60.103

60.23 **Acetamide - sodium bromide complex**
$2C_2H_5O$, Na^+, Br^-
P.Piret, L.Rodrique, Y.Gobillon, M.van Meerssche
Acta Cryst., **20,** 482, 1966
Residue 1 also classified in 1

60.24 **Hydrazine - bis(ethanol) complex**
$2C_2H_6O$, H_4N_2
R.Liminga *Acta Chem. Scand.*, **21,** 1206, 1967
Residue 1 also classified in 5

60.25 **Ethylenediamine - lithium bromide complex**
$2C_2H_8N_2$, Li^+, Br^-
F.Durant, P.Piret, M.van Meerssche *Acta Cryst.*, **23,** 780, 1967
Residue 1 also classified in 3

60.26 **bis(Ethylenediamine) - lithium chloride complex**
$2C_2H_8N_2$, Li^+, Cl^-
F.Durant, P.Piret, M.van Meerssche *Acta Cryst.*, **23,** 780, 1967
Residue 1 also classified in 3

60.27 **bis(Ethylenediamine) - lithium chloride complex**
$2C_2H_8N_2$, Li^+, Cl^-
S.Jamet-Delcroix *J. Chim. Phys. Phys.-Chim. Biol.*, **64,** 601, 1967
Residue 1 also classified in 3

60.28 **1,3,5 - Trinitrohexahydrotriazine - tetramethylene - sulfone complex**
$C_3H_6N_6O_6$, $C_4H_8O_2S$
B.Rerat, J.Berthou, A.Laurent, C.Rerat
C. R. Acad. Sci., Fr., C, **267,** 760, 1968
Residue 1 also classified in 33; residue 2 classified in 60, 39

60.29 **Acetone - bromine (at −30 °C)**
C_3H_6O, $2Br$
O.Hassel, K.O.Stromme *Acta Chem. Scand.*, **13,** 275, 1959
Residue 1 also classified in 5
See also *Int. Distances,* M 98s; *Structure Reports,* **23,** 524, 1959

60.C **Quinol - acetone**
C_3H_6O, $C_6H_6O_2$
For complete entry see 60.88

60.30   **Acetone - sodium iodide**
$3C_3H_6O$, Na^+ , I^-
P.Piret, Y.Gobillon, M.van Meerssche   *Bull. Soc. Chim. Fr.*, 205, 1963
Residue 1 also classified in 5

60.31   **N - Methylacetamide - sodium perchlorate complex**
$2C_3H_7NO$, Na^+ , ClO_4^-
D.J.Haas   *Thesis, New York,* 1965
Residue 1 also classified in 1

60.32   **Tri(dimethylformide) - sodium iodide**
$3C_3H_7NO$, Na^+ , I^-
Y.Gobillon, P.Piret, M.van Meerssche   *Bull. Soc. Chim. Fr.*, 551, 1962
Residue 1 also classified in 1

60.33   **N - Methylacetamide - lithium chloride complex**
$4C_3H_7NO$, Li^+ , Cl^-
D.J.Haas   *Nature,* **201,** 64, 1964
Residue 1 also classified in 1

60.34   **N - Methylacetamide - potassium iodide - potassium tri - iodide complex**
$6C_3H_7NO$, $2K^+$, I^- , I_3^-
K.Toman, J.Honzl, J.Jecny   *Acta Cryst.*, **18,** 673, 1965
Residue 1 also classified in 1

60.35   **Trimethylamine iodomonochloride**
C_3H_9N , ClI
O.Hassel, H.Hope   *Acta Chem. Scand.*, **14,** 391, 1960
Residue 1 also classified in 3

60.36   **Trimethylamine - iodine (at −20 °C)**
C_3H_9N , I_2
K.O.Stromme   *Acta Chem. Scand.*, **13,** 268, 1959
Residue 1 also classified in 3
See also *Int. Distances,* M 101s; *Structure Reports,* **23,** 568, 1959

60.37   **Trimethylamine - sulfur dioxide complex**
C_3H_9N , O_2S
D.van der Helm, J.D.Childs, S.D.Christian   *J. Chem. Soc. (D),* 887, 1969
Residue 1 also classified in 3

60.38   **Pyrazine - tetrabromoethylene complex**
$C_4H_4N_2$, C_2Br_4
T.Dahl, O.Hassel   *Acta Chem. Scand.*, **22,** 2851, 1968
Residue 1 also classified in 33; residue 2 classified in 60, 5

60.39 **Pyrazine - tetraiodoethylene complex**
$C_4H_4N_2$, C_2I_4
T.Dahl, O.Hassel *Acta Chem. Scand.*, **22**, 2851, 1968
Residue 1 also classified in 33; residue 2 classified in 60, 5

60.C **Acridine - cytosine complex monohydrate**
$C_4H_5N_3O$, $C_{13}H_9N$, H_2O
For complete entry see 60.132

60.40 **bis(Diacetamide) - sodium bromide complex**
$2C_4H_7NO_2$, Br^- , Na^+
J.P.Roux, J.C.A.Boeyens *Acta Cryst. (B)*, **25**, 1700, 1969
Residue 1 also classified in 1

60.41 **bis(Diacetamide) - potassium iodide complex**
$2C_4H_7NO_2$, I^- , K^+
J.P.Roux, J.C.A.Boeyens *Acta Cryst. (B)*, **25**, 2395, 1969
Residue 1 also classified in 1

60.42 **1 - Oxa - 4 - selenacyclohexane iodine monochloride complex**
C_4H_8OSe , ClI
C.Knobler, J.D.McCullough *Inorg. Chem.*, **7**, 365, 1968
Residue 1 also classified in 42

60.43 **1,4 - Oxaselenane - iodine complex**
C_4H_8OSe , I_2
H.Maddox, J.D.McCullough *Inorg. Chem.*, **5**, 522, 1966
Residue 1 also classified in 42

60.44 **1,4 - Dioxan - bromine**
$C_4H_8O_2$, 2Br
O.Hassel, J.Hvoslef *Acta Chem. Scand.*, **8**, 873, 1954
Residue 1 also classified in 38
See also *Int. Distances*, M 171; *Structure Reports*, **18**, 637, 1954

60.45 **1,4 - Dioxan - mercuric chloride**
$C_4H_8O_2$, Cl_2Hg
O.Hassel, J.Hvoslef *Acta Chem. Scand.*, **8**, 1953, 1954
Residue 1 also classified in 38
See also *Int. Distances*, M 171; *Structure Reports*, **18**, 638, 1954

60.46   **1,4 - Dioxan - lithium chloride complex**
$C_4H_8O_2$, Li
F.Durant, Y.Gobillon, P.Piret, M.van Meerssche
Bull. Soc. Chim. Belges, **75,** 52, 1966
Residue 1 also classified in 38

60.47   **1,4 - Dioxane - lithium chloride complex monohydrate**
$C_4H_8O_2$, Li^+ , Cl^- , H_2O
F.Durant, M.Griffe   *Bull. Soc. Chim. Belges,* **77,** 557, 1968
Residue 1 also classified in 38

60.48   **1,4 - Dioxan - dinitrogen tetroxide complex**
$C_4H_8O_2$, N_2O_4
P.Groth, O.Hassel   *Acta Chem. Scand.,* **19,** 120, 1965
Residue 1 also classified in 38

60.49   **1,4 - Dioxan sulphuric acid**
$C_4H_8O_2$, H_2O_4S
O.Hassel, C.Romming   *Acta Chem. Scand.,* **14,** 398, 1960
Residue 1 also classified in 38

60.C   **Oxalyl bromide - 1,4 - dioxan complex**
$C_4H_8O_2$, $C_2Br_2O_2$
For complete entry see 60.17

60.C   **Oxalyl chloride - 1,4 - dioxan complex**
$C_4H_8O_2$, $C_2Cl_2O_2$
For complete entry see 60.18

60.50   **1,4 - Dioxan - di - iodoacetylene complex**
$C_4H_8O_2$, C_2I_2
P.Gagnaux, B.P.Susz   *Helv. Chim. Acta,* **43,** 948, 1960
Residue 1 also classified in 38; residue 2 classified in 5, 60

60.51   **Silver perchlorate - dioxan**
$3C_4H_8O_2$, Ag^+ , ClO_4^-
R.J.Prosen, K.N.Trueblood   *Acta Cryst.,* **9,** 741, 1956
Residue 1 also classified in 84
See also *Int. Distances,* M 171s; *Structure Reports,* **20,** 620, 1956

60.C   **1,3,5 - Trinitrohexahydrotriazine - tetramethylene - sulfone complex**
$C_4H_8O_2S$, $C_3H_6N_6O_6$
For complete entry see 60.28

60.52 **1,4 - Dithiane - iodine complex**
$C_4H_8S_2$, $2I_2$
G.Y.Chao, J.D.McCullough *Acta Cryst.*, **13**, 727, 1960
Residue 1 also classified in 39

60.53 **1,4 - Dithiane - iodoform**
$C_4H_8S_2$, CHI_3
T.Bjorvatten, O.Hassel *Acta Chem. Scand.*, **15**, 1429, 1961
Residue 1 also classified in 39; residue 2 classified in 60, 5

60.C **Di - iodoacetylene - 1,4 - dithiane complex**
$C_4H_8S_2$, C_2I_2
For complete entry see 60.19

60.54 **1,4 - Dithiane - antimony tri - iodide complex**
$2C_4H_8S_2$, I_3Sb
T.Bjorvatten *Acta Chem. Scand.*, **20,** 1863, 1966
Residue 1 also classified in 39

60.55 **Tetrahydroselenophene - iodine complex (refined as space group no. 62)**
C_4H_8Se , I_2
H.Hope, J.D.McCullough *Acta Cryst.*, **17,** 712, 1964
Residue 1 also classified in 39

60.56 **Tetrahydroselenophene - iodine complex (refined as space group no.33)**
C_4H_8Se , I_2
H.Hope, J.D.McCullough *Acta Cryst.*, **17,** 712, 1964
Residue 1 also classified in 39

60.57 **1,4 - Diselenane - iodine complex**
$C_4H_8Se_2$, $2I_2$
G.Y.Chao, J.D.McCullough *Acta Cryst.*, **14,** 940, 1961
Residue 1 also classified in 39

60.C **Iodoform - 1,4 - diselenane**
$C_4H_8Se_2$, $2CHI_3$
For complete entry see 60.4

60.C **·Di - iodoacetylene - 1,4 - diselenane complex**
$C_4H_8Se_2$, C_2I_2
For complete entry see 60.20

60.58 **1,4 - Diselenane - tetraiodoethylene complex**
$C_4H_8Se_2$, C_2I_4
T.Dahl, O.Hassel *Acta Chem. Scand.*, **19,** 2000, 1965
Residue 1 also classified in 39; residue 2 classified in 60

60.C Guanidinium chloride - N,N - dimethylacetamide complex
C_4H_9NO , $3CH_6N_3{}^+$, $3Cl^-$
For complete entry see 60.16

60.59 Morpholine - β - iodophenylacetylene complex
C_4H_9NO , C_8H_5I
R.H.Baughman *J. Org. Chem.*, **29**, 964, 1964
Residue 1 also classified in 40; residue 2 classified in 60, 19

60.60 Piperazine silver iodide
$0.5C_4H_{10}N_2$, Ag^+ , I^-
G.B.Ansell, W.G.Finnegan *J. Chem. Soc. (D)*, 1300, 1969
Residue 1 also classified in 33

60.61 Diethylether - bromodichloromethane complex (form ii)
$C_4H_{10}O$, $CHBrCl_2$
P.Andersen, T.Thurmann-Moe *Acta Chem. Scand.*, **18**, 433, 1964
Residue 1 also classified in 5; residue 2 classified in 60, 5

60.62 Piperazinium bis(dichloroiodide)
$C_4H_{12}N_2{}^{2+}$, $2Cl_2I^-$
C.Romming *Acta Chem. Scand.*, **12**, 668, 1958
Residue 1 also classified in 33
See also *Int. Distances*, M 117s; *Structure Reports*, **22**, 768, 1958

60.63 Pyridine - iodine monochloride
C_5H_5ClIN
O.Hassel, C.Romming *Acta Chem. Scand.*, **10**, 696, 1956
Also classified in 33
See also *Int. Distances*, M 120s; *Structure Reports*, **20**, 604, 1956

60.64 Pyridine - lithium chloride
C_5H_5N , Li^+ , Cl^-
F.Durant, J.Verbist, M.van Meerssche
Bull. Soc. Chim. Belges, **75**, 806, 1966
Residue 1 also classified in 33

60.65 Pyridine - cyanogen iodide complex
C_5H_5N , CIN
T.Dahl, O.Hassel, K.Sky *Acta Chem. Scand.*, **21**, 592, 1967
Residue 1 also classified in 33; residue 2 classified in 60, 7

60.66 **Pyridine - lithium chloride**
$2C_5H_5N$, Li^+ , Cl^- , H_2O
F.Durant, P.Piret, M.van Meerssche *Acta Cryst.*, **22**, 52, 1967
Residue 1 also classified in 33

60.67 **Cystylglycine - sodium iodide**
$2C_5H_{10}N_2O_3S$, Na^+ , I^-
H.B.Dryer *Acta Cryst.*, **4**, 42, 1951
Residue 1 also classified in 48
See also *Int. Distances*, M 186; *Structure Reports*, **15**, 407, 1951

60.68 **L - Cysteine ethyl ester hydrochloride - urea complex**
$C_5H_{12}NO_2S^+$, $CH_4N_2O_2$, Cl^-
D.J.Haas *Acta Cryst.*, **19**, 860, 1965
Residue 1 also classified in 48; residue 2 classified in 8, 60

60.69 **Hexabromobenzene - 1,2,4,5 - tetrabromobenzene complex**
C_6Br_6 , $C_6H_2Br_4$
G.Gafner, F.H.Herbstein *J. Chem. Soc.*, 5290, 1964
Residue 1 also classified in 19; residue 2 classified in 60, 19

60.70 **Chloranil - bis(8 - hydroxyquinolinato) palladium(ii)**
C_6Cl_4O , $C_{18}H_{12}N_2O_2Pd$
B.Kamenar, C.K.Prout, J.D.Wright *J. Chem. Soc.*, 4851, 1965
Residue 1 also classified in 18; residue 2 classified in 60, 83, 84

60.C **8 - Hydroxyquinoline - chloranil complex**
$C_6Cl_4O_2$, $2C_9H_7NO$
For complete entry see 60.107

60.C **N,N,N′,N′ - Tetramethyl - p - diaminobenzene - chloranil complex**
$C_6Cl_4O_2$, $C_{10}H_{16}N_2$
For complete entry see 60.122

60.71 **Chloranil - hexamethylbenzene complex**
$C_6Cl_4O_2$, $C_{12}H_{18}$
N.D.Jones, R.E.Marsh *Acta Cryst.*, **15**, 809, 1962
Residue 1 also classified in 18; residue 2 classified in 60, 19

60.C **Perylene - fluoranil complex**
$C_6F_4O_2$, $C_{20}H_{12}$
For complete entry see 60.152

60.72 Tetracyanoethylene - naphthalene complex
C_6N_4 , $C_{10}H_8$
R.M.Williams, S.C.Wallwork *Acta Cryst.*, **22,** 899, 1967
Residue 1 also classified in 37; residue 2 classified in 60, 7

60.C Ferrocene - tetracyanoethylene complex
C_6N_4 , $C_{10}H_{10}Fe$
For complete entry see 60.119

60.C Pyrene - tetracyanoethylene complex (projections)
C_6N_4 , $C_{16}H_{10}$
For complete entry see 60.141

60.C Pyrene - tetracyanoethylene complex
C_6N_4 , $C_{16}H_{10}$
For complete entry see 60.142

60.C Perylene - tetracyanoethylene complex
C_6N_4 , $C_{20}H_{12}$
For complete entry see 60.153

60.73 Benzotrifurazan - triethyl phosphate complex
$C_6N_6O_3$, $C_6H_{15}OP$
T.S.Cameron, C.K.Prout *Chem. Communic.*, 684, 1968
Residue 1 also classified in 40; residue 2 classified in 60, 46

60.74 Benzotrifuroxan - 13,14 - dithiatricyclo(8,2,1,1(4,7)) tetradeca - 4,6,10,12 - tetraene
$C_6N_6O_6$, $C_{12}H_{12}S_2$
B.Kamenar, C.K.Prout *J. Chem. Soc.*, 4838, 1965
Residue 1 also classified in 40, 10; residue 2 classified in 60, 39

60.C Hexabromobenzene - 1,2,4,5 - tetrabromobenzene complex
$C_6H_2Br_4$, C_6Br_6
For complete entry see 60.69

60.75 1 - Imino - 4 - (oxime - N - oxide) - benzdifurazan triphenyl phosphine oxide complex
$C_6H_2N_6O_4$, $C_{18}H_{15}OP$
A.S.Bailey, T.S.Cameron, J.M.Evans, C.K.Prout
Chem. Communic., 664, 1966
Residue 1 also classified in 40, 10; residue 2 classified in 60, 64

60.76 **Picryl azide - bis(8 - hydroxyquinolinato) copper(ii)**
$2C_6H_2N_6O_6$, $C_{18}H_{12}CuN_2O_2$
A.S.Bailey, C.K.Prout *J. Chem. Soc.*, 4867, 1965
Residue 1 also classified in 15, 9

60.C **p - Iodoaniline - s - trinitrobenzene complex**
$C_6H_3N_3O_6$, C_6H_6IN
For complete entry see 60.86

60.C **Indole - s - trinitrobenzene complex**
$C_6H_3N_3O_6$, C_8H_7N
For complete entry see 60.102

60.C **Skatole - s - trinitrobenzene complex (at −140 ° C)**
$C_6H_3N_3O_6$, C_9H_9N
For complete entry see 60.108

60.C **2,4,6 - Tri(dimethylamino) - 1,3,5 - triazine - s - trinitrobenzene complex**
$C_6H_3N_3O_6$, $C_9H_{18}N_6$
For complete entry see 60.113

60.C **Azulene - s - trinitrobenzene complex**
$C_6H_3N_3O_6$, $C_{10}H_8$
For complete entry see 60.115

60.C **Azulene - s - trinitrobenzene complex**
$C_6H_3N_3O_6$, $C_{10}H_8$
For complete entry see 60.116

60.C **Tricarbonyl chromium anisole - 1,3,5 - trinitrobenzene complex**
$C_6H_3N_3O_6$, $C_{10}H_8CrO_4$
For complete entry see 60.118

60.C **Phenothiazine - s - trinitrobenzene complex**
$C_6H_3N_3O_6$, $C_{12}H_9NS$
For complete entry see 60.129

60.C **Anthracene - s - trinitrobenzene**
$C_6H_3N_3O_6$, $C_{14}H_{10}$
For complete entry see 60.133

60.C **Anthracene - s - trinitrobenzene (at −100 ° C)**
$C_6H_3N_3O_6$, $C_{14}H_{10}$
For complete entry see 60.134

60.C   Acepleiadylene - s - trinitrobenzene complex
$C_6H_3N_3O_6$, $C_{16}H_{10}$
For complete entry see 60.143

60.77   Picric acid - 1 - bromo - 2 - aminonaphthalene complex
$C_6H_3N_3O_7$, $C_{10}H_8BrN$
E.Carstensen-Oeser, S.Gottlicher, G.Habermehl
Chem. Ber., **101,** 1648, 1968
Residue 1 also classified in 15, 17; residue 2 classified in 60, 24

60.78   p - Benzoquinone - p - bromophenol complex
$C_6H_4O_2$, C_6H_5BrO
G.G.Shipley, S.C.Wallwork   *Acta Cryst.*, **22,** 593, 1967
Residue 1 also classified in 18; residue 2 classified in 60, 17

60.79   p - Benzoquinone - di(p - bromophenol) complex
$C_6H_4O_2$, $2C_6H_5BrO$
G.G.Shipley, S.C.Wallwork   *Acta Cryst.*, **22,** 585, 1967
Residue 1 also classified in 18; residue 2 classified in 60, 17

60.80   p - Benzoquinone - p - chlorophenol complex
$C_6H_4O_2$, C_6H_5ClO
G.G.Shipley, S.C.Wallwork   *Acta Cryst.*, **22,** 593, 1967
Residue 1 also classified in 18; residue 2 classified in 60, 17

60.81   p - Benzoquinone - di(p - chlorophenol) complex
$C_6H_4O_2$, $2C_6H_5ClO$
G.G.Shipley, S.C.Wallwork   *Acta Cryst.*, **22,** 585, 1967
Residue 1 also classified in 18; residue 2 classified in 60, 17

60.C   p - Benzoquinone - p - bromophenol complex
C_6H_5BrO , $C_6H_4O_2$
For complete entry see 60.78

60.C   Testosterone - p - bromophenol complex
C_6H_5BrO , $C_{19}H_{28}O_2$
For complete entry see 60.151

60.C   p - Benzoquinone - di(p - bromophenol) complex
$2C_6H_5BrO$, $C_6H_4O_2$
For complete entry see 60.79

60.C   p - Benzoquinone - p - chlorophenol complex
C_6H_5ClO , $C_6H_4O_2$
For complete entry see 60.80

60.C p - Benzoquinone - di(p - chlorophenol) complex
$2C_6H_5ClO$, $C_6H_4O_2$
For complete entry see 60.81

60.82 Benzenediazonium chloride - acetic acid complex (low temp.study)
$C_6H_5N_2^+$, $C_2H_4O_2$, Cl^-
C.Romming, T.Tjornhom *Acta Chem. Scand.*, **22**, 2934, 1968
Residue 1 also classified in 9; residue 2 classified in 60, 1

60.83 Benzene - dialuminium bromide complex
C_6H_6 , Al_2Br_6
D.D.Eley, J.H.Taylor, S.C.Wallwork *J. Chem. Soc.*, 3867, 1961
Residue 1 also classified in 19

60.84 Benzene - bromine (at −45 °C)
C_6H_6 , Br_2
O.Hassel, K.O.Stromme *Acta Chem. Scand.*, **12**, 1146, 1958
Residue 1 also classified in 19
See also *Int. Distances*, M 129s; *Structure Reports*, **22**, 689, 1958

60.85 Benzene - dinitrogen tetroxide complex
C_6H_6 , N_2O_4
K.O.Stromme *Acta Cryst. (B)*, **24**, 1607, 1968
Residue 1 also classified in 19

60.86 p - Iodoaniline - s - trinitrobenzene complex
C_6H_6IN , $C_6H_3N_3O_6$
H.M.Powell, G.Huse, P.W.Cooke *J. Chem. Soc.*, 153, 1943
Residue 1 also classified in 16; residue 2 classified in 60, 15
See also *Int. Distances*, M 232; *Structure Reports*, **9**, 372, 1943

60.87 p - Nitroaniline cadmium chloride monohydrate
$2C_6H_6N_2O_2$, Cd_2Cl_4 , $2H_2O$
G.F.Wolodina, A.W.Ablov *Dokl. Akad. Nauk S. S. S. R.*, **182**, 105, 1968
Residue 1 also classified in 15, 16

60.88 Quinol - acetone
$C_6H_6O_2$, C_3H_6O
J.D.Lee, S.C.Wallwork *Acta Cryst.*, **12**, 210, 1959
Residue 1 also classified in 17; residue 2 classified in 60, 5
See also *Int. Distances*, M 152s; *Structure Reports*, **23**, 650, 1959

60.89 Aniline - antimony trichloride complex
C_6H_7N
R.Hulme, J.C.Scruton   *J. Chem. Soc. (A)*, 2448, 1968
Also classified in 16

60.90 4 - Picoline - iodine complex
C_6H_7N , I_2
O.Hassel, C.Romming, T.Tufte   *Acta Chem. Scand.*, **15**, 967, 1961
Residue 1 also classified in 33

60.91 Cyclohexane - 1,4 - dione - mercuric chloride
$(C_6H_8Cl_2HgO_2)_n$
P.Groth, O.Hassel   *Acta Chem. Scand.*, **18**, 1327, 1964
Also classified in 21

60.92 Cyclohexane - 1,4 - dione - di - iodoacetylene
$C_6H_8O_2$, C_2I_2
P.Groth, O.Hassel   *Acta Chem. Scand.*, **19**, 1733, 1965
Residue 1 also classified in 21; residue 2 classified in 60, 5

60.93 Pentamethylenetetrazole - iodine monochloride complex (photographic data)
$C_6H_{10}N_4$, ClI
N.C.Baenziger, A.D.Nelson, A.Tulinsky, J.H.Bloor, A.I.Popov
J. Amer. Chem. Soc., **89**, 6463, 1967
Residue 1 also classified in 35

60.94 Pentamethylenetetrazole - iodine monochloride complex (diffractometer data)
$C_6H_{10}N_4$, ClI
N.C.Baenziger, A.D.Nelson, A.Tulinsky, J.H.Bloor, A.I.Popov
J. Amer. Chem. Soc., **89**, 6463, 1967
Residue 1 also classified in 35

60.95 Glycylglycylglycine - calcium chloride complex trihydrate
$C_6H_{11}N_3O_4$, Ca^{2+} , $2Cl^-$, $3H_2O$
D.van der Helm, T.V.Willoughby   *Acta Cryst. (B)*, **25**, 2317, 1969
Residue 1 also classified in 48

60.96 Hexamethylenetetramine - iodoform complex
$C_6H_{12}N_4$, CHI_3
T.Dahl, O.Hassel   *Acta Chem. Scand.*, **22**, 2036, 1968
Residue 1 also classified in 37; residue 2 classified in 60, 5

60.97 **Hexamethylenetetramine - calcium bromide complex decahydrate (triclinic form)**
$2C_6H_{12}N_4$, Ca^{2+} , $2Br^-$, $10H_2O$
L.Mazzarella, A.L.Kovacs, P.de Santis, A.M.Liquori
Acta Cryst., **22,** 65, 1967
Residue 1 also classified in 37

60.C **Benzotrifurazan - triethyl phosphate complex**
$C_6H_{15}OP$, $C_6N_6O_3$
For complete entry see 60.73

60.C **Phenothiazine - 3,5 - dinitrobenzoic acid complex**
$C_7H_4N_2O_6$, $C_{12}H_9NS$
For complete entry see 60.130

60.C **Caffeine - 5 - chlorosalicylic acid complex**
$C_7H_5ClO_3$, $C_8H_{10}N_4O_2$
For complete entry see 60.104

60.98 **Thallium(i) benzoate - tetra(thiourea) complex**
$C_7H_5O_2^-$, $4CH_4N_2S$, Tl^+
L.H.W.Verhoef, J.C.A.Boeyens *Acta Cryst. (B)*, **25,** 617, 1969
Residue 1 also classified in 14; residue 2 classified in 60, 8

60.99 **Benzamide - hydrogen tri - iodide complex**
$2C_7H_7NO$, HI_3
J.M.Reddy, K.Knox, M.B.Robin *J. Chem. Phys.*, **40,** 1082, 1964
Residue 1 also classified in 13

60.100 **2,6 - Lutidine - urea complex**
C_7H_9N , CH_4N_2O
J.D.Lee, S.C.Wallwork *Acta Cryst.*, **19,** 311, 1965
Residue 1 also classified in 33; residue 2 classified in 60, 8

60.101 **1,2,4,5 - Tetracyanobenzene - hexamethylbenzene complex**
$C_8H_2N_4$, $C_{12}H_{18}$
N.Niimura, Y.Ohashi, Y.Saito *Bull. Chem. Soc. Jap.*, **41,** 1815, 1968
Residue 1 also classified in 7; residue 2 classified in 60, 19

60.C **Morpholine - β - iodophenylacetylene complex**
C_8H_5I , C_4H_9NO
For complete entry see 60.59

60.102 **Indole - s - trinitrobenzene complex**
C_8H_7N , $C_6H_3N_3O_6$
A.W.Hanson *Acta Cryst.*, **17**, 559, 1964
Residue 1 also classified in 35; residue 2 classified in 60, 15

60.103 **Ethyl picarte - cesium ethylate complex**
$C_8H_7N_3O_7$, $C_2H_5O^-$, Cs^+
G.L.Casalone, R.Destro, C.M.Gramaccioli, C.Mariani, A.Mugnoli,
M.Simonetta *Acta Cryst.*, **21**, A107, 1966
Residue 1 also classified in 17, 15; residue 2 classified in 60, 5

60.C **Carbon tetrabromide - p - xylene complex**
C_8H_{10} , CBr_4
For complete entry see 60.1

60.104 **Caffeine - 5 - chlorosalicylic acid complex**
$C_8H_{10}N_4O_2$, $C_7H_5ClO_3$
E.Shefter *J. Pharm. Sci.*, **57**, 1163, 1968
Residue 1 also classified in 44, 58; residue 2 classified in 60, 13, 17

60.105 **p - Iodo - N,N - dimethylaniline - hydrochloride - hydroiodide - iodine - complex (refined as space group no. 15)**
$2C_8H_{11}IN^+$, Cl^- , I_3^-
R.Allmann *Z. Kristallogr.*, **117**, 184, 1962
Residue 1 also classified in 16

60.106 **p - Iodo - N,N - dimethylaniline - hydrochloride - hydroiodide - iodine - complex (refined as space group no. 9)**
$2C_8H_{11}IN^+$, Cl^- , I_3^-
R.Allmann *Z. Kristallogr.*, **117**, 184, 1962
Residue 1 also classified in 16

60.C **bis(Diacetatobromo rhodium(iii)) - bisguanidine**
$C_8H_{12}Br_2O_8Rh_2$, $2CH_5N_3$
For complete entry see 81.56

60.C **bis(Diacetatochloro rhodium(iii)) - bisguanidine**
$C_8H_{12}Cl_2O_8Rh_2$, $2CH_5N_3$
For complete entry see 81.58

60.C **1,3 - Iodoform - quinoline**
$3C_9H_7N$, CHI_3
For complete entry see 60.3

60.107 **8 - Hydroxyquinoline - chloranil complex**
$2C_9H_7NO$, $C_6Cl_4O_2$
C.K.Prout, A.G.Wheeler *J. Chem. Soc. (A)*, 469, 1967
Residue 1 also classified in 35; residue 2 classified in 60, 18

60.108 **Skatole - s - trinitrobenzene complex (at −140 °C)**
C_9H_9N , $C_6H_3N_3O_6$
A.W.Hanson *Acta Cryst.*, **17**, 559, 1964
Residue 1 also classified in 35; residue 2 classified in 60, 15

60.109 **5 - (2 - Methylmercapto - 4 - methyl - 4,5 - dihydro - 5 - thiadiazolidene) - 3 - ethyl - rhodanine - iodine complex**
$2C_9H_{11}N_3OS_4$, I_2
M.Bois D'Enghien-Peteau, J.Meunier-Piret, M.van Meerssche
J. Chim. Phys. Phys.-Chim. Biol., **65**, 1221, 1968
Residue 1 also classified in 41

60.110 **1,3,7,9 - Tetramethyluric acid - pyrene complex**
$C_9H_{12}N_4O_3$, $C_{16}H_{10}$
A.Damiani, P.de Santis, E.Giglio, A.M.Liquori, R.Puliti, A.Ripamonti
Acta Cryst., **19**, 340, 1965
Residue 1 also classified in 44; residue 2 classified in 60, 29

60.111 **1,3,7,9 - Tetramethyluric acid - 3,4 - benzpyrene complex**
$2C_9H_{12}N_4O_3$, $C_{20}H_{12}$
A.Damiani, E.Giglio, A.M.Liquori, A.Ripamonti
Acta Cryst., **23**, 675, 1967
Residue 1 also classified in 44; residue 2 classified in 60, 30

60.112 **1,3,7,9 - Tetramethyluric acid - coronene complex**
$2C_9H_{12}N_4O_3$, $C_{24}H_{12}$
A.Damiani, P.de Santis, E.Giglio, A.M.Liquori, R.Puliti, A.Ripamonti
Acta Cryst., **16**, A57, 1963
Residue 1 also classified in 44; residue 2 classified in 60, 30

60.113 **2,4,6 - Tri(dimethylamino) - 1,3,5 - triazine - s - trinitrobenzene complex**
$C_9H_{18}N_6$, $C_6H_3N_3O_6$
R.M.Williams, S.C.Wallwork *Acta Cryst.*, **21**, 406, 1966
Residue 1 also classified in 33; residue 2 classified in 60, 15

60.C **Naphthalene - 1,2,4,5 - tetracyanobenzene complex**
$C_{10}H_2N_4$, $C_{10}H_8$
For complete entry see 60.117

60.C N,N,N′,N′ - Tetramethyl - p - phenylenediamine - 1,2,4,5 - tetracyanobenzene complex
$C_{10}H_2N_4$, $C_{10}H_{16}N_2$
For complete entry see 60.123

60.C Palladium(ii) 8 - hydroxyquinolinate - 1,2,4,5 - tetracyanobenzene complex (disordered)
$C_{10}H_2N_4$, $C_{18}H_{12}N_2O_2Pd$
For complete entry see 60.148

60.C Copper(ii) 8 - hydroxyquinolinate - bis(1,2,4,5 - tetracyanobenzene) complex (photographic data)
$2C_{10}H_2N_4$, $C_{18}H_{12}CuN_2O_2$
For complete entry see 60.146

60.C Copper(ii) 8 - hydroxyquinolinate - bis(1,2,4,5 - tetracyanobenzene) complex (diffractometer data)
$2C_{10}H_2N_4$, $C_{18}H_{12}CuN_2O_2$
For complete entry see 60.147

60.C Anthracene - pyromellitic dianhydride complex
$C_{10}H_2O_6$, $C_{14}H_{10}$
For complete entry see 60.135

60.C Pyrene - pyromellitic dianhydride
$C_{10}H_2O_6$, $C_{16}H_{10}$
For complete entry see 60.144

60.C Perylene - pyromellitic dianhydride complex
$C_{10}H_2O_6$, $C_{20}H_{12}$
For complete entry see 60.154

60.114 Naphthalene - antimony trichloride complex
$C_{10}H_8$, $2Cl_3Sb$
R.Hulme, J.T.Szymanski *Acta Cryst. (B)*, **25**, 753, 1969
Residue 1 also classified in 24

60.C Tetracyanoethylene - naphthalene complex
$C_{10}H_8$, C_6N_4
For complete entry see 60.72

60.115 Azulene - s - trinitrobenzene complex
$C_{10}H_8$, $C_6H_3N_3O_6$
A.W.Hanson *Acta Cryst.*, **19**, 19, 1965
Residue 1 also classified in 27; residue 2 classified in 60, 15

20

60.116 Azulene - s - trinitrobenzene complex
$C_{10}H_8$, $C_6H_3N_3O_6$
D.S.Brown, S.C.Wallwork *Acta Cryst.*, **19,** 149, 1965
Residue 1 also classified in 27; residue 2 classified in 60, 15

60.117 Naphthalene - 1,2,4,5 - tetracyanobenzene complex
$C_{10}H_8$, $C_{10}H_2N_4$
S.Kumakura, F.Iwasaki, Y.Saito *Bull. Chem. Soc. Jap.*, **40,** 1826, 1967
Residue 1 also classified in 24; residue 2 classified in 60, 7

60.C Picric acid - 1 - bromo - 2 - aminonaphthalene complex
$C_{10}H_8BrN$, $C_6H_3N_3O_7$
For complete entry see 60.77

60.118 Tricarbonyl chromium anisole - 1,3,5 - trinitrobenzene complex
$C_{10}H_8CrO_4$, $C_6H_3N_3O_6$
O.L.Carter, A.T.McPhail, G.A.Sim *J. Chem. Soc. (A)*, 822, 1966
Residue 1 also classified in 74; residue 2 classified in 60, 15

60.119 Ferrocene - tetracyanoethylene complex
$C_{10}H_{10}Fe$, C_6N_4
E.Adman, M.Rosenblum, S.Sullivan, T.N.Margulis
J. Amer. Chem. Soc., **89,** 4540, 1967
Residue 1 also classified in 73; residue 2 classified in 60, 7

60.120 Dipyridine - iodine heptaiodide
$C_{10}H_{10}I_4N_2^+$, I_3^- , $2I_2$
O.Hassel, H.Hope *Acta Chem. Scand.*, **15,** 407, 1961
Residue 1 also classified in 33

60.121 Mesitaldehyde - perchloric acid complex
$2C_{10}H_{12}O$, $HClO_4$
C.D.Fisher, L.H.Jensen, W.M.Schubert *J. Amer. Chem. Soc.*, **87,** 33, 1965
Residue 1 also classified in 19

60.122 N,N,N',N' - Tetramethyl - p - diaminobenzene - chloranil complex
$C_{10}H_{16}N_2$, $C_6Cl_4O_2$
J.L.de Boer, A.Vos *Acta Cryst. (B)*, **24,** 720, 1968
Residue 1 also classified in 16; residue 2 classified in 60, 18

60.123 N,N,N',N' - Tetramethyl - p - phenylenediamine - 1,2,4,5 - tetracyanobenzene complex
$C_{10}H_{16}N_2$, $C_{10}H_2N_4$
Y.Ohashi, H.Iwasaki, Y.Saito *Bull. Chem. Soc. Jap.*, **40,** 1789, 1967
Residue 1 also classified in 16; residue 2 classified in 60, 7

60.124 **7,7,8,8 - Tetracyanoquinodimethan - N,N,N′,N′ - tetramethyl - p - phenylenediamine complex**
$C_{10}H_{16}N_2$, $C_{12}H_4N_4$
A.W.Hanson *Acta Cryst.*, **19,** 610, 1965
Residue 1 also classified in 7; residue 2 classified in 60, 16

60.C **bis(7,7,8,8 - Tetracyanoquinodimethan) - N,N,N′,N′ - tetramethyl - p - phenylenediamine complex**
$C_{10}H_{16}N_2$, $2C_{12}H_4N_4$
For complete entry see 60.127

60.C **7,7,8,8 - Tetracyanoquinodimethan - N,N,N′,N′ - tetramethyl - p - phenylenediamine complex**
$C_{12}H_4N_4$, $C_{10}H_{16}N_2$
For complete entry see 60.124

60.125 **7,7,8,8 - Tetracyanoquinodimethane - anthracene complex**
$C_{12}H_4N_4$, $C_{14}H_{10}$
R.M.Williams, S.C.Wallwork *Acta Cryst. (B),* **24,** 168, 1968
Residue 1 also classified in 7; residue 2 classified in 60, 26

60.126 **7,7,8,8 - Tetracyanoquinodimethane - copper(ii) 8 - hydroxyquinolate complex**
$C_{12}H_4N_4$, $C_{18}H_{12}CuN_2O_2$
R.M.Williams, S.C.Wallwork *Acta Cryst.*, **23,** 448, 1967
Residue 1 also classified in 7; residue 2 classified in 60, 83, 84

60.C **Ditoluenechromium - bis(7,7,8,8 - tetracyanoquinodimethane) complex**
$C_{12}H_4N_4^-$, $C_{14}H_{16}Cr^+$, $C_{12}H_4N_4$
For complete entry see 60.140

60.127 **bis(7,7,8,8 - Tetracyanoquinodimethan) - N,N,N′,N′ - tetramethyl - p - phenylenediamine complex**
$2C_{12}H_4N_4$, $C_{10}H_{16}N_2$
A.W.Hanson *Acta Cryst. (B),* **24,** 768, 1968
Residue 1 also classified in 7; residue 2 classified in 60, 16

60.128 **Phenazine - iodine complex**
$C_{12}H_8N_2$, I_2
T.Uchida *Bull. Chem. Soc. Jap.*, **40,** 2244, 1967
Residue 1 also classified in 36

60.129 **Phenothiazine - s - trinitrobenzene complex**
$C_{12}H_9NS$, $C_6H_3N_3O_6$
C.J.Fritchie
Amer. Cryst. Assoc., Abstr. Papers (Winter Meeting), 50, 1967
Residue 1 also classified in 41; residue 2 classified in 60, 15

60.130 **Phenothiazine - 3,5 - dinitrobenzoic acid complex**
$C_{12}H_9NS$, $C_7H_4N_2O_6$
C.J.Fritchie, B.L.Trus *Chem. Communic.*, 833, 1968
Residue 1 also classified in 41; residue 2 classified in 60, 13, 15

60.131 **5 - Ethyl - 5 - phenylbarbituric acid - 8 - bromo - 9 - ethyladenine complex**
$C_{12}H_{12}N_2O_3$, $2C_7H_8BrN_5$
S.H.Kim, A.Rich *Proc. Nation..Acad. Sci. U. S. A.*, **60**, 402, 1968
Residue 1 also classified in 43, 60, 44

60.C **Benzotrifuroxan - 13,14 - dithiatricyclo(8,2,1,1(4,7)) tetradeca - 4,6,10,12 - tetraene**
$C_{12}H_{12}S_2$, $C_6N_6O_6$
For complete entry see 60.74

60.C **Chloranil - hexamethylbenzene complex**
$C_{12}H_{18}$, $C_6Cl_4O_2$
For complete entry see 60.71

60.C **1,2,4,5 - Tetracyanobenzene - hexamethylbenzene complex**
$C_{12}H_{18}$, $C_8H_2N_4$
For complete entry see 60.101

60.C **Hexahelicene - 4 - bromo - 2,5,7 - trinitrofluorenone complex**
Phenanthro (3,4 - c) phenanthrene - 4 - bromo - 2,5,7 - trinitro fluorenone complex)
$C_{13}H_4BrN_3O_7$, $C_{26}H_{16}$
For complete entry see 60.156

60.132 **Acridine - cytosine complex monohydrate**
$C_{13}H_9N$, $C_4H_5N_3O$, H_2O
E.Shefter *Science,* **160,** 1351, 1968
Residue 1 also classified in 36; residue 2 classified in 60, 44

60.133 **Anthracene - s - trinitrobenzene**
$C_{14}H_{10}$, $C_6H_3N_3O_6$
D.S.Brown, S.C.Wallwork, A.Wilson *Acta Cryst.*, **17**, 168, 1964
Residue 1 also classified in 26; residue 2 classified in 60, 15

60.134 Anthracene - s - trinitrobenzene (at -100 °C)
$C_{14}H_{10}$, $C_6H_3N_3O_6$
D.S.Brown, S.C.Wallwork, A.Wilson *Acta Cryst.*, **17**, 168, 1964
Residue 1 also classified in 26; residue 2 classified in 60, 15

60.135 Anthracene - pyromellitic dianhydride complex
$C_{14}H_{10}$, $C_{10}H_2O_6$
J.C.A.Boeyens, F.H.Herbstein *J. Phys. Chem.*, **69**, 2160, 1965
Residue 1 also classified in 26; residue 2 classified in 60, 13, 38

60.C 7,7,8,8 - Tetracyanoquinodimethane - anthracene complex
$C_{14}H_{10}$, $C_{12}H_4N_4$
For complete entry see 60.125

60.136 Stilbene - antimony trichloride
$C_{14}H_{12}$, $4Cl_3Sb$
R.Hulme, M.B.Hursthouse *Acta Cryst.*, **21**, A143, 1966
Residue 1 also classified in 19

60.137 Dibenzyl - antimony trichloride (1 - 2)
$C_{14}H_{14}$, $2Cl_3Sb$
R.Hulme, M.B.Hursthouse *Acta Cryst.*, **21**, A143, 1966
Residue 1 also classified in 19

60.138 Dibenzyl - antimony trichloride (1 - 4)
$C_{14}H_{14}$, $4Cl_3Sb$
R.Hulme, M.B.Hursthouse *Acta Cryst.*, **21**, A143, 1966
Residue 1 also classified in 19

60.139 Benzyl sulfide - iodine
$C_{14}H_{14}S$, I_2
C.Romming *Acta Chem. Scand.*, **14**, 2145, 1960
Residue 1 also classified in 11, 19

60.140 Ditoluenechromium - bis(7,7,8,8 - tetracyanoquinodimethane) complex
$C_{14}H_{16}Cr^+$, $C_{12}H_4N_4^-$, $C_{12}H_4N_4$
R.P.Shibaeva, L.O.Atovmyan, M.N.Orfanova
J. Chem. Soc. (D), 1494, 1969
Residue 1 also classified in 74; residue 2 classified in 60, 7

60.141 Pyrene - tetracyanoethylene complex (projections)
$C_{16}H_{10}$, C_6N_4
H.Kuroda, I.Ikemoto, H.Akamatu *Bull. Chem. Soc. Jap.*, **39**, 547, 1966
Residue 1 also classified in 29; residue 2 classified in 60, 7

60.142 **Pyrene - tetracyanoethylene complex**
$C_{16}H_{10}$, C_6N_4
I.Ikemoto, H.Kuroda   *Acta Cryst. (B)*, **24**, 383, 1968
Residue 1 also classified in 29; residue 2 classified in 60, 7

60.143 **Acepleiadylene - s - trinitrobenzene complex**
$C_{16}H_{10}$, $C_6H_3N_3O_6$
A.W.Hanson   *Acta Cryst.*, **21**, 97, 1966
Residue 1 also classified in 29; residue 2 classified in 60, 15

60.C **1,3,7,9 - Tetramethyluric acid - pyrene complex**
$C_{16}H_{10}$, $C_9H_{12}N_4O_3$
For complete entry see 60.110

60.144 **Pyrene - pyromellitic dianhydride**
$C_{16}H_{10}$, $C_{10}H_2O_6$
F.H.Herbstein, J.S.Snyman   *Acta Cryst.*, **21**, A115, 1966
Residue 1 also classified in 29; residue 2 classified in 60, 13, 38

60.145 **N,N - Ethylenebis(salicylaldiminato) copper(ii) - sodium perchlorate complex p - xylene solvate**
$2C_{16}H_{14}CuN_2O_2$, Na^+, ClO_4^-, C_8H_{10}
G.H.W.Milburn, M.R.Truter, B.L.Vickery   *Chem. Communic.*, 1188, 1968
Residue 1 also classified in 78

60.146 **Copper(ii) 8 - hydroxyquinolinate - bis(1,2,4,5 - tetracyanobenzene) complex (photographic data)**
$C_{18}H_{12}CuN_2O_2$, $2C_{10}H_2N_4$
P.Murray-Rust, J.D.Wright   *J. Chem. Soc. (A)*, 247, 1968
Residue 1 also classified in 83, 84; residue 2 classified in 60, 7

60.147 **Copper(ii) 8 - hydroxyquinolinate - bis(1,2,4,5 - tetracyanobenzene) complex (diffractometer data)**
$C_{18}H_{12}CuN_2O_2$, $2C_{10}H_2N_4$
P.Murray-Rust, J.D.Wright   *J. Chem. Soc. (A)*, 247, 1968
Residue 1 also classified in 83, 84; residue 2 classified in 60, 7

60.C **7,7,8,8 - Tetracyanoquinodimethane - copper(ii) 8 - hydroxyquinolate complex**
$C_{18}H_{12}CuN_2O_2$, $C_{12}H_4N_4$
For complete entry see 60.126

60.C **Chloranil - bis(8 - hydroxyquinolinato) palladium(ii)**
$C_{18}H_{12}N_2O_2Pd$, C_6Cl_4O
For complete entry see 60.70

60.148 **Palladium(ii) 8 - hydroxyquinolinate - 1,2,4,5 - tetracyanobenzene complex (disordered)**
$C_{18}H_{12}N_2O_2Pd$, $C_{10}H_2N_4$
B.Kamenar, C.K.Prout, J.D.Wright *J. Chem. Soc. (A),* 661, 1966
Residue 1 also classified in 83, 84; residue 2 classified in 60, 7

60.C **1 - Imino - 4 - (oxime - N - oxide) - benzdifurazan triphenyl phosphine oxide complex**
$C_{18}H_{15}OP$, $C_6H_2N_6O_4$
For complete entry see 60.75

60.149 **Tetra(triphenylphosphine oxide) - tetrakis - (μ(3) - hydroxomolybdenum dicarbonyl nitrosyl) adduct**
$4C_{18}H_{15}OP$, $C_8H_4Mo_4N_4O_{16}$
V.Albano, P.Bellon, G.Ciani, M.Manassero *J. Chem. Soc. (D),* 1242, 1969
Residue 1 also classified in 64

60.150 **Triphenylphosphine sulfide - iodine complex**
$2C_{18}H_{15}PS$, $3I_2$
W.W.Schweikert, E.A.Meyers *J. Phys. Chem.,* **72,** 1561, 1968
Residue 1 also classified in 64, 11

60.151 **Testosterone - p - bromophenol complex**
$C_{19}H_{28}O_2$, C_6H_5BrO
A.Cooper, G.Kartha, E.M.Gopalakrishna, D.A.Norton
Acta Cryst. (B), **25,** 2409, 1969
Residue 1 also classified in 51; residue 2 classified in 60, 17

60.152 **Perylene - fluoranil complex**
$C_{20}H_{12}$, $C_6F_4O_2$
A.W.Hanson *Acta Cryst.,* **16,** 1147, 1963
Residue 1 also classified in 30; residue 2 classified in 60, 18

60.153 **Perylene - tetracyanoethylene complex**
$C_{20}H_{12}$, C_6N_4
I.Ikemoto, H.Kuroda *Bull. Chem. Soc. Jap.,* **40,** 2009, 1967
Residue 1 also classified in 30; residue 2 classified in 60, 7

60.C **1,3,7,9 - Tetramethyluric acid - 3,4 - benzpyrene complex**
$C_{20}H_{12}$, $2C_9H_{12}N_4O_3$
For complete entry see 60.111

60.154 **Perylene - pyromellitic dianhydride complex**
$C_{20}H_{12}$, $C_{10}H_2O_6$
J.C.A.Boeyens, F.H.Herbstein *J. Phys. Chem.*, **69,** 2160, 1965
Residue 1 also classified in 30; residue 2 classified in 60, 13, 38

60.C **1,3,7,9 - Tetramethyluric acid - coronene complex**
$C_{24}H_{12}$, $2C_9H_{12}N_4O_3$
For complete entry see 60.112

60.155 **bis(Diphenylglyoximato) nickel(ii) - iodine complex**
$2C_{24}H_{22}N_4NiO_4$, I_2
A.S.Foust, R.H.Soderberg *J. Amer. Chem. Soc.*, **89,** 5507, 1967
Residue 1 also classified in 83

60.156 **Hexahelicene - 4 - bromo - 2,5,7 - trinitrofluorenone complex**
Phenanthro (3,4 - c) phenanthrene - 4 - bromo - 2,5,7 - trinitro fluorenone
complex)
$C_{26}H_{16}$, $C_{13}H_4BrN_3O_7$
I.R.Mackay, J.M.Robertson, J.G.Sime *J. Chem. Soc. (D),* 1470, 1969
Residue 1 also classified in 30; residue 2 classified in 60, 28

CLATHRATES

61.1   **Ethylene oxide hydrate**
C_2H_4O , $7.2H_2O$
R.K.McMullan, G.A.Jeffrey   *J. Chem. Phys.*, **42,** 2725, 1965
Residue 1 also classified in 38

61.2   **Trimethylamine hydrate**
C_3H_9N , $10.25H_2O$
D.Panke   *J. Chem. Phys.*, **48,** 2990, 1968
Residue 1 also classified in 3

61.3   **Tetrahydrofuran hydrogen sulfide hydrate**
$8C_4H_8O$, $7.33H_2S$, $136H_2O$
T.C.W.Mak, R.K.McMullan   *J. Chem. Phys.*, **42,** 2732, 1965
Residue 1 also classified in 38

61.4   **Diethylamine hydrate**
$C_4H_{11}N$, $8.67H_2O$
T.H.Jordan, T.C.W.Mak   *J. Chem. Phys.*, **47,** 1222, 1967
Residue 1 also classified in 3

61.5   **t - Butylamine hydrate**
$C_4H_{11}N$, $9.75H_2O$
R.K.McMullan, G.A.Jeffrey, T.H.Jordan
J. Chem. Phys., **47,** 1229, 1967
Residue 1 also classified in 3

61.6   **Tetramethylammonium fluoride tetrahydrate**
$C_4H_{12}N^+$, F^- , $4H_2O$
W.J.McLean, G.A.Jeffrey   *J. Chem. Phys.*, **47,** 414, 1967
Residue 1 also classified in 3

61.7   **Tetramethylammonium hydroxide pentahydrate**

$C_4H_{12}N^+$, HO^-, $5H_2O$

R.K.McMullan, T.C.W.Mak, G.A.Jeffrey

J. Chem. Phys., **44,** 2338, 1966

Residue 1 also classified in 3

61.8   **Tetramethylammonium sulfate tetrahydrate (at −26 °C)**

$2C_4H_{12}N^+$, O_4S^{2-}, $4H_2O$

W.J.McLean, G.A.Jeffrey   *J. Chem. Phys.*, **49,** 4556, 1968

Residue 1 also classified in 3

61.9   **Quinol - sulphur dioxide**

$C_6H_6O_2$, O_2S

D.E.Palin, H.M.Powell   *J. Chem. Soc.*, 208, 1947

Residue 1 also classified in 17

See also *Int. Distances*, M 199; *Structure Reports*, **11,** 645, 1947

61.10   **Hexamethylenetetramine hexahydrate**

$C_6H_{12}N_4$, $6H_2O$

T.C.W.Mak   *J. Chem. Phys.*, **43,** 2799, 1965

Residue 1 also classified in 37

61.C   **Tetra - n - butyl ammonium benzoate hydrate**

$C_7H_5O_2^-$, $C_{16}H_{36}N^+$, $39.5H_2O$

For complete entry see 61.14

61.11   **Tri - n - butyl sulfonium fluoride hydrate (cubic form)**

$C_{12}H_{27}S^+$, F^-, $20H_2O$

G.A.Jeffrey, R.K.McMullan   *J. Chem. Phys.*, **37,** 2231, 1962

Residue 1 also classified in 11

61.12   **Tri - n - butyl sulfonium fluoride hydrate (monoclinic form)**

$C_{12}H_{27}S^+$, F^-, $23H_2O$

P.T.Beurskens, G.A.Jeffrey   *J. Chem. Phys.*, **40,** 2800, 1964

Residue 1 also classified in 11

61.13   **Tetra - n - butyl ammonium fluoride hydrate**

$C_{16}H_{36}N^+$, F^-, $32.8H_2O$

R.K.McMullan, M.Bonamico, G.A.Jeffrey

J. Chem. Phys., **39,** 3295, 1963

Residue 1 also classified in 3

61.14   **Tetra - n - butyl ammonium benzoate hydrate**
$C_{16}H_{36}N^+$, $C_7H_5O_2^-$, $39.5H_2O$
M.Bonamico, G.A.Jeffrey, R.K.McMullan
J. Chem. Phys., **37**, 2219, 1962
Residue 1 also classified in 3; residue 2 classified in 13, 61

61.C   **tris(o - Phenylenedioxy) phosphonitrile trimer bromobenzene inclusion compound**
$C_{18}H_{12}N_3O_6P_3$, $0.5C_6H_5Br$
For complete entry see 64.49

61.C   **tris(o - Phenylenedioxy) phosphonitrile trimer benzene inclusion compound**
$C_{18}H_{12}N_3O_6P_3$, $0.5C_6H_6$
For complete entry see 64.50

61.15   **Triphenylphosphine - lithium iodide complex**
$5C_{18}H_{15}OP$, ILi
Y.M.G.Yasin, O.J.R.Hodder, H.M.Powell   *Chem. Communic.*, 705, 1966
Residue 1 also classified in 64

61.16   **4 - p - Hydroxy - 2,2,4 - trimethylthiachroman - ethanol**
$3C_{18}H_{20}OS$, C_2H_6O
D.D.MacNicol, H.H.Mills, F.B.Wilson   *J. Chem. Soc, (D)*, 1332, 1969
Residue 1 also classified in 39

61.17   **trans,anti,trans,anti,trans - Perhydrotriphenylene - chloroform**
$C_{18}H_{30}$, $0.5CHCl_3$
G.Allegra, M.Farina, A.Immirzi, A.Colombo, U.Rossi, R.Broggi, G.Natta
J. Chem. Soc. (B), 1020, 1967
Residue 1 also classified in 29

61.18   **trans - Perhydrotriphenylene cyclohexane solvate**
$C_{18}H_{30}$, $0.39C_6H_{12}$
A.Immirzi, G.Allegra   *Atti Accad. Nazion. Lincei, R. C.,*
Cl. Sci. Fis. Mat. Nat., **43**, 181, 1967
Residue 1 also classified in 29

61.19   **trans,anti,trans,anti,trans - Perhydrotriphenylene - n - heptane**
$C_{18}H_{30}$, $0.225C_7H_{16}$
G.Allegra, M.Farina, A.Immirzi, A.Colombo, U.Rossi, R.Broggi, G.Natta
J. Chem. Soc. (B), 1020, 1967
Residue 1 also classified in 29

61.20 **trans,anti,trans,anti,trans - Perhydrotriphenylene - n - heptane**
$C_{18}H_{30}$, $0.225C_7H_{16}$
G.Allegra, M.Farina, A.Colombo, G.Casagrande-Tettamanti, U.Rossi,
G.Natta *J. Chem. Soc. (B)*, 1028, 1967
Residue 1 also classified in 29

61.21 **Copper(ii) bis - 1 - ephedrine - benzene**
$C_{20}H_{26}CuN_2O_2$, $0.67C_6H_6$
Y.Amano, K.Osaki, T.Watanabe *Bull. Chem. Soc. Jap.*, **37,** 1363, 1964
Residue 1 also classified in 83, 84

61.22 **Tetra iso - amyl ammonium fluoride hydrate**
$C_{20}H_{44}N^+$, F^- , $38H_2O$
D.Feil, G.A.Jeffrey *J. Chem. Phys.*, **35,** 1863, 1961
Residue 1 also classified in 3

BORON COMPOUNDS

62.1   **Decachloro - 1,7 - dicarbaclovododecaborane**
$C_2H_2B_{10}Cl_{10}$
J.A.Potenza, W.N.Lipscomb
Proc. Nation. Acad. Sci. U. S. A., **56,** 1917, 1966

62.2   **Diboron tetrachloride - ethylene addition compound (at −70 ° C)**
1,2 - bis - Dichloroborane - ethane
$C_2H_4B_2Cl_4$
E.B.Moore, W.N.Lipscomb *Acta Cryst.*, **9,** 668, 1956
See also *Int. Distances,* M 79s; *Structure Reports,* **20,** 528, 1956

62.3   **Octachloro - 1,2 - dicarbaclovododecaborane**
$C_2H_4B_{10}Cl_8$
J.A.Potenza, W.N.Lipscomb *Inorg. Chem.*, **3,** 1673, 1964

62.4   **Octachloro - 1,2 - dicarbaclovododecaborane (refinement)**
$C_2H_4B_{10}Cl_8$
G.S.Pawley *Acta Cryst.*, **20,** 631, 1966

62.5   **Dimethylsulfoxide - boron trifluoride**
$C_2H_6BF_3O_5$
E.L.McGandy *Dissert. Abstr.*, **22,** 754, 1961
Also classified in 11

62.6   **Carbon monoxide - borane - monomethylamine complex**
$C_2H_7BNO^-$, CH_6N^+
R.W.Parry, C.E.Nordman, J.C.Carter, G.Terhaar
Adv. Chem. Ser., 302, 1964
Residue 2 classified in 3

62.7   **2,3 - Dicarbahexaborane(8)**
$C_2H_8B_4$
F.P.Boer, W.E.Streib, W.N.Lipscomb *Inorg. Chem.*, **3,** 1666, 1964

62.8 **2,3 - Dicarbahexaborane(8) (refinement)**
$C_2H_8B_4$
G.S.Pawley *Acta Cryst.*, **20**, 631, 1966

62.9 **Tribromo - o - carborane**
$C_2H_9B_{10}Br_3$
J.A.Potenza, W.N.Lipscomb *Inorg. Chem.*, **5**, 1478, 1966

62.10 **Dibromo - o - carborane**
$C_2H_{10}B_{10}Br_2$
J.A.Potenza, W.N.Lipscomb *Inorg. Chem.*, **5**, 1471, 1966

62.11 **9,10 - Dibromo - 1,7 - dicarbadodecaborane(12)**
$C_2H_{10}B_{10}Br_2$
H.Beall, W.N.Lipscomb *Inorg. Chem.*, **6**, 874, 1967

62.12 **Iodo - o - dicarbadodecaborane**
$C_2H_{11}B_{10}I$
V.G.Andrianov, V.I.Stanko, Yu.T.Struchkov, A.I.Klimova
Zh. Strukt. Khim., **8**, 707, 1967

62.13 **2,3 - Dimethylpentaborane**
$C_2H_{13}B_5$
L.B.Friedman, W.N.Lipscomb *Inorg. Chem.*, **5**, 1752, 1966

62.14 **Iodo - m - dicarbadodecaborane**
$C_2H_{13}B_{10}I$
V.G.Andrianov, V.I.Stanko, Yu.T.Struchkov
Zh. Strukt. Khim., **8**, 558, 1967

62.15 **Acetonitrile - tridecahydrononaborane**
$C_2H_{16}B_9N$
F.E.Wang, P.G.Simpson, W.N.Lipscomb
J. Chem. Phys., **35**, 1335, 1961
Also classified in 7

62.16 **1 - Ethyldecaborane**
$C_2H_{18}B_{10}$
A.Perloff *Acta Cryst.*, **17**, 332, 1964

62.17 **Trimethylamine boron trichloride**
$C_3H_9BCl_3N$
H.Hess *Acta Cryst. (B)*, **25**, 2338, 1969

62.18   **B - Trimethylborazole**
$C_3H_9B_3N_3$
K.Anzenhofer   *Molec. Phys.*, **11**, 495, 1966

62.19   **Trimethylamine triborane**
$C_3H_{16}B_3N$
H.G.Norment   *Acta Cryst.*, **14**, 1216, 1961

62.20   **Dimethylamino - boron dichloride dimer**
$C_4H_{12}B_2Cl_4N_2$
H.Hess   *Z. Kristallogr.*, **118**, 361, 1963

62.21   **Dimethylaminoboron difluoride**
$C_4H_{12}B_2F_4N_2$
A.C.Hazell   *J. Chem. Soc. (A)*, 1392, 1966

62.22   **2,3 - Dimethyl - 2,3 - dicarbahexaborane**
$C_4H_{12}B_4$
F.P.Boer, W.E.Streib, W.N.Lipscomb   *Inorg. Chem.*, **3**, 1666, 1964

62.23   **1,7 - Dimethyl - 1,7 - dicarbaclovo - octaborane(8)**
$C_4H_{12}B_6$
H.Hart, W.N.Lipscomb   *Inorg. Chem.*, **7**, 1070, 1968

62.24   **8,9,10,12 - Tetrabromo - C,C' - dimethyl - o - carborane**
$C_4H_{12}B_{10}Br_4$
J.A.Potenza, W.N.Lipscomb   *Inorg. Chem.*, **5**, 1483, 1966

62.25   **1,6 - Dimethyl - 1,6 - dicarbaclovononaborane(9)**
$C_4H_{13}B_7$
T.F.Koetzle, F.E.Scarbrough, W.N.Lipscomb   *Inorg. Chem.*, **7**, 1076, 1968

62.26   **1,2 - bis(Bromomethyl) - 1,2 - dicarbaclovododecaborane**
$C_4H_{14}B_{10}Br_2$
D.Voet, W.N.Lipscomb   *Inorg. Chem.*, **3**, 1679, 1964

62.27   **1,2 - bis(Bromomethyl) - 1,2 - dicarbaclovododecaborane (refinement)**
$C_4H_{14}B_{10}Br_2$
G.S.Pawley   *Acta Cryst.*, **20**, 631, 1966

62.28   **6,9 - Dimethylcarborane**
$C_4H_{15}B_9$
C.Tsai, W.E.Streib   *J. Amer. Chem. Soc.*, **88**, 4513, 1966

62.29 **bis(Dimethylamine)boronium chloride**
$C_4H_{16}BN_2^+$, Cl^-
Z.V.Zvonkova, Z.P.Povet'eva, V.M.Vozzennikov, V.P.Gluskova, V.I.Jakovenko, A.N.Khvatkina *Acta Cryst.*, **21**, A155, 1966
Residue 1 also classified in 3

62.30 **Dimethylaminoborine dimer**
$C_4H_{16}B_2N_2$
P.J.Schapiro *Dissert. Abstr.*, **22**, 2607, 1962

62.C **Tetramethylammonium bis(π - 5,9,10 - tribromo - (1) - 2,3 - dicarbollyl) cobalt(iii)**
$C_4H_{16}B_{18}Br_6Co^-$, $C_4H_{12}N^+$
For complete entry see 3.31

62.31 **6,8 - Dimethyl - 6,8 - dicarbanonaborane**
$C_4H_{17}B_7$
D.Voet, W.N.Lipscomb *Inorg. Chem.*, **6**, 113, 1967

62.32 **Tetramethylbiphosphine - bis(monoborane) (high temperature form)**
$C_4H_{18}B_2P_2$
H.L.Carrell, J.Donohue *Acta Cryst. (B)*, **24**, 699, 1968
Also classified in 64

62.33 **Cesium bis(1,2 - dicarbollyl) cobaltate**
$C_4H_{22}B_{18}Co^-$, Cs^+
A.Zalkin, T.E.Hopkins, D.H.Templeton *Inorg. Chem.*, **6**, 1911, 1967
Residue 1 also classified in 71

62.C **Triphenylmethylphosphonium bis((3) - 1,2 - dicarbollyl) cuprate(iii)**
$C_4H_{22}B_{18}Cu^-$, $C_{19}H_{18}P^+$
For complete entry see 64.56

62.34 **Tetraethylammonium bis - (3) - 1,2 - dicarbollyl cuprate(ii)**
$C_4H_{22}B_{18}Cu^{2-}$, $2C_8H_{20}N^+$
R.M.Wing *J. Amer. Chem. Soc.*, **89**, 5599, 1967
Residue 1 also classified in 71; residue 2 classified in 3

62.35 **bis(o - Dodecacarborane)**
$C_4H_{22}B_{20}$
L.H.Hall, A.Perloff, F.A.Maver, S.Block
J. Chem. Phys., **43**, 3911, 1965

62.36 **(1) - 1',5 - bis(Acetonitrile)ollyl di - icosahedralborane acetonitrile solvate**
$C_4H_{22}B_{20}N_2$, C_2H_3N
J.H.Enemark, L.B.Friedman, W.N.Lipscomb *Inorg. Chem.*, **5**, 2165, 1966

62.37   **3 - Ethylamine - 5,6 - μ - ethyl - amino - octaborane(12)**
Ethylammonium - ethylimmonium undecahydro - octaborane
$C_4H_{23}B_8N_2$
R.Lewin, P.G.Simpson, W.N.Lipscomb   *J. Chem. Phys.*, **39**, 1532, 1963

62.38   **bis(Dimethylsulfide) - dodecahydrodecaborane**
$C_4H_{24}B_{10}S_2$
D.E.Sands, A.Zalkin   *Acta Cryst.*, **15**, 410, 1962
Also classified in 11

62.39   **Iron(ii) P - methyl - (3) - 1,7 - carbaphosphollide**
$C_4H_{26}B_{18}FeP_2$
L.J.Todd, I.C.Paul, J.L.Little, P.S.Welcker, C.R.Peterson
J. Amer. Chem. Soc., **90**, 4489, 1968
Also classified in 64, 71

62.40   **meso - Pentane - 2,4 - diol borate**
$C_5H_{11}BO_3$
S.Kuribayashi   *Bull. Chem. Soc. Jap.*, **39**, 2784, 1966

62.41   **Cesium π - (1) - 2,3 - dicarbollyl rhenium tricarbonyl**
$C_5H_{11}B_9O_3Re^-$, Cs^+
A.Zalkin, T.E.Hopkins, D.H.Templeton   *Inorg. Chem.*, **5**, 1189, 1966
Residue 1 also classified in 71

62.42   **p - Bromophenylboronic acid**
$C_6H_6BBrO_2$
Z.V.Zvonkova, V.P.Gluskova   *Kristallografija*, **3**, 559, 1958
See also *Int. Distances*, M 128s; *Structure Reports*, **22**, 691, 1958

62.43   **1,8,10,9 - Triazaboradecalin**
$C_6H_{14}BN_3$
G.J.Bullen   *J. Chem. Soc. (A)*, 404, 1969

62.44   **Dimethylaminoborine trimer**
$C_6H_{24}B_3N_3$
L.M.Trefonas, F.S.Mathews, W.N.Lipscomb   *Acta Cryst.*, **14**, 273, 1961

62.C π - Cyclopentadienyl - π - (1)2,3 - dicarbollyl iron(iii)
$C_7H_{16}B_9Fe$
For complete entry see 73.5

62.45   **Dimethylphosphinoborine tetramer**
$C_8H_{32}B_4P_4$
P.Goldstein, R.A.Jacobson   *J. Amer. Chem. Soc.*, **84**, 2457, 1962
Also classified in 64

62.46 **tris - 1,3,5 - (Dimethylamino) - 1,3,5 - triboracyclohexane**
$C_9H_{24}B_3N_3$
H.Hess *Acta Cryst. (B)*, **25**, 2334, 1969

62.C **Diphenyliodonium fluoroborate**
$C_{12}H_{10}I^+$, BF_4^-
For complete entry see 19.38

62.47 **Hexaethyl borazine**
$C_{12}H_{30}B_3N_3$
M.A.Viswamitra, S.N.Vaidya *Z. Kristallogr.*, **121**, 472, 1965

62.C **Triethylamine - (tricobalt enneacarbonyl carbon) - oxyborane**
$C_{16}H_{17}BCo_3NO_{10}$
For complete entry see 71.40

62.48 **3,6 - Di - tert - butyl - 1,2,4,5 - tetra - azadiborane dimer**
$C_{16}H_{44}B_4N_8$
P.C.Thomas, I.C.Paul *Chem. Communic.*, 1130, 1968

62.49 **Tri - (1,3,2 - benzodioxaborol - 2 - yl)amine**
$C_{18}H_{12}B_3NO_6$
G.J.Bullen, P.R.Mallinson *Chem. Communic.*, 1076, 1967

62.50 **Tetra - B - isothiocyanato tetra - N - t - butylborazocine**
$C_{20}H_{36}B_4N_8S_4$
P.T.Clarke, H.M.Powell *J. Chem. Soc. (B)*, 1172, 1966

62.51 **Rubidium tetraphenylborate**
$C_{24}H_{20}B^-$, Rb^+
Ya.Ozol, S.Vimba, A.Ievins *Kristallografija*, **7**, 362, 1962

62.C **Cobalt(ii) chloride tetraphenylborate tris(o - diphenylphosphinophenyl) phosphine**
$C_{24}H_{20}B^-$, $C_{54}H_{42}ClCoP_4^+$
For complete entry see 86.107

62.52 **B,B - Di - iodo - P,P - diphenylphosphinoborine**
$C_{24}H_{20}B_2I_4P_2$
G.J.Bullen, P.R.Mallinson *J. Chem. Soc. (D)*, 132, 1969
Also classified in 64

62.53 **1,3 - Di - p - chlorophenyl - 2 - triethylcarbinyl - 4 - ethyl - 5,5 - diethyl - 1,3 - diaza - 2,4 - diborolidine**
$C_{26}H_{38}B_2Cl_2N_2$
C.Tsai, W.E.Streib *Tetrahedron Letters*, 669, 1968

SILICON COMPOUNDS

63.1   Cyclotetra(methylenedichlorosilicon)
$C_4H_8Cl_8Si_4$
E.Krahe   *Thesis, Munster*, 1965

63.2   **Diethylsilanediol**
$C_4H_{12}O_2Si$
M.Kakudo, T.Watase   *J. Chem. Phys.*, **21**, 167, 1953
See also *Int. Distances*, M 175; *Structure Reports*, **17**, 679, 1953

63.3   **Tetra(methylsilicon) hexasulfide**
$C_4H_{12}S_6Si_4$
J.C.J.Bart, J.J.Daly   *Chem. Communic.*, 1207, 1968
Also classified in 11

63.4   **Diallylsilanediol**
$C_6H_{12}O_2Si$
N.Kasai, M.Kakudo   *Bull. Chem. Soc. Jap.*, **27**, 605, 1954
See also *Int. Distances*, M 206; *Structure Reports*, **18**, 681, 1954

63.C   **Dibromotrimethylsiloxy aluminium**
$C_6H_{18}Al_2Br_4O_2$
For complete entry see 68.11

63.5   **Hexamethylcyclotrisiloxane**
$C_6H_{18}O_3Si_3$
G.Pyronel   *Atti Accad. Nazion. Lincei, R. C.,*
Cl. Sci. Fis. Mat. Nat., **16**, 231, 1954
See also *Int. Distances*, M 209; *Structure Reports*, **18**, 682, 1954

63.C   **Pentabromotrialuminium methylsiloxane**
$C_8H_{24}Al_3Br_5O_6Si_4$
For complete entry see 68.16

63.6   bis - Tetramethyl - disilanilene - dioxide
$C_8H_{24}O_2Si_4$
T.Takano, N.Kasai, M.Kakudo   *Bull. Chem. Soc. Jap.*, **36**, 585, 1963

63.7   Octamethylcyclotetrasiloxane (at −50 ° C)
$C_8H_{24}O_4Si_4$
H.Steinfink, B.Post, I.Fankuchen   *Acta Cryst.*, **8**, 420, 1955
See also *Int. Distances*, M 220; *Structure Reports*, **19**, 549, 1955

63.8   Octamethylspiro - 5,5 - pentasiloxane
$C_8H_{24}O_6Si_5$
W.L.Roth, D.Harker   *Acta Cryst.*, **1**, 34, 1948
See also *Int. Distances*, M 220; *Structure Reports*, **11**, 620, 1948

63.9   Octamethyl - octa(siloxane)
$C_8H_{24}O_8Si_8$
T.Higuchi, A.Shimada   *Bull. Chem. Soc. Jap.*, **39**, 1316, 1966

63.10   Octa - (methylsilsesquioxane)
$C_8H_{24}O_{12}Si_8$
K.Larsson   *Ark. Kemi*, **16**, 203, 1960

63.11   Octamethylcyclo - tetrasilazane
$C_8H_{28}N_4Si_4$
G.S.Smith, L.E.Alexander   *Acta Cryst.*, **16**, 1015, 1963

63.12   Tetrafluorobis(pyridine)silicon
$C_{10}H_{10}F_4N_2Si$
V.A.Bain, R.C.G.Killean, M.Webster   *Acta Cryst. (B)*, **25**, 156, 1969

63.13   p - bis(Dimethylhydroxysilyl)benzene
$C_{10}H_{18}O_2Si_2$
L.E.Alexander, M.G.Northolt, R.Engmann
J. Phys. Chem., **71**, 4298, 1967

63.14   Tetramethyl - NN′ - bis(trimethylsilyl) cyclodisilazane
$C_{10}H_{30}N_2Si_4$
P.J.Wheatley   *J. Chem. Soc.*, 1721, 1962

63.15   Dimethylsilylamine pentamer
$C_{10}H_{45}N_5Si_5$
R.Rudman, W.C.Hamilton, S.Novick, T.D.Goldfarb
J. Amer. Chem. Soc., **89**, 5157, 1967

63.16 **Phenyl - (2,2',2'' - nitriloethoxy)silane**
$C_{12}H_{17}NO_3Si$
J.W.Turley, F.P.Boer *J. Amer. Chem. Soc.*, **90,** 4026, 1968

63.17 **2,2,4,4,6,6,8,8 - Octamethyl - 2,4,6,8 - tetrasilabicyclo(3.3.0)oct - 1(5) - ene**
$C_{12}H_{28}Si_4$
H.G.von Schnering, E.Krahe, G.Fritz
Z. Anorg. Allg. Chem., **365,** 113, 1969

63.18 **Tricyclic methyldisilanylenesiloxane**
$C_{12}H_{36}O_6Si_8$
T.Higuchi, A.Shimada *Bull. Chem. Soc. Jap.*, **40,** 752, 1967

63.19 **2,6 - Di - p - bromophenyl - 1,1 - dimethyl - 1 - sila - 2,4,6 - triazacyclohexane - 3,5 - dione acetone solvate**
$C_{16}H_{15}BrN_3O_2Si$, C_3H_6O
J.J.Daly, W.Fink *J. Chem. Soc.*, 4958, 1964

63.20 **(–) - α - Naphthylphenylmethylsilylfluoride (absolute configuration)**
$C_{17}H_{15}FSi$
T.Ashida, R.Pepinsky, Y.Okaya *Acta Cryst.*, **16,** A48, 1963

63.21 **(+) - α - Naphthylphenylmethylsilylfluoride**
$C_{17}H_{15}F\dot{S}i$
Y.Okaya, T.Ashida *Acta Cryst.*, **20,** 461, 1966

63.22 **(–) - α - Naphthylphenylmethylsilane**
$C_{17}H_{16}Si$
Y.Okaya, T.Ashida *Acta Cryst.*, **20,** 461, 1966

63.C **Tetramethylammonium bis(o - phenylenedioxy) - phenylsiliconate**
$C_{18}H_{13}O_4Si^-$, $C_4H_{12}N^+$
For complete entry see 3.32

63.23 **hexakis(Trimethylsilyl) - 2,4 - diamino - 1,3,2,4 - diazadiboretin**
$C_{18}H_{54}B_2N_4Si_6$
H.Hess *Acta Cryst. (B),* **25,** 2342, 1969

63.24 **tris(Hexamethyldisilylamine) iron(iii)**
$C_{18}H_{54}FeN_3Si_6$
D.C.Bradley, M.B.Hursthouse, P.F.Rodesiler *J. Chem. Soc. (D)*, 14, 1969
Also classified in 83

63.C **1,1' - bis(Pentamethyldisilanyl) - ferrocene**
$C_{20}H_{38}FeSi_4$
For complete entry see 73.94

63.25 **Triphenyl trimethyl disiloxane**
$C_{21}H_{24}OSi_2$
H.Kandler, E.Wolfel   *Acta Cryst.*, **16**, A74, 1963

63.26 **Phenyl(2,2′,2″ - nitrilotriphenoxy)silane**
$C_{24}H_{17}NO_3Si$
F.P.Boer, J.W.Turley, J.J.Flynn   *J. Amer. Chem. Soc.*, **90**, 5102, 1968

63.27 **1,1,4,4 - Tetramethyl - 2,3,5,6 - tetraphenyl - 1,4 - disilicocyclohexa - 2,5 - diene**
$C_{32}H_{32}Si_2$
N.G.Bokii, Yu.T.Struchkov   *Zh. Strukt. Khim.*, **6**, 571, 1965

63.28 **1,1,4,4 - Tetramethyl - 2,3,5,6 - tetraphenyl - 1,4 - disilacyclohexadiene**
$C_{32}H_{32}Si_2$
M.E.Vol'pin, V.G.Dulova, Yu.T.Struchkov, N.K.Bokiy, D.N.Kursanov
J. Organometal. Chem., **8**, 87, 1967

63.29 **1 - Methyldiphenylsilyl - 3 - (1,1,3 - trimethyl - 3,3 - diphenylsilazanyl) - 2,2,4,4 - tetramethyl cyclodisilazane**
$C_{32}H_{45}N_3Si_5$
G.Chioccola, J.J.Daly   *J. Chem. Soc. (A)*, 1658, 1968

63.30 **N - Ethyl - 2,2,4,4,6,6 - hexaphenyl - 3 - azacyclotrisiloxane**
$C_{38}H_{35}NO_2Si_3$
W.Fink, P.J.Wheatley   *J. Chem. Soc. (A)*, 1517, 1967

PHOSPHORUS COMPOUNDS

64.1 **2,4 - Dithio - 2,4 - dichloro - 1,3 - dimethyl - cyclodiphosphazan**
$C_2H_6Cl_2N_2P_2S_2$
J.Weiss, G.Hartmann *Z. Naturforsch., B,* **21,** 891, 1966

64.2 **N - Methyltrichlorophosphineimine dimer**
$C_2H_6Cl_6N_2P_2$
H.Hess, D.Forst *Z. Anorg. Allg. Chem.,* **342,** 240, 1966

64.3 **N - Methyltrichlorophosphineimine dimer**
$C_2H_6Cl_6N_2P_2$
L.G.Hoard, R.A.Jacobson *J. Chem. Soc. (A),* 1203, 1966

64.4 **Potassium - O - O - dimethyl - phosphordithioate**
$C_2H_6O_2PS_2^-$, K^+
P.Coppens, C.H.MacGillavry, S.G.Hovenkamp, H.Douwes
Acta Cryst., **15,** 765, 1962

64.5 **Methyl metadithiophosphonate**
$C_2H_6P_2S_4$
J.J.Daly *J. Chem. Soc.,* 4065, 1964

64.6 **Dimethylphosphinic acid**
$C_2H_7O_2P$
F.Giordano, A.Ripamonti *Acta Cryst.,* **22,** 678, 1967

64.7 **2 - Aminoethylphosphonic acid**
$C_2H_8NO_3P$
Y.Okaya *Acta Cryst.,* **20,** 712, 1966
Also classified in 3

64.8 **N,N' - Dimethyl - N - (thiophosphordiamido) - thiophosphortriamide**
$C_2H_{13}N_5P_2S_2$
M.L.Ziegler, J.Weiss *Z. Anorg. Allg. Chem.,* **361,** 136, 1968

64.9 **(Triaminophosphine) - (di(methylimino) aminophosphine) nitride**
$C_2H_{16}N_7P_2^+$, I^-
M.L.Ziegler $Z. Anorg. Allg. Chem.$, 362, 257, 1968

64.C **Calcium 1,3 - diphosphorylimidazole**
$C_3H_3N_2O_6P_2^{3-}$, $1.5Ca^{2+}$, $6H_2O$
For complete entry see 32.9

64.C **Trimethylphosphine oxide antimony(v) chloride**
$C_3H_9Cl_5OPSb$
For complete entry see 66.6

64.10 **Nitrilotrimethylene triphosphonic acid**
$C_3H_{12}NO_9P_3$
J.J.Daly, P.J.Wheatley $J. Chem. Soc. (A)$, 212, 1967
Also classified in 3

64.11 **Tetra(trifluoromethyl)cyclotetraphosphine**
$C_4F_{12}P_4$
G.J.Palenik, J.Donohue $Acta Cryst.$, 15, 564, 1962

64.12 **Disodium N - phosphorylcreatinine hydrate**
$C_4H_8N_3O_5P^{2-}$, $2Na^+$, $4.5H_2O$
J.R.Herriott, W.E.Love $Acta Cryst. (B)$, 24, 1014, 1968
Residue 1 also classified in 48

64.13 **Tetramethyl diphosphine disulfide**
$C_4H_{12}P_2S_2$
C.Pedone, A.Sirigu $J. Chem. Phys.$, 47, 339, 1967

64.C **Tetramethylbiphosphine - bis(monoborane) (high temperature form)**
$C_4H_{18}B_2P_2$
For complete entry see 62.32

64.C **Iron(ii) P - methyl - (3) - 1,7 - carbaphosphollide**
$C_4H_{26}B_{18}FeP_2$
For complete entry see 62.39

64.14 **pentakis - (Trifluoromethyl) - cyclopentaphosphine, (at -100 ° C)**
$C_5F_{15}P_5$
C.J.Spencer, W.N.Lipscomb $Acta Cryst.$, 14, 250, 1961

64.15 **2α - Bromo - 5β - bromomethyl - 5α - methyl - 2β - oxo - 11,3,2 - dioxaphosphorinane**
$C_5H_9Br_2O_3P$
T.A.Beineke $Acta Cryst. (B)$, 25, 413, 1969

64.16   Cyclohexane - 1a,3a,5a - thiophosphonic acid ester
$C_6H_9O_3PS$
P.Andersen, K.E.Hjortaas   *Acta Chem. Scand.*, **14**, 829, 1960

64.17   3 - α - Oxo - 3 - β - hydrido - 7 - β - hydroxy - 2,4 - dioxa - 3 -
phosphabicyclo(3.3.1)nonane
$C_6H_{11}O_4P$
D.M.Nimrod   *Dissert. Abstr. (B)*, **28**, 3672, 1968

64.18   Triethylphosphine sulfide
$C_6H_{15}PS$
M.van Meerssche, A.Leonard   *Acta Cryst.*, **12**, 1053, 1959
See also *Int. Distances*, M 134s; *Structure Reports*, **23**, 607, 1959

64.19   Triethylphosphine selenide
$C_6H_{15}PSe$
M.van Meerssche, A.Leonard   *Bull. Soc. Chim. Belges*, **69**, 45, 1960

64.20   2,2,2,4,6,8,8,8 - Octachloro - 1,3,5,7,9,10 - hexamethyl - 1,3,5,7,9,10 -
hexaza - 2,4,6,8 - tetraphospha(2,4,6,8 - P(v)) dispiro(3.1.3.1) - decane
$C_6H_{18}Cl_8N_6P_4$
J.Weiss, G.Hartmann   *Z. Anorg. Allg. Chem.*, **351**, 152, 1967

64.21   2,4,6 - Trimethoxy - 1,3,5 - trimethyl - 2,4,6 - trioxo - cyclotriphosphazane
$C_6H_{18}N_3O_6P_3$
G.B.Ansell, G.J.Bullen   *J. Chem. Soc. (A)*, 3026, 1968

64.22   Dimethylphosphinoborine trimer
$C_6H_{24}B_3P_3$
W.C.Hamilton   *Acta Cryst.*, **8**, 199, 1955
See also *Int. Distances*, M 210; *Structure Reports*, **19**, 545, 1955

64.23   Carbon disulfide - triethylphosphine complex
$C_7H_{15}PS_2$
T.N.Margulis, D.H.Templeton   *J. Chem. Phys.*, **36**, 2311, 1962

64.24   bis(Cyclotetramethylene) diphosphine disulfide
$C_8H_{16}P_2S_2$
J.D.Lee, G.W.Goodacre   *Acta Cryst. (B)*, **25**, 2127, 1969

64.25   2,3,4 - Trimethylpentane - 2,4 - phosphinic acid monohydrate
$C_8H_{17}O_2P$, H_2O
D.D.Swank, C.N.Caughlan   *Chem. Communic.*, 1051, 1968

64.26 **Tetraethyl diphosphine disulfide**
$C_8H_{20}P_2S_2$
S.N.Dutta, M.M.Woolfson *Acta Cryst.*, **14,** 178, 1961

64.27 **bis(Diethylthiophosphoryl) diselenide**
$C_8H_{20}P_2S_2Se_2$
S.Husebye *Acta Chem. Scand.*, **29,** 51, 1966

64.28 **Octamethoxycyclotetraphosphazatetraene**
$C_8H_{24}N_4O_8P_4$
G.B.Ansell, G.J.Bullen *Chem. Communic.*, 430, 1966

64.29 **Octamethylcyclotetraphosphonitrile (at 140 °K)**
$C_8H_{24}N_4P_4$
M.W.Dougill *J. Chem. Soc.,*.   1961

64.C **Octamethylphosphonitrilium copper(ii) trichloride**
$C_8H_{25}Cl_3CuN_4P_4$
For complete entry see 83.66

64.30 **Octamethylphosphonitrilium tetrachlorocobaltate**
$2C_8H_{25}N_4P_4{}^+$, Cl_4Co^{2-}
J.Trotter, S.H.Whitlow, N.L.Paddock *J. Chem. Soc. (D),* 695, 1969

64.C **Dimethylphosphinoborine tetramer**
$C_8H_{32}B_4P_4$
For complete entry see 62.45

64.31 **Methyl phenyl pholanium iodide**
$C_{11}H_{16}P^+$, I^-
E.Alver, B.H.Holtedahl *Acta Chem. Scand.,* **21,** 359, 1967

64.32 **2,2 - Diphenyl - 4,4,6,6 - tetrachloro - cyclotriphosphazatriene**
$C_{12}H_{10}Cl_4N_3P_3$
N.V.Mani, F.R.Ahmed, W.H.Barnes *Acta Cryst.*, **19,** 693, 1965

64.33 **1,1 - Diphenylphosphonitrilic fluoride trimer**
$C_{12}H_{10}F_4N_3P_3$
C.W.Allen, I.C.Paul, T.Moeller *J. Amer. Chem. Soc.*, **89,** 6361, 1967

64.34 **Octamethyltetra - amino - 2,3 - diphosphinyl buta - 1,3 - diene**
$C_{12}H_{28}N_4O_2$
L.Born *Acta Cryst. (B),* **25,** 1460, 1969

64.C **Lead(ii) O,O - di - isopropyl phosphorodithioate**
$C_{12}H_{28}O_4P_2PbS_4$
For complete entry see 69.26

64.35   **1,1,3,3 - Tetra(isopropylamino) - 5,5 - dichloro - cyclotriphosphatriazene**
$C_{12}H_{32}Cl_2N_7P_3$
N.V.Mani, A.J.Wagner   *Chem. Communic.*, 658, 1968

64.36   **9,10 - Dihydro - 9 - hydroxy - 9 - phosphaphenanthrene - 9 - oxide**
$C_{13}H_{11}O_2P$
P.J.Wheatley   *J. Chem. Soc.*, 3733, 1962

64.37   **Fluorodiphenyl - N - methylphosphine imide**
$C_{13}H_{13}FNP$
G.W.Adamson, J.C.J.Bart   *J. Chem. Soc. (D)*, 1036, 1969

64.38   **Methyl diphenylthiophosphinite**
$C_{13}H_{13}OPS$
G.Lepicard, D.de Saint-Giniez-Liebig, A.Laurent, C.Rerat
Acta Cryst. (B), **25**, 617, 1969

64.39   **Methyl diphenylselenophosphinite**
$C_{13}H_{13}OPSe$
G.Lepicard, D.de Saint-Giniez-Liebig, A.Laurent, C.Rerat
Acta Cryst. (B), **25**, 617, 1969

64.40   **2,6 - Dimethyl - 4 - phenylphosphabenzene**
$C_{13}H_{13}P$
J.C.J.Bart, J.J.Daly   *Angew. Chem.*, **80**, 843, 1968

64.C   **1,2,4 Triazole - 5 - methylamino - 3 - phenyl - 1 - bis(dimethylamido) phosphate**
$C_{13}H_{21}N_6OP$
For complete entry see 32.35

64.C   **1,2,4 - Triazole - 5 - phenylamino - 3 - methyl - 2 - bis(dimethylamido) phosphate**
$C_{13}H_{21}N_6OP$
For complete entry see 32.37

64.41   **2,2,4,4 - Tetrafluoro - 1,3 - dimethyl - 2,4 - diphenyl - 1,3 - diaza - 2,4 - diphosphacyclobutane**
$C_{14}H_{16}F_4N_2P_2$
J.W.Cox, E.R.Corey   *Chem. Communic.*, 123, 1967

64.42   **N,N - Dimethyldiphenylphosphinamide**
$C_{14}H_{16}NOP$
M.-ul-Haque, C.N.Caughlan   *Chem. Communic.*, 921, 1966

64.43   **1,2 - Dimethyl - 1,2 - diphenyldiphosphine disulfide (meso form)**
$C_{14}H_{16}P_2S_2$
P.J.Wheatley   *J. Chem. Soc.*, 523, 1960

64.44   **2,2,3,4,4 - Pentamethyl - 1 - phenylphosphetan - 1 - oxide**
$C_{14}H_{21}OP$
M.-ul-Haque, C.N.Caughlan   *Chem. Communic.*, 1228, 1968

64.C   **1,2,4 - Triazole - 5 - phenylamino - 3 - ethyl - 2 - bis(dimethylamido)**
phosphate
$C_{14}H_{23}N_6OP$
For complete entry see 32.39

64.C   **1,2,4 Triazole - 5 - phenylamino - 3 - ethyl - 1 - bis(dimethylamido)**
phosphate
$C_{14}H_{23}N_6OP$
For complete entry see 32.40

64.C   **1,2,4 - Triazole - 5 - phenylamino - 3 - isopropyl - 2 - bis(dimethylamido)**
phosphate
$C_{15}H_{25}N_6OP$
For complete entry see 32.43

64.C   **1,2,4 - Triazole - 5 - phenylamino - 3 - isopropyl - 1 - bis(dimethylamido)**
phosphate hemihydrate
$C_{15}H_{25}N_6OP$, $0.5H_2O$
For complete entry see 32.44

64.45   **Hexadecamethoxycyclo - octaphosphonitrile**
$C_{16}H_{48}N_8O_{16}P_8$
N.L.Paddock, J.Trotter, S.H.Whitlow   *J. Chem. Soc. (A)*, 2227, 1968

64.46   **octakis(Dimethylamino) cyclotetraphosphazatetraene**
$C_{16}H_{48}N_{12}P_4$
G.J.Bullen   *J. Chem. Soc.*, 3193, 1962

64.47   **6 - Chloro - 5,6,7,12 - tetrahydro - 2,5,7,10 - tetramethyl**
dibenzo(d,g)(1,3,2)diazaphosphocine - 6 - oxide
$C_{17}H_{20}ClN_2OP$
C.Y.Cheng, R.A.Shaw, T.S.Cameron, C.K.Prout
Chem. Communic., 616, 1968
Also classified in 42

64.48 (+) - Methyl - n - propyl - phenyl - benzyl - phosphonium bromide (absolute configuration)

$C_{17}H_{22}BrP$

A.F.Peerdeman, J.P.C.Holst, L.Horner, H.Winkler

Tetrahedron Letters, 811, 1965

64.49 tris(o - Phenylenedioxy) phosphonitrile trimer bromobenzene inclusion compound

$C_{18}H_{12}N_3O_6P_3$, $0.5C_6H_5Br$

L.A.Siegel, J.H.van den Hende   *J. Chem. Soc. (A)*, 817, 1967

Residue 1 also classified in 61

64.50 tris(o - Phenylenedioxy) phosphonitrile trimer benzene inclusion compound

$C_{18}H_{12}N_3O_6P_3$, $0.5C_6H_6$

L.A.Siegel, J.H.van den Hende   *J. Chem. Soc. (A)*, 817, 1967

Residue 1 also classified in 61

64.51 p - Bromophenyldiphenylphosphine

$C_{18}H_{14}BrP$

H.J.Kuhn, K.Plieth   *Naturwissenschaften*, **53**, 359, 1966

64.52 Chlorotriphenylphosphonium bis(cis - 1,2 - bis(trifluoromethyl)ethene - 1,2 - dithiolato) gold

$C_{18}H_{15}ClP^+$, $C_8AuF_{12}S_4^-$

J.H.Enemark, J.A.Ibers   *Inorg. Chem.*, **7**, 2636, 1968

Residue 2 classified in 85

64.C 1 - Imino - 4 - (oxime - N - oxide) - benzdifurazan triphenyl phosphine oxide complex

$C_{18}H_{15}OP$, $C_6H_2N_6O_4$

For complete entry see 60.75

64.C Tetra(triphenylphosphine oxide) - tetrakis - (μ(3) - hydroxomolybdenum dicarbonyl nitrosyl) adduct

$4C_{18}H_{15}OP$, $C_8H_4Mo_4N_4O_{16}$

For complete entry see 60.149

64.C Triphenylphosphine - lithium iodide complex

$5C_{18}H_{15}OP$, ILi

For complete entry see 61.15

64.53 Triphenyl phosphorus

$C_{18}H_{15}P$

J.J.Daly   *J. Chem. Soc.*, 3799, 1964

64.C   **Triphenylphosphine sulfide - iodine complex**
$2C_{18}H_{15}PS$, $3I_2$
For complete entry see 60.150

64.54   **bis(N - Ethylbenzthiazole(2)) - phosphamethine cyanine perchlorate**
$C_{18}H_{18}N_2PS_2$, ClO_4
R.Allmann   *Chem. Ber.*, **99**, 1332, 1966
Residue 1 also classified in 41

64.C   **bis(Triphenylphosphine)methyl tetrachloroferrate**
$C_{19}H_{16}P^+$, Cl_4Fe^-
For complete entry see 12.15

64.55   **Methylenetriphenylphosphorane**
$C_{19}H_{17}P$
J.C.J.Bart   *Angew. Chem.*, **80**, 697, 1968

64.56   **Triphenylmethylphosphonium bis((3) - 1,2 - dicarbollyl) cuprate(iii)**
$C_{19}H_{18}P^+$, $C_4H_{22}B_{18}Cu^-$
R.M.Wing   *J. Amer. Chem. Soc.*, **90**, 4828, 1968
Residue 2 classified in 62

64.C   **Triphenylmethylphosphonium bis(1,2 - dicyanoethylene - 1,2 - dithiolate) nickel(iii)**
$C_{19}H_{18}P^+$, $C_8N_4NiS_4^-$
For complete entry see 85.43

64.57   **Triphenylphosphoranylideneketene**
$C_{20}H_{15}OP$
J.J.Daly, P.J.Wheatley   *J. Chem. Soc. (A)*, 1703, 1966

64.58   **Triphenylphosphoranylidenethioketene**
$C_{20}H_{15}PS$
J.J.Daly   *J. Chem. Soc. (A)*, 1913, 1967

64.C   **Zinc n - butylphenylphosphinate (monoclinic form)**
$(C_{20}H_{28}O_4P_2Zn)_n$
For complete entry see 84.50

64.C   **Zinc n - butylphenylphosphinate (orthorhombic form)**
$(C_{20}H_{28}O_4P_2Zn)_n$
For complete entry see 84.51

64.59 1 - Ethyl - 2 - (1 - ethyl - 1,2 - dihydroquinolin - 2 -
ylidenephosphino)quinolinium perchlorate
$C_{22}H_{32}N_2P^+$, ClO_4^-
I.Kawada, R.Allmann   *Angew. Chem.*, **80**, 40, 1968
Residue 1 also classified in 35

64.60 2,2,2 - Tri - isopropoxy - 4,5 - (2',2'' - biphenyleno) - 1,3,2 -
dioxaphospholene (monoclinic form)
$C_{23}H_{29}O_5P$
R.D.Spratley, W.C.Hamilton, J.Ladell
J. Amer. Chem. Soc., **89**, 2272, 1967
Also classified in 28

64.61 2,2,2 - Tri - isopropoxy - 4,5 - (2',2'' - biphenyleno) - 1,3,2 -
dioxaphospholene (orthorhombic form)
$C_{23}H_{29}O_5P$
W.C.Hamilton, S.J.La Placa, F.Ramirez, C.P.Smith
J. Amer. Chem. Soc., **89**, 2268, 1967
Also classified in 28

64.C 1 - Triphenylphosphino - 2 - oxo - benzdifuroxan benzene solvate
$C_{24}H_{15}N_4O_5P$, C_6H_6
For complete entry see 40.31

64.62 tris(Phenylethynyl)phosphine
$C_{24}H_{15}P$
D.Mootz, G.Sassmannshausen   *Z. Anorg. Allg. Chem.*, **355**, 200, 1967

64.63 1,2,3 - Triphenyl - 1,2,3 - triphosphaindane
$C_{24}H_{19}P_3$
J.J.Daly   *J. Chem. Soc. (A)*, 1020, 1966

64.C B,B - Di - iodo - P,P - diphenylphosphinoborine
$C_{24}H_{20}B_2I_4P_2$
For complete entry see 62.52

64.64 2,2 - Dichloro - 4,4,6,6 - tetraphenyl cyclotriphosphazatriene
$C_{24}H_{20}Cl_2N_3P_3$
N.V.Mani, F.R.Ahmed, W.H.Barnes   *Acta Cryst.*, **21**, 375, 1966

64.65 2,4,6,8 - Tetrachloro - 2,4,6,8 - tetraphenyl cyclotetraphosphazatetraene
$C_{24}H_{20}Cl_4N_4P_4$
G.J.Bullen, P.R.Mallinson, A.H.Burr   *J. Chem. Soc. (D)*, 691, 1969

64.66 **Tetraphenylphosphonium bis(tetracyano quinodimethanide)**
$C_{24}H_{20}P^+$, $C_{24}H_8N_8^-$
P.Goldstein, K.Seff, K.N.Trueblood *Acta Cryst. (B)*, **24**, 778, 1968
Residue 2 classified in 7

64.67 **Imino bis(aminodiphenylphosphorus) chloride**
$C_{24}H_{24}N_3P_2^+$, Cl^-
J.W.Cow, E.R.Corey *J. Chem. Soc. (D)*, 205, 1969

64.68 **2 - t - Butyl - 4 - (4 - methoxyphenyl) 5,6 - dihydronaphtho(1,2 - b)phosphorin**
$C_{24}H_{25}OP$
W.Fischer, E.Hellner, A.Chatzidakis, K.Dimroth
Tetrahedron Letters, 6227, 1968

64.69 **Tetracyclohexylcyclotetraphosphine (form i)**
$C_{24}H_{44}P_4$
J.C.J.Bart *Acta Cryst. (B)*, **25**, 762, 1969

64.70 **Dodeca(dimethylamino) cyclohexaphosphazahexaene**
Phosphonitrilic dimethylamide hexamer
$C_{24}H_{72}N_{18}P_6$
A.J.Wagner, A.Vos *Acta Cryst. (B)*, **24**, 1423, 1968

64.71 **1,1 - Dimethyl - 2,4,6 - triphenyl phosphabenzene**
$C_{25}H_{23}P$
J.J.Daly, G.Markl *J. Chem. Soc. (D)*, 1057, 1969

64.72 **Benzoyl (triphenylphosphoranylidene)methyl chloride**
$C_{26}H_{20}ClOP$
F.S.Stephens *J. Chem. Soc.*, 5658, 1965

64.73 **Benzoyl (triphenylphosphoranylidene)methyl iodide**
$C_{26}H_{20}IOP$
F.S.Stephens *J. Chem. Soc.*, 5640, 1965

64.74 **bis(Diphenylphosphino)acetylene**
$C_{26}H_{20}P_2$
J.C.J.Bart *Acta Cryst. (B)*, **25**, 489, 1969

64.75 **p - Tolyl triphenylphosphoranylidenemethyl sulfone**
$C_{26}H_{23}O_2PS$
P.J.Wheatley *J. Chem. Soc.*, 5785, 1965
Also classified in 11

64.76   bis(Diphenylphosphino)ethylamine ethyl iodide
$C_{28}H_{30}NP_2{}^+$, I^-
D.S.Payne, J.A.A.Mokuolu, J.C.Speakman   *Chem. Communic.*, 599, 1965

64.77   2,4,6,8 - Tetramethylamino - 2,4,6,8 - tetraphenyl
cyclotetraphosphazatetraene (γ form)
$C_{28}H_{36}N_8P_4$
G.J.Bullen, P.R.Mallinson, A.H.Burr   *J. Chem. Soc. (D),* 691, 1969

64.78   N - p - Bromophenyl triphenylphosphine imide - dimethyl acetylene
dicarboxylate adduct
$C_{30}H_{25}BrNO_4P$
T.C.W.Mak, J.Trotter   *Acta Cryst.*, **18,** 81, 1965

64.79   Pentaphenyl phosphorus
$C_{30}H_{25}P$
P.J.Wheatley   *J. Chem. Soc.*, 2206, 1964

64.80   Phosphobenzene pentamer
$C_{30}H_{25}P_5$
J.J.Daly   *J. Chem. Soc.*, 6147, 1964

64.81   2,2,4,4,6,6 - Hexaphenylcyclotriphosphazatriene
$C_{36}H_{30}N_3P_3$
F.R.Ahmed, P.Singh, W.H.Barnes   *Acta Cryst. (B),* **25,** 316, 1969

64.82   Phosphobenzene hexamer (trigonal form)
$C_{36}H_{30}P_6$
J.J.Daly   *J. Chem. Soc.*, 4789, 1965

64.83   Phosphobenzene hexamer (triclinic form)
$C_{36}H_{30}P_6$
J.J.Daly   *J. Chem. Soc. (A),* 428, 1966

64.84   4,4 - bis(Trifluoromethyl) - 2,2,2 - triphenyl - 3 -
(triphenylphosphoranylidene) - 1,2 - oxaphosphetane
$C_{40}H_{30}F_6OP_2$
G.Chioccola, J.J.Daly   *J. Chem. Soc. (A),* 568, 1968

64.C π - Cyclopentadienyl - bis(triphenylphosphite) iron iodide
$C_{41}H_{35}FeIO_6P_2$
For complete entry see 73.120

64.C   bis(Uranyl acetate triphenylphosphine oxide)
$C_{44}H_{42}O_{14}P_2U_2$
For complete entry see 81.90

64.85   **1,3,5 - Trinitro - 2,4,6 - tri(triphenylphosphoranylideneamino) benzene toluene solvate**

$C_{60}H_{45}N_6O_6P_3$, $0.5C_7H_8$

T.S.Cameron, C.K.Prout   *Chem. Communic.*, 455, 1967

ARSENIC COMPOUNDS

65.1   **Di - iodomethylarsine (at 5 ° C)**
CH_3AsI_2
N.Camerman, J.Trotter   *Acta Cryst.*, **16**, 922, 1963

65.2   **Cacodylic acid**
$C_2H_7AsO_2$
J.Trotter, T.Zobel   *J. Chem. Soc.*, 4466, 1965

65.3   **Methyldicyanoarsine**
$C_3H_3AsN_2$
E.O.Schlemper, D.Britton   *Acta Cryst.*, **20**, 777, 1966

65.4   **Cyanodimethylarsine**
C_3H_6AsN
N.Camerman, J.Trotter   *Canad. J. Chem.*, **41**, 460, 1963

65.5   **1,3,5 - Trichloro - 2,4,6 - trimethyl - tri(arsenonitrile)**
$C_3H_9As_3Cl_3N_3$
J.Weiss, W.Eisenhuth   *Z. Naturforsch., B*, **22**, 454, 1967

65.6   **Aluminium tetrahydroborate - trimethylarsine**
$C_3H_{21}AlAsB_3$
N.A.Bailey, P.H.Bird, M.G.H.Wallbridge   *Inorg. Chem.*, **7**, 1575, 1968

65.7   **Tetramethylarsonium bromide**
$C_4H_{12}As^+, Br^-$
E.Collins, D.J.Sutor, F.G.Mann   *J. Chem. Soc.*, 4051, 1963

65.8   **Dimethylarsino dimethyldithioarsinate**
$C_4H_{12}As_2S_2$
N.Camerman, J.Trotter   *J. Chem. Soc.*, 219, 1964

65.9   **Arsenomethane**
$C_5H_{15}As_5$
J.H.Burns, J.Waser   *J. Amer. Chem. Soc.*, **79**, 859, 1957
See also *Int. Distances*, M 124s; *Structure Reports*, **21**, 511, 1957

65.10 o - Phenylenediarsine oxychloride
$C_6H_4As_2Cl_2O$
W.R.Cullen, J.Trotter   *Canad. J. Chem.*, **40**, 1113, 1962

65.11 Arsenobenzene
C_6H_5As
K.Hedberg, E.W.Hughes, J.Waser   *Acta Cryst.*, **14**, 369, 1961

65.12 Phenylarsonic acid
$C_6H_7AsO_3$
A.Shimada   *Bull. Chem. Soc. Jap.*, **33**, 301, 1960

65.13 Phenylarsonic acid
$C_6H_7AsO_3$
Yu.T.Struchkov   *Izvest. Akad. Nauk S. S. S. R., Ser. Khim.*, 1962, 1960

65.14 m - Aminobenzene arsonic acid
$C_6H_8AsNO_3$
A.Shimada   *Bull. Chem. Soc. Jap.*, **35**, 1600, 1962
Also classified in 16

65.15 Hexa(methylamine) tetra - arsenic
$C_6H_{18}As_4N_6$
J.Weiss, W.Eisenhuth   *Z. Anorg. Allg. Chem.*, **350**, 9, 1967

65.16 Arsenious xanthate
$C_9H_{15}AsO_3S_6$
G.Carrai, G.Gottardi   *Z. Kristallogr.*, **113**, 373, 1960

65.17 Potassium di - o - phenylenedioxyarsenate(iii)
$C_{12}H_8AsO_4^-$, K^+
A.C.Skapski   *Chem. Communic.*, 10, 1966

65.18 10 - Bromo - 5,10 - dihydrophenarsazine
$C_{12}H_9AsBrN$
M.Fukuyo, K.Nakatsu, A.Shimada   *Bull. Chem. Soc. Jap.*, **39**, 1614, 1966

65.19 10 - Chloro - 5,10 - dihydrophenarsazine
$C_{12}H_9AsClN$
A.Camerman, J.Trotter   *J. Chem. Soc.*, 730, 1965

65.20 10 - Chloro - 5,10 - dihydrophenarsazine
$C_{12}H_9AsClN$
M.Fukuyo, K.Nakatsu, A.Shimada   *Bull. Chem. Soc. Jap.*, **39**, 1614, 1966

65.21   **Bromodiphenylarsine**
$C_{12}H_{10}AsBr$
J.Trotter   *J. Chem. Soc.*, 2567, 1962

65.22   **Chlorodiphenyl arsine**
$C_{12}H_{10}AsCl$
J.Trotter   *Canad. J. Chem.*, **40,** 1590, 1962

65.23   **Iodo - diphenylarsine**
$C_{12}H_{10}AsI$
J.Trotter   *Canad. J. Chem.*, **41,** 191, 1963

65.24   **5,10 - Dihydro - 5,10 - dimethylarsanthrene**
$C_{14}H_{14}As_2$
O.Kennard, F.G.Mann, D.G.Watson, J.K.Fawcett, K.A.Kerr
Chem. Communic., 269, 1968

65.25   **5,10 - Dihydro - 5,10 - dimethylarsanthren dibromide**
$C_{14}H_{14}As_2Br_2$
D.J.Sutor, F.R.Harper   *Acta Cryst.*, **12,** 585, 1959
See also *Int. Distances,* M 180s; *Structure Reports,* **23,** 705, 1959

65.26   **5,10 - Dihydro - 5,10 - dimethylarsanthren di - iodide**
$C_{14}H_{14}As_2I_2$
D.J.Sutor, F.R.Harper   *Acta Cryst.*, **12,** 585, 1959
See also *Int. Distances,* M 180s; *Structure Reports,* **23,** 706, 1959

65.27   **Arsenic(iii) N,N - diethyldithiocarbamate**
$C_{15}H_{30}AsN_3S_6$
M.Colapietro, A.Domenicano, L.Scaramuzza, A.Vaciago
Chem. Communic., 302, 1968

65.28   **Phenylarsine bis(diethyldithiocarbamate)**
$C_{16}H_{25}AsN_2S_4$
R.Bally   *Acta Cryst.*, **23,** 295, 1967

65.29   **9 - Phenyl - 9 - arsafluorene**
$C_{18}H_{13}As$
D.Sartain, M.R.Truter   *J. Chem. Soc.*, 4414, 1963

65.30   **Triphenylarsine oxide monohydrate**
$C_{18}H_{15}AsO$, H_2O
G.Ferguson, E.W.Macaulay   *J. Chem. Soc. (A),* 1, 1969

65.31 **Triphenylarsine hydroxybromide**
$C_{18}H_{16}AsBrO$
G.Ferguson, E.W.Macauly *Chem. Communic.*, 1288, 1968

65.32 **Triphenylarsine hydroxychloride**
$C_{18}H_{16}AsClO$
G.Ferguson, E.W.Macaulay *Chem. Communic.*, 1288, 1968

65.C **Triphenylmethylarsonium bis(toluene - 3,4 - dithiolato) cobaltate ethanol solvate**
$C_{19}H_{18}As^+$, $C_{14}H_{12}CoS_4^-$, $0.5C_2H_6O$
For complete entry see 85.73

65.33 **bis(Triphenylmethylarsonium) tetrachloro nickel(ii)**
$2C_{19}H_{18}As^+$, Cl_4Ni^{2-}
P.Pauling *Inorg. Chem.*, **5**, 1498, 1966

65.34 **Tri - p - tolylarsine**
$C_{21}H_{21}As$
J.Trotter *Canad. J. Chem.*, **41**, 14, 1963

65.35 **tris - (p - Nitrophenylethynyl)arsine**
$C_{24}H_{12}AsN_3O_6$
D.Mootz, W.Look *Z. Anorg. Allg. Chem.*, **356**, 244, 1968
Also classified in 15

65.36 **Tetraphenylarsonium iodide**
$C_{24}H_{20}As^+$, I^-
R.C.L.Mooney *J. Amer. Chem. Soc.*, **62**, 2955, 1940
See also *Int. Distances*, M 252; *Structure Reports*, **8**, 312, 1940

65.37 **Tetraphenylarsonium hydrogen dinitrate**
$C_{24}H_{20}As^+$, $HN_2O_6^-$
B.D.Faithful, S.C.Wallwork *Chem. Communic.*, 1211, 1967

65.38 **Tetraphenylarsonium oxotetrabromoaquomolybdate**
$C_{24}H_{20}As^+$, $H_2Br_4MoO_2^-$
J.G.Scane *Acta Cryst.*, **23**, 85, 1967

65.39 **Tetraphenylarsonium cis - diaquotetrachlororuthenate monohydrate**
$C_{24}H_{20}As^+$, $H_2Cl_4ORu^-$, H_2O
T.E.Hopkins, A.Zalkin, D.H.Templeton, M.G.Adamson
Inorg. Chem., **5**, 1428, 1966

65.C   **Tetraphenylarsonium oxo - tetrabromo - acetonitrile rhenate(v)**
$C_{24}H_{20}As^+$, $C_2H_3Br_4NORe^-$
For complete entry see 83.1

65.40   **Tetraphenylarsonium 3 - fluoro - 1,1,4,5,5 - pentacyano - 2 - azapentadienide**
$C_{24}H_{20}As^+$, $C_9FN_6^-$
G.J.Palenik   *Acta Cryst.*, **20,** 471, 1966
Residue 2 classified in 7

65.41   **Tetraphenylarsonium rhenium carbonyl hydride**
$C_{24}H_{20}As^+$, $C_{12}H_2O_{12}Re_3^-$
M.R.Churchill, P.H.Bird, H.D.Kaesz, R.Bau, B.Fontal
J. Amer. Chem. Soc., **90,** 7135, 1968

65.42   **Tetraphenylarsonium tri - iodo(triphenylphosphine) nickelate(ii)**
$C_{24}H_{20}As^+$, $C_{18}H_{15}I_3NiP^-$
R.P.Taylor, D.H.Templeton, W.DeW.Horrocks Junior
Inorg. Chem., **7,** 2629, 1968
Residue 2 classified in 86

65.43   **bis(Tetraphenylarsonium) tri - μ - chloro - octachlorotrirhenate monohydrate**
$2C_{24}H_{20}As^+$, $Cl_{11}Re_3^{2-}$, H_2O
B.R.Penfold, W.T.Robinson   *Inorg. Chem.*, **5,** 1758, 1966

65.44   **Tetraphenylarsonium tetranitratocobaltate(v)**
$2C_{24}H_{20}As^+$, $CoN_4O_{12}^{2-}$
J.G.Bergman, F.A.Cotton   *Inorg. Chem.*, **5,** 1208, 1966

65.45   **Tetraphenylarsonium penta - azido iron(iii)**
$2C_{24}H_{20}As^+$, FeN_{15}^{2-}
J.Drummond, J.S.Wood   *J. Chem. Soc. (D),* 1373, 1969

65.46   **Tetraphenylarsonium bis(trithiocarbonato) nickelate(ii)**
$2C_{24}H_{20}As^+$, $C_2NiS_6^{2-}$
J.S.McKechnie, S.L.Miesel, I.C.Paul   *Chem. Communic.*, 152, 1967
Residue 2 classified in 85

65.47   **Tetraphenylarsonium bis(N - cyanodithio) nickelate(ii)**
$2C_{24}H_{20}As^+$, $C_4N_4NiS_4^{2-}$
F.A.Cotton, C.B.Harris   *Inorg. Chem.*, **7,** 2140, 1968
Residue 2 classified in 85

65.48 **bis(Diphenylarsenic) oxide**
$C_{24}H_{20}As_2O$
W.R.Cullen, J.Trotter   *Canad. J. Chem.*, **41,** 2983, 1963

65.49 **Tri - p - xylylarsine**
$C_{24}H_{27}As$
J.Trotter   *Acta Cryst.*, **16,** 1187, 1963

65.50 **Di - isothiocyanato - (N,N - bis(2 - diethylaminoethyl) - 2 - diphenylarsinoethylamine) nickel(ii)**
$C_{30}H_{42}AsN_5NiS_2$
M.di Vaira, L.Sacconi   *J. Chem. Soc. (D),* 10, 1969
Also classified in 83

65.51 **Di(bis(triphenylarsine oxide)hydronium) dimercury hexabromide**
$2C_{36}H_3As_2O_2{}^+$, $Br_6Hg_2{}^{2-}$
G.S.Harris, F.Inglis, J.McKechnie, K.K.Cheung, G.Ferguson
Chem. Communic., 442, 1967

65.52 **Diphenylarsenonitrile tetramer**
$C_{48}H_{40}As_4N_4$
M.D.Glick   *Dissert. Abstr.*, **25,** 5546, 1965

ANTIMONY AND BISMUTH COMPOUNDS

66.1   2 - Chloro - 1,3 - dithia - 2 - stibacyclopentane
       $C_2H_4ClS_2Sb$
       M.A.Bush, P.F.Lindley, P.Woodward   *J. Chem. Soc. (A)*, 221, 1967

66.2   **Dimethylformamide - antimony(v) chloride complex**
       $C_3H_7Cl_5NOSb$
       L.Brun, C.-I.Branden   *Acta Cryst.*, **20**, 749, 1966

66.3   **Trimethylstibine di - iodide**
       $C_3H_9BI_2S$
       A.F.Wells   *Z. Kristallogr.*, **99**, 367, 1938
       See also *Int. Distances*, M 156; *Strukturbericht*, **6**, 219, 1938

66.4   **Trimethylstibine dibromide**
       $C_3H_9Br_2Sb$
       A.F.Wells   *Z. Kristallogr.*, **99**, 367, 1938
       See also *Int. Distances*, M 155; *Strukturbericht*, **6**, 219, 1938

66.5   **Trimethylstibine dichloride**
       $C_3H_9Cl_2Sb$
       A.F.Wells   *Z. Kristallogr.*, **99**, 367, 1938
       See also *Int. Distances*, M 156; *Strukturbericht*, **6**, 219, 1938

66.6   **Trimethylphosphine oxide antimony(v) chloride**
       $C_3H_9Cl_5OPSb$
       C.-I.Branden, I.Lindqvist   *Acta Chem. Scand.*, **15**, 167, 1961
       Also classified in 64

66.7   **Antimony hydrogen bis(thioglycollate)**
       $C_4H_5O_4S_2Sb$
       I.Hansson   *Acta Chem. Scand.*, **22**, 509, 1968

66.8   **Di - μ - ethoxy - bis(tetrachloroantimony)**
       $C_4H_{10}Cl_8O_2Sb_2$
       H.Preiss   *Z. Anorg. Allg. Chem.*, **362**, 24, 1968

66.C Tetramethylstibonium tetrakis(trimethylsiloxy) aluminate
$C_4H_{12}Sb^+$, $C_{12}H_{36}AlO_4Si_4{}^-$
For complete entry see 68.21

66.9 trans - trans - trans - Dichlorotri - 2 - chlorovinylstibine
$C_6H_6Cl_5Sb$
Yu.T.Struchkov, T.L.Khotsyanova
Dokl. Akad. Nauk S. S. S. R., **91**, 565, 1953
See also *Int. Distances*, M 197; *Structure Reports*, **17**, 671, 1953

66.10 Pentachloro antimony - benzoyl chloride
$C_7H_5Cl_6OSb$
J.-M.Le Carpentier, B.Chevrier, R.Weiss
Bull. Soc. Fr. Mineral. Cristallogr., **91**, 544, 1968

66.11 Ammonium antimonyl D - tartrate trihydrate
$C_8H_4O_{12}Sb_2{}^{2-}$, $2H_4N^+$, $3H_2O$
G.A.Kiosse, N.I.Golovastikov, A.V.Ablov, N.V.Belov
Dokl. Akad. Nauk S. S. S. R., **177**, 329, 1967

66.12 Ammonium antimonyl DL - tartrate tetrahydrate
$C_8H_8O_{12}Sb_2{}^{2-}$, $2H_4N^+$, $4H_2O$
G.A.Kiosse, N.I.Golovastikov, N.V.Belov
Dokl. Akad. Nauk S. S. S. R., **155**, 545, 1964

66.13 Antimony(iii) xanthate
$C_9H_{15}O_3S_6Sb$
G.Gottardi *Z. Kristallogr.*, **115**, 451, 1961

66.14 Monoaquo diphenyltrichlorostibine
$C_{12}H_{12}Cl_3OSb$
T.N.Polynova, M.A.Porai-Koshits *Zh. Strukt. Khim.*, **8**, 112, 1967

66.15 Triphenylbismuth
$C_{18}H_{15}Bi$
J.Wetzel *Z. Kristallogr.*, **104**, 305, 1942
See also *Int. Distances*, M 247; *Structure Reports*, **9**, 388, 1942

66.16 Triphenylbismuth
$C_{18}H_{15}Bi$
D.M.Hawley, G.Ferguson *J. Chem. Soc. (A)*, 2059, 1968

66.17 Triphenylbismuth dichloride
$C_{18}H_{15}BiCl_2$
D.M.Hawley, G.Ferguson *J. Chem. Soc. (A)*, 2539, 1968

66.18   **Triphenyldichlorostibine**
$C_{18}H_{15}Cl_2Sb$
T.N.Polynova, M.A.Porai-Koshits   *Zh. Strukt. Khim.*, **7,** 742, 1966

66.19   **Dimethoxyphenylantimony**
$C_{20}H_{21}O_2Sb$
K.Shen, W.E.McEwen, S.J.La Placa, W.C.Hamilton, A.P.Wolf
J. Amer. Chem. Soc., **90,** 1718, 1968

66.20   **Tetraphenyl antimony hydroxide**
$C_{24}H_{21}OSb$
A.L.Beauchamp, M.J.Bennett, F.A.Cotton
J. Amer. Chem. Soc., **91,** 297, 1969

66.21   **Methoxytetraphenylantimony**
$C_{25}H_{23}OSb$
K.Shen, W.E.McEwen, S.J.La Placa, W.C.Hamilton, A.P.Wolf
J. Amer. Chem. Soc., **90,** 1718, 1968

66.22   **Pentaphenyl antimony**
$C_{30}H_{25}Sb$
P.J.Wheatley   *J. Chem. Soc.*, 3718, 1964

66.23   **Pentaphenylantimony**
$C_{30}H_{25}Sb$
A.L.Beauchamp, M.J.Bennett, F.A.Cotton
J. Amer. Chem. Soc., **90,** 6675, 1968

66.24   **tris - 3 - Sulfanilamido - 6 - methoxypyridazine - bismuth chloride**
$C_{33}H_{36}BiCl_3N_{12}O_9S_3$
L.Cavalca, M.Nardelli, G.Fava, G.Giraldi   *Acta Cryst.*, **16,** A69, 1963

66.C   **Ferrous phenanthroline antimony d - tartrate octahydrate**
$C_{36}H_{24}FeN_6{}^{2+}$, $C_8H_4O_{12}Sb_2{}^{2-}$, $8H_2O$
For complete entry see 83.210

GROUPS IA AND IIA COMPOUNDS

67.C   **Strontium cyanamide**
$CN_2{}^{2-}$, Sr^{2+}
For complete entry see 7.1

67.1   **Dimethylberyllium**
$(C_2H_6Be_2)_n$
A.I.Snow, R.E.Rundle   *Acta Cryst.*, **4**, 348, 1951
See also *Int. Distances*, M 136; *Structure Reports*, **15**, 415, 1951

67.2   **bis(Ethylacetato) calcium difluorophosphate**
$C_8H_{16}CuF_4O_8P_2$
H.Grunze, K.-H.Jost, G.-U.Wolf   *Z. Anorg. Allg. Chem.*, **365**, 294, 1969

67.C   **Lithium aluminium tetraethyl**
$C_8H_{20}AlLi$
For complete entry see 68.15

67.3   **Magnesium bromide diethyletherate**
$C_8H_{20}Br_2MgO_2$
H.Schibilla, M.S. le Bihan   *Acta Cryst.*, **23**, 332, 1967

67.4   **Ethyl lithium**
$C_8H_{20}Li_4$
H.Dietrich   *Acta Cryst.*, **16**, 681, 1963

67.5   **Sodium hydridodiethylberyllate dietherate**
$C_8H_{22}Be_2{}^{2-}$, $2C_4H_{10}NaO^+$
G.W.Adamson, H.M.M.Shearer   *Chem. Communic.*, 240, 1965

67.6   **Barium - 2 - dicyanomethylene - 1,1,3,3 - tetracyanopropane hexahydrate**
$C_{10}N_6{}^{2-}$, Ba^{2+}, $6H_2O$
D.A.Bekoe, P.K.Gantzel, K.N.Trueblood   *Acta Cryst.*, **16**, A62, 1963
Residue 1 also classified in 7, 12

67.7   **Calcium hexacyanoisobutene hexahydrate**
Calcium 2 - dicyanomethylene - 1,1,3,3 - tetracyanopropane hexahydrate
$C_{10}N_6{}^{2-}$, Ca^{2+} , $6H_2O$
D.A.Bekoe, P.K.Gantzel, K.N.Trueblood   *Acta Cryst.*, **22**, 657, 1967
Residue 1 also classified in 7, 12

67.8   **bis - Acetylacetone beryllium**
$C_{10}H_{14}BeO_4$
V.Amirthalingam, V.M.Padmanabhan, J.Shankar
Acta Cryst., **13**, 201, 1960

67.9   **Diaquo bis(acetylacetonato) magnesium(ii)**
$C_{10}H_{21}MgO_6$
B.Morosin   *Acta Cryst.*, **22**, 315, 1967

67.10   **Ethyl magnesium bromide dietherate**
$C_{10}H_{25}BrMgO_2$
L.J.Guggenberger, R.E.Rundle   *J. Amer. Chem. Soc.*, **90**, 5375, 1968

67.11   **Tetraberyllium mono - oxyhexa - acetate**
Basic beryllium acetate
$C_{12}H_{18}Be_4O_{13}$
A.Tulinsky, C.R.Worthington, E.Pignataro   *Acta Cryst.*, **12**, 623+, 1959
See also *Int. Distances,* M 170s; *Structure Reports,* **23**, 534, 1959

67.12   **Magnesium bis(dimethyl aluminium dimethoxide) dioxane complex**
$(C_{12}H_{32}Al_2MgO_6)_n$
J.L.Atwood, G.D.Stucky   *J. Organometal. Chem.*, **13**, 53, 1968
Also classified in 68

67.13   **bis(Dimethylamino) beryllium trimer**
$C_{12}H_{36}Be_3N_6$
J.L.Atwood, G.D.Stucky   *Chem. Communic.*, 1169, 1967

67.14   **Phenyl magnesium bromide dietherate**
$C_{14}H_{25}BrMgO_2$
G.Stucky, R.E.Rundle   *J. Amer. Chem. Soc.*, **86**, 4825, 1964

67.15   **Ethylmagnesium bromide - triethylamine complex**
$C_{16}H_{40}Br_2Mg_2N_2$
J.Toney, G.D.Stucky   *Chem. Communic.*, 1168, 1967

67.16   **bis(t - Butoxymagnesium bromide - diethylether)**
$C_{16}H_{40}Br_2Mg_2O_4$
P.T.Moseley, H.M.M.Shearer   *Chem. Communic.*, 279, 1968

67.17 **Magnesium oxybromide - diethylether complex**
$C_{16}H_{40}Br_2Mg_4O_5$
G.Stucky, R.E.Rundle   *J. Amer. Chem. Soc.*, **86**, 4821, 1964

67.18 **bis - Calcium strontium hexapropionate (paraelectric form)**
$C_{18}H_{30}Ca_2O_{12}Sr$
H.Maruyama, Y.Tomile, I.Mizutani, Y.Yamazaki, Y.Uesu, N.Yamada,
J.Kobayashi   *J. Phys. Soc. Jap.*, **23**, 899, 1967

67.19 **bis - Calcium strontium hexapropionate (ferroelectric form, at −50 °C)**
$C_{18}H_{30}Ca_2O_{12}Sr$
I.Mizutani, Y.Yamazaki, Y.Uesu, N.Yamada, J.Kobayashi, H.Maruyama,
Y.Tomiie   *J. Phys. Soc. Jap.*, **23**, 900, 1967

67.20 **Calcium hexa - antipyrine perchlorate**
$C_{66}H_{72}CaN_{12}O_6^{2+}$, $2ClO_4^-$
M.Vijayan, M.A.Viswamitra   *Acta Cryst. (B)*, **24**, 1067, 1968
Residue 1 also classified in 32

67.21 **Magnesium hexa - antipyrine perchlorate**
$C_{66}H_{72}MgN_{12}O_6^{2+}$, $2ClO_4^-$
M.Vijayan, M.A.Viswamitra   *Acta Cryst.*, **23**, 1000, 1967
Residue 1 also classified in 32

GROUP III COMPOUNDS

68.1 **Methyl aluminium chloride dimer**
$C_2H_6Al_2Cl_4$
G.Allegra, G.Perego, A.Immirzi *Makromol. Chem.*, **61**, 69, 1963

68.2 **Trichlorotrimethylamine aluminium(iii)**
$C_3H_9AlCl_3N$
D.F.Grant, R.C.G.Killean, J.L.Lawrence *Acta Cryst. (B)*, **25**, 377, 1969

68.3 **Trimethyl indium**
C_3H_9In
E.L.Amma, R.E.Rundle *J. Amer. Chem. Soc.*, **80**, 4141, 1958
See also *Int. Distances*, M 101s; *Structure Reports*, **22**, 571, 1958

68.4 **Trimethylamine gallane**
$C_3H_{12}GaN$
D.F.Shriver, C.E.Nordman *Inorg. Chem.*, **2**, 1298, 1963

68.5 **Aluminium tetrahydroborate - trimethylamine**
$C_3H_{21}AlB_3N$
N.A.Bailey, P.H.Bird, M.G.H.Wallbridge *Inorg. Chem.*, **7**, 1575, 1968

68.6 **Aluminium tetrahydroborate - trimethylamine (at −160 ° C)**
$C_3H_{21}AlB_3N$
N.A.Bailey, P.H.Bird, M.G.H.Wallbridge *Inorg. Chem.*, **7**, 1575, 1968

68.7 **Thallium(i) perchlorate - tetra(thiourea) complex**
$C_4H_{16}N_8S_4$, ClO_4Tl
J.C.A.Boeyens, F.H.Herbstein *Inorg. Chem.*, **6**, 1408, 1967
Residue 1 also classified in 8

68.8 **Thallium(i) nitrate - tetra(thiourea) complex**
$C_4H_{16}N_8S_4$, NO_3Tl
J.C.A.Boeyens, F.H.Herbstein *Inorg. Chem.*, **6**, 1408, 1967
Residue 1 also classified in 8

68.9 **Cyclopentadienyl indium**
$(C_5H_5In)_n$
E.Frasson, F.Menegus, C.Panattoni *Nature*, **199**, 1087, 1963

68.10 **Trimethylaluminium**
$C_6H_{18}Al_2$
R.G.Vranka, E.L.Amma *J. Amer. Chem. Soc.*, **89**, 3121, 1967

68.11 **Dibromotrimethylsiloxy aluminium**
$C_6H_{18}Al_2Br_4O_2$
M.Bonamico, G.Dessy *J. Chem. Soc. (A)*, 1786, 1967
Also classified in 63

68.12 **Aluminium hydride - N,N,N′,N′ - tetramethyl ethylenediamine adduct**
$C_6H_{19}AlN_2$
G.J.Palenik *Acta Cryst.*, **17**, 1573, 1964

68.13 **Aluminium hydride - trimethylamine complex**
$C_6H_{21}AlN$
C.W.Heitsch, C.E.Nordman, R.W.Parry *Inorg. Chem.*, **2**, 508, 1963

68.14 **Benzoyl chloride - aluminium chloride complex**
$C_7H_5AlCl_4O$
S.E.Rasmussen, N.C.Broch *Acta Chem. Scand.*, **20**, 1351, 1966

68.15 **Lithium aluminium tetraethyl**
$C_8H_{20}AlLi$
R.L.Gerteis, R.E.Dickerson, T.L.Brown *Inorg. Chem.*, **3**, 872, 1964
Also classified in 67

68.16 **Pentabromotrialuminium methylsiloxane**
$C_8H_{24}Al_3Br_5O_6Si_4$
M.Bonamico, G.Dessy *J. Chem. Soc. (A)*, 291, 1968
Also classified in 63

68.17 **Hydrogen aquoethylenediamine tetra - acetato gallium(iii)**
$C_{10}H_{15}GaN_2O_9$
J.L.Hoard, C.H.L.Kennard, G.S.Smith *Inorg. Chem.*, **2**, 1316, 1963

68.18 **bis(Trimethyl aluminium) dioxanate**
$C_{10}H_{26}Al_2O_2$
J.L.Atwood, G.D.Stucky *J. Amer. Chem. Soc.*, **89**, 5362, 1967

68.19 **Diethyl (salicylaldehydato) thallium(iii)**
$C_{11}H_{15}O_2Tl$
G.H.W.Milburn, M.R.Truter *J. Chem. Soc. (A)*, 648, 1967

68.20   **Aluminium triethyl - potassium fluoride**
$C_{12}H_{30}AlF^-$, K^+
G.Allegra, G.Perego   *Acta Cryst.*, **16**, 185, 1963

68.C   **Magnesium bis(dimethyl aluminium dimethoxide) dioxane complex**
$(C_{12}H_{32}Al_2MgO_6)_n$
For complete entry see 67.12

68.21   **Tetramethylstibonium tetrakis(trimethylsiloxy) aluminate**
$C_{12}H_{36}AlO_4Si_4^-$, $C_4H_{12}Sb^+$
P.J.Wheatley   *J. Chem. Soc.*, 3200, 1963
Residue 1 also classified in 3; residue 2 classified in 66

68.22   **Dimethyl thallium perchlorate 1,10 - phenanthroline complex**
$C_{14}H_{14}ClN_2O_4Tl$
T.L.Blundell, H.M.Powell   *Chem. Communic.*, 54, 1967
Also classified in 36

68.23 μ - **Diphenylamino** - μ - **methyl - tetramethyl - dialuminium**
$C_{17}H_{25}Al_2N$
V.R.Magnuson, G.D.Stucky   *J. Amer. Chem. Soc.*, **90**, 3269, 1968

68.24   **Triphenyl gallium**
$C_{18}H_{15}Ga$
J.F.Malone, W.S.McDonald   *J. Chem. Soc. (D)*, 591, 1969

68.25   **Triphenyl indium**
$C_{18}H_{15}In$
J.F.Malone, W.S.McDonald   *J. Chem. Soc. (D)*, 591, 1969

68.26   **Dichloro bis(2,2' - bipyridyl)gallium tetrachlorogallate**
$C_{20}H_{16}Cl_2Ga^+$, Cl_4Ga^-
R.Restivo, G.J.Palenik   *J. Chem. Soc. (D)*, 867, 1969

68.C   **Dicyclopentadienyl titanium diethyl aluminium dimer**
$C_{28}H_{40}Al_2Ti_2$
For complete entry see 73.111

68.27   **Dimethylaluminumoxybenzylidenaniline dimer**
$C_{30}H_{32}Al_2N_2O_2$
Y.Kai, N.Yasuoka, N.Kasai, M.Kakudo, H.Yasuda, H.Tani
Chem. Communic., 1332, 1968

68.28   **bis(Dimethyl aluminium N - phenyl - N - acetyl - benzamide) complex**
$C_{34}H_{38}Al_2N_2O_4$
Y.Kai, N.Yasuoka, N.Kasai, M.Kakudo, H.Yasuda, H.Tani
J. Chem. Soc. (D), 575, 1969

68.29   **Triphenyl aluminium**
$C_{36}H_{30}Al_2$
J.F.Malone, W.S.McDonald   *Chem. Communic.*, 444, 1967

68.30   **Triphenyl aluminium**
$C_{36}H_{30}Al_2$
H.D.McBride   *Dissert. Abstr. (B)*, **27**, 3891, 1967

68.31   **Diphenyl aluminium nitride tetramer**
$C_{48}H_{40}Al_4N_4$
T.R.R.McDonald, W.S.McDonald   *Proc. Chem. Soc.*, 382, 1963

68.32   **Diphenyl aluminium - phenyl - p - bromophenylketimine dimer benzene solvate**
$C_{50}H_{38}Al_2Br_2N_2$, $2C_6H_6$
W.S.McDonald   *Acta Cryst. (B)*, **25**, 1385, 1969

GERMANIUM, TIN, LEAD COMPOUNDS

69.C 2,2',2' - Terpyridyl(dimethyl)chlorotin(iv) dimethyl (trichloro)stannate(iv)
$C_2H_6Cl_3Sn^-$, $C_{17}H_{17}ClN_3Sn^+$
For complete entry see 69.32

69.1 Dimethyl tin difluoride
$C_2H_6F_2Sn$
E.O.Schlemper, W.C.Hamilton *Inorg. Chem.*, **5**, 995, 1966

69.2 Trimethyl tin fluoride
C_3H_9FSn
H.C.Clark, R.J.O'Brien, J.Trotter *J. Chem. Soc.*, 2332, 1964

69.3 Trimethyl tin hydroxide
$C_3H_{10}OSn$
N.Kasai, K.Yasuda, R.Okawara *J. Organometal. Chem.*, **3**, 172, 1965

69.4 1,1,4,4 - Tetrachloro - 1,4 - digermaniacyclohexa - 2,5 - diene
$C_4H_4Cl_4Ge_2$
N.G.Bokii, Yu.T.Struchkov *Zh. Strukt. Khim.*, **9**, 838, 1968

69.5 1,1,4,4 - Tetraiodo - 1,4 - digermacyclohexa - 2,5 - diene
$C_4H_4Ge_2I_4$
N.G.Bokii, Yu.T.Struchkov *Zh. Strukt. Khim.*, **8**, 122, 1967

69.6 Trimethyl germanium cyanide
C_4H_9GeN
E.O.Schlemper, D.Britton *Inorg. Chem.*, **5**, 511, 1966

69.7 Trimethyltin isothiocyanate
C_4H_9NSSn
R.A.Forder, G.M.Sheldrick *J. Chem. Soc. (A)*, 1125, 1969

69.8 Trimethyl tin cyanide
C_4H_9NSn
E.O.Schlemper, D.Britton *Inorg. Chem.*, **5**, 507, 1966

69.9 **Tetra(methyltin) hexasulfide**
$C_4H_{12}S_6Sn_4$
C.Dorfelt, A.Janek, D.Kobelt, E.F.Paulus, H.Sherer
J. Organometal. Chem., **14**, 22, 1968

69.10 **Tin(iv) chloride glutaronitrile**
$(C_5H_6Cl_4N_2Sn)_n$
D.M.Barnhart, C.N.Caughlan, M.-ul-Haque   *Inorg. Chem.*, **7**, 1135, 1968
Also classified in 7

69.11 **Lead(ii) thiourea acetate**
$C_5H_{10}N_2O_4PbS$
M.Nardelli, G.Fava, G.Branchi   *Acta Cryst.*, **13**, 898, 1960

69.12 **Lead ethylxanthate**
$C_6H_{10}O_2PbS_4$
H.Hagihara, S.Yamashita   *Acta Cryst.*, **21**, 350, 1966

69.13 **cis - Dichloro - cis - bis(dimethylsulfoxide) - trans - dimethyltin (iv)**
$C_6H_{18}Cl_2O_2S_2Sn$
N.W.Isaacs, C.H.L.Kennard, W.Kitching   *Chem. Communic.*, 820, 1968

69.14 **Trimethyl tin - pentacarbonyl manganese**
$C_8H_9MnO_5Sn$
R.F.Bryan   *J. Chem. Soc. (A)*, 696, 1968
Also classified in 71

69.15 **Lead(iv) acetate**
$C_8H_{12}O_4Pb$
B.Kamenar   *Acta Cryst.*, **16**, A34, 1963

69.16 **Chloro(trimethyl)pyridine tin**
$C_8H_{14}ClNSn$
R.Hulme   *J. Chem. Soc.*, 1524, 1963

69.17 **Tin(iv) chloride tetrahydrothiophene complex**
$C_8H_{16}Cl_4S_2Sn$
I.R.Beattie, R.Hulme, L.Rule   *J. Chem. Soc.*, 1581, 1965

69.18 **1,1,4,4 - Tetramethyl - 1,4 - digermacyclohexa - 2,5 - diene**
$C_8H_{16}Ge_2$
N.G.Bokii, G.N.Zakharova, Yu.T.Struchkov
Zh. Strukt. Khim., **8**, 501, 1967

69.19 1 - Ethylgermatrane
$C_8H_{17}GeNO_3$
Ya.Ya.Blejdelis, A.A.Kemme, L.O.Atovmyan, R.P.Shibaeva
Khim. Geterocikl. Svedn. Latv. S. S. S. R., **4**, 184, 1968

69.20 bis(Acetato dimethyl stannyl)oxide
$C_8H_{18}O_5Sn_2$
Z.V.Zvonkova, Z.P.Povet'eva, V.M.Vozzennikov, V.P.Gluskova,
V.I.Jakovenko, A.N.Khvatkina *Acta Cryst.*, **21**, A155, 1966

69.21 Dicyclopentadienyl lead (form 1, orthorhombic)
$(C_{10}H_{10}Pb)_n$
C.Panattoni, G.Bombieri, U.Croatto *Acta Cryst.*, **21**, 823, 1966

69.22 Lead(ii) n - butylxanthate
$C_{10}H_{18}O_2PbS_4$
H.Hagihara, Y.Watanabe, S.Yamashita *Acta Cryst. (B)*, **24**, 960, 1968

69.23 Diphenyl tin m - xylene solvate
$C_{12}H_{10}Sn$, C_8H_{10}
D.H.Olson, R.E.Rundle, inorG.CheM.

69.24 Dichlorodimethyl bis(pyridine - N - oxide) tin(iv)
$C_{12}H_{16}Cl_2N_2O_2Sn$
E.A.Blom, B.R.Penfold, W.T.Robinson *J. Chem. Soc. (A)*, 913, 1969

69.25 tris - μ - (Dimethyl germanium) - di(iron tricarbonyl)
$C_{12}H_{18}Fe_2Ge_3O_6$
E.H.Brooks, M.Elder, W.A.G.Graham, D.Hall
J. Amer. Chem. Soc., **90**, 3587, 1968

69.26 Lead(ii) O,O - di - isopropyl phosphorodithioate
$C_{12}H_{28}O_4P_2PbS_4$
S.L.Lawton, G.T.Kokotailo *Nature*, **221**, 550, 1969
Also classified in 64

69.27 Lead isopropylxanthate - pyridine complex
$C_{13}H_{19}NO_2PbS_4$
H.Hagihara, N.Yoshida, Y.Watanabe *Acta Cryst. (B)*, **25**, 1775, 1969

69.C bis(Dicarbonyl - π - cyclopentadienyl iron) dichlorogermane
$C_{14}H_{10}Cl_2Fe_2Ge$
For complete entry see 73.47

69.C   Dichlorobis(dicarbonyl - π - cyclopentadienyl iron) tin(iv)
$C_{14}H_{10}Cl_2Fe_2O_4Sn$
For complete entry see 73.48

69.C   bis(Dicarbonyl - π - cyclopentadienyl iron) tin dinitrate
$C_{14}H_{10}Fe_2N_2O_4Sn$
For complete entry see 73.49

69.28 μ - Chloro - (2,2' - dipyridyl tricarbonylmolybdenum) - methyldichlorotin
$C_{14}H_{11}Cl_3MoN_2O_3Sn$
M.Elder, W.A.G.Graham, D.Hall, R.Kummer
J. Amer. Chem. Soc., **90**, 2189, 1968
Also classified in 83

69.C   bis - (Dicarbonyl - π - cyclopentadienyl - iron) - dimethyl - lead
$C_{16}H_{16}Fe_2O_4Pb$
For complete entry see 73.68

69.C   bis(Dicarbonyl - π - cyclopentadienyl iron) dimethyl tin
$C_{16}H_{16}Fe_2O_4Sn$
For complete entry see 73.69

69.29   bis(Iron tetracarbonyl - diethylgermanium)
$C_{16}H_{20}Fe_2Ge_2O_8$
J.-C.Zimmer, M.Huber   *C. R. Acad. Sci., Fr., C,* **267,** 1685, 1968

69.30   Mandelic acid - germanium dioxide complex dihydrate
$C_{16}H_{20}GeO_{10}$
C.Sterling   *J. Inorg. Nucl. Chem.*, **29,** 1211, 1967

69.31   bis(1,2 - Diethoxycarbonyl - ethyl) tin(iv) dibromide (form i)
$C_{16}H_{26}Br_2O_8Sn$
M.Yoshida, T.Ueki, N.Yasuoka, N.Kasai, M.Kakudo, I.Omae, S.Kikkawa,
S.Matsuda   *Bull. Chem. Soc. Jap.*, **41,** 1113, 1968

69.32   2,2',2' - Terpyridyl(dimethyl)chlorotin(iv) dimethyl (trichloro)stannate(iv)
$C_{17}H_{17}ClN_3Sn^+$, $C_2H_6Cl_3Sn^-$
F.W.B.Einstein, B.R.Penfold   *J. Chem. Soc. (A),* 3019, 1968
Residue 2 classified in 69

69.33   Lead 8 - mercaptoquinolinate
$C_{18}H_{12}N_2PbS_2$
E.A.Shugam, V.M.Agre, Yu.A.Bankowskii, E.A.Lukasha
Zh. Strukt. Khim., **8,** 171, 1967

69.34 **Tetrachloro - 1,4 - bistriethyl - stannyloxybenzene**
$C_{18}H_{30}Cl_4O_2Sn_2$
P.J.Wheatley *J. Chem. Soc.*, 5027, 1961

69.35 **Tetramethyl tritin tetra(iron tetracarbonyl)**
$C_{20}H_{12}Fe_4O_{16}Sn_3$
R.M.Sweet, C.J.Fritchie Junior, R.A.Schunn *Inorg. Chem.*, **6**, 749, 1967
Also classified in 71

69.36 **Acetyltriphenylgermane**
$C_{20}H_{18}GeO$
R.W.Harrison, J.Trotter *J. Chem. Soc. (A)*, 258, 1968

69.37 **Dimethyl tin bis(8 - hydroxyquinolinate)**
$C_{20}H_{18}N_2O_2Sn$
E.O.Schlemper *Inorg. Chem.*, **6**, 2012, 1967

69.38 **Lead diethyldithiocarbamate**
$C_{20}H_{40}N_4Pb_2S_8$
Z.V.Zvonkova, A.N.Khvatkina, N.S.Ivanova
Kristallografija, **12**, 1065, 1967

69.39 **Tetracarbonyl cobalt - diphenyl tin - manganese pentacarbonyl**
$C_{21}H_{10}CoMnO_9Sn$
B.P.Biryukov, O.P.Solodova, Yu.T.Struchkov
Zh. Strukt. Khim., **9**, 228, 1968

69.40 **Triphenyl tin pentacarbonyl manganese**
$C_{23}H_{15}MnO_5Sn$
H.P.Weber, R.F.Bryan *Acta Cryst.*, **22**, 822, 1967
Also classified in 71

69.41 **Tribenzyl tin acetate**
$C_{23}H_{24}O_2Sn$
N.W.Alcock, R.E.Timms *J. Chem. Soc. (A)*, 1873, 1968

69.42 **Diphenyldichloro lead**
$(C_{24}H_{20}Cl_4Pb_2)_n$
V.Busetti, M.Mammi, A.Del Pra *Acta Cryst.*, **16**, A71, 1963

69.C **bis(Dicarbonyl - π - cyclopentadienyl iron) - dicyclopentadienyl tin**
$C_{24}H_{20}Fe_2O_4Sn$
For complete entry see 73.103

69.43   **Tri(cyclo - octa - 1,5 - diene - platinum - di(tin trichloride)**
$C_{24}H_{36}Cl_6Pt_3Sn_2$
L.J.Guggenberger   *Chem. Communic.*, 512, 1968
Also classified in 74

69.C   **bis(π - Cyclopentadienyl dicarbonyl iron) di(phenylsulfonyl) tin**
$C_{26}H_{20}Fe_2O_8S_2Sn$
For complete entry see 73.110

69.44   **1,1,4,4 - Tetraphenyl - 1,4 - digermanacyclohexa - 2,5 - diene**
$C_{28}H_{24}Ge_2$
M.E.Vol'pin, V.G.Dulova, Yu.T.Struchkov, N.K.Bokiy, D.N.Kursanov
J. Organometal. Chem., **8**, 87, 1967

69.45   **1,1,4,4 - Tetraphenyl - 1,4 - digermanacyclohexa - 2,5 - diene**
$C_{28}H_{24}Ge_2$
N.G.Bokii, Yu.T.Struchkov   *Zh. Strukt. Khim.*, **8**, 122, 1967

69.46   **(4 - Bromo - 1,2,3,4 - tetraphenyl - cis,cis - 1,3 - butadienyl)dimethyl tin bromide**
$C_{30}H_{26}Br_2Sn$
F.P.Boer, J.J.Flynn, H.H.Freedman, S.V.McKinley, V.R.Sandel
J. Amer. Chem. Soc., **89**, 5068, 1967

69.47   **Tricyclohexyl tin acetate**
$C_{30}H_{36}O_2Sn$
N.W.Alcock, R.E.Timms   *J. Chem. Soc. (A)*, 1876, 1968

69.48   **Triphenylgermanium manganese pentacarbonyl**
$C_{35}H_{15}GeMnO_5$
B.T.Kilbourn, T.L.Blundell, H.M.Powell   *Chem. Communic.*, 444, 1965
Also classified in 11

69.49   **Lead hexa - antipyrine perchlorate**
$C_{66}H_{72}N_{12}O_6Pb^{2+}, 2ClO_4^-$
M.Vijayan, M.A.Viswamitra   *Acta Cryst.*, **21**, 522, 1966

TELLURIUM COMPOUNDS

70.1 α - **Dimethyltellurium dichloride**
$C_2H_6Cl_2Te$
G.D.Christofferson, J.D.McCullough, R.A.Sparks
Acta Cryst., **11,** 782, 1958
See also *Int. Distances,* M 87s; *Structure Reports,* **22,** 563, 1958

70.2 **Tellurium dimethanethiosulfonate**
$C_2H_6O_4S_4Te$
O.Foss, E.H.Vihovde *Acta Chem. Scand.*, **8,** 1032, 1954
See also *Int. Distances,* M 139; *Structure Reports,* **18,** 678, 1954

70.3 **bis(Thiourea) tellurium(ii) bromide**
$C_2H_8Br_2N_4S_2Te$
O.Foss, K.Johnsen, K.Maartmann-Moe, K.Maroy
Acta Chem. Scand., **20,** 113, 1966

70.4 **bis(Thiourea) tellurium(ii) chloride**
$C_2H_8Cl_2N_4S_2Te$
O.Foss, K.Johnsen, K.Maartmann-Moe, K.Maroy
Acta Chem. Scand., **20,** 113, 1966

70.5 **Trimethyltellurium(iv) methyltellurium(iv) tetra - iodide**
β - Dimethyltellurium di - iodide
$C_3H_9Te^+$, $CH_3I_4Te^-$
F.Einstein, J.Trotter, C.S.Williston *J. Chem. Soc. (A),* 2018, 1967

70.6 **Tellurium(ii) dimethyldithiophosphate**
$C_4H_{12}O_4P_2S_4Te$
S.Husebye *Acta Chem. Scand.*, **20,** 24, 1966

70.7 **Tellurium dimethanethiosulfonate thiourea complex**
$C_4H_{14}N_4O_4S_6Te$
O.Foss, K.Maroy, S.Husebye *Acta Chem. Scand.*, **19,** 2361, 1965

70.8   Tetrathiourea tellurium(ii) dichloride
$C_4H_{16}N_8S_4Te^{2+}$, $2Cl^-$
K.Fosheim, O.Foss, A.Scheie, S.Solheimsnes
Acta Chem. Scand., **19**, 2336, 1965

70.9   Tetrathiourea tellurium(ii) dichloride dihydrate
$C_4H_{16}N_8S_4Te^{2+}$, $2Cl^-$, $2H_2O$
K.Fosheim, O.Foss, A.Scheie, S.Solheimsnes
Acta Chem. Scand., **19**, 2336, 1965

70.10   Tellurium di(ethylxanthate)
$C_6H_{10}O_2S_4Te$
S.Husebye   *Acta Chem. Scand.*, **21**, 42, 1967

70.11   Tellurium dibromide ethylenethiourea complex
$C_6H_{12}Br_2N_4S_2Te$
O.Foss, H.M.Kjoge, K.Maroy   *Acta Chem. Scand.*, **19**, 2349, 1965

70.12   Tellurium di - iodide ethylenethiourea complex
$C_6H_{12}I_2N_4S_2Te$
O.Foss, H.M.Kjoge, K.Maroy   *Acta Chem. Scand.*, **19**, 2349, 1965

70.13   Trithiourea tellurium(ii) hydrogen difluoride
$C_6H_{24}N_{12}S_6Te_2^{4+}$, $4HF_2^-$
O.Foss, S.Hauge   *Acta Chem. Scand.*, **19**, 2395, 1965

70.14   Benzenetellurenyl bromide thiourea complex
$C_7H_9BrN_2STe$
O.Foss, S.Husebye   *Acta Chem. Scand.*, **20**, 132, 1966

70.15   Benzenetellurenyl chloride thiourea complex
$C_7H_9ClN_2S_2Te$
O.Foss, S.Husebye   *Acta Chem. Scand.*, **20**, 132, 1966

70.16   Phenyl bis(thiourea) tellurium chloride
$C_8H_{13}N_4S_2Te^+$, Cl^-
O.Foss, K.Maroy   *Acta Chem. Scand.*, **20**, 123, 1966

70.17   trans - Tetrabromo bis(tetramethylthiourea) tellurium(iv)
$C_{10}H_{24}Br_4N_4S_2Te$
S.Husebye, J.W.George   *Inorg. Chem.*, **8**, 313, 1969

70.18   trans - Tetrachloro bis(tetramethylthiourea) tellurium(iv)
$C_{10}H_{24}Cl_4N_4S_2Te$
S.Husebye, J.W.George   *Inorg. Chem.*, **8**, 313, 1969

70.19   Di - p - chlorodiphenyl tellurium di - iodide
$C_{12}H_8Cl_2I_2Te$
G.Y.Chao, J.D.McCullough   *Acta Cryst.*, **15**, 887, 1962

70.20   pp' - Dichlorodiphenyl ditelluride
$C_{12}H_8Cl_2Te_2$
F.H.Kruse, R.E.Marsh, J.D.McCullough   *Acta Cryst.*, **10**, 201, 1957
See also *Int. Distances*, M 167s; *Structure Reports*, **21**, 595, 1957

70.21   Tellurium(iv) catecholate
$C_{12}H_8O_4Te$
O.Lindqvist   *Acta Chem. Scand.*, **21**, 1473, 1967

70.22   Diphenyl tellurium dibromide
$C_{12}H_{10}Br_2Fe$
G.D.Christofferson, J.D.McCullough   *Acta Cryst.*, **11**, 249, 1958
See also *Int. Distances*, M 168s; *Structure Reports*, **22**, 704, 1958

70.23   Tellurium dibenzenethiosulfonate
$C_{12}H_{10}O_4S_4Te$
P.Oyum, O.Foss   *Acta Chem. Scand.*, **10**, 279, 1956
See also *Int. Distances*, M 169s; *Structure Reports*, **20**, 583, 1956

70.24   Tellurium ditoluene - p - thiolsulphonate
$C_{14}H_{14}O_4S_4Te$
O.Foss, P.Oyum   *Acta Chem. Scand.*, **9**, 1014, 1955
Also classified in 11
See also *Int. Distances*, M 241; *Structure Reports*, **19**, 574, 1955

70.25   Di - p - tolyl telluride
$C_{14}H_{14}Te$
W.R.Blackmore, S.C.Abrahams   *Acta Cryst.*, **8**, 317, 1955
See also *Int. Distances*, M 241; *Structure Reports*, **19**, 575, 1955

TRANSITION METAL-C COMPOUNDS

71.1 **Trichloromethyl mercury bromide**
$CBrCl_3Hg$
T.A.Babushkina, E.V.Bryukhova, F.K.Velichko, V.I.Pakhomov,
G.K.Semin *Zh. Strukt. Khim.*, **9**, 207, 1968

71.2 **trans - β - Chlorovinyl mercury chloride**
$C_2H_2Cl_2Hg$
V.I.Pakhomov, A.I.Kitaigorodskij *Zh. Strukt. Khim.*, **7**, 860, 1966

71.3 **Methoxycarbonyl mercuric chloride**
$C_2H_3ClHgO_2$
T.C.W.Mak, J.Trotter *J. Chem. Soc.*, 3243, 1962

71.4 **Methyl mercury(ii) cyanide (neutron study)**
C_2H_3HgN
J.C.Mills, H.S.Preston, C.H.L.Kennard
J. Organometal. Chem., **14**, 33, 1968

71.5 **Methylmercury(ii) cyanide**
C_2H_3HgN
J.C.Mills, H.S.Preston, C.H.L.Kennard
J. Organometal. Chem., **14**, 33, 1968

71.6 **Ethyl zinc iodide**
$(C_2H_5IZn)_n$
P.T.Moseley, H.M.M.Shearer *Chem. Communic.*, 876, 1966

71.7 **Ethyl(pentammine) rhodium(iii) bromide**
$C_2H_{20}N_5Rh^{2+}, 2Br^-$
A.C.Skapski, P.G.H.Troughton *J. Chem. Soc. (D)*, 666, 1969

71.C **Tetramethylammonium bis(π - 5,9,10 - tribromo - (1) - 2,3 - dicarbollyl) cobalt(iii)**
$C_4H_{16}B_{18}Br_6Co^-, C_4H_{12}N^+$
For complete entry see 3.31

71.C   **Cesium bis(1,2 - dicarbollyl) cobaltate**
$C_4H_{22}B_{18}Co^-$, Cs^+
For complete entry see 62.33

71.C   **Tetraethylammonium bis - (3) - 1,2 - dicarbollyl cuprate(ii)**
$C_4H_{22}B_{18}Cu^{2-}$, $2C_8H_{20}N^+$
For complete entry see 62.34

71.C   **Iron(ii) P - methyl - (3) - 1,7 - carbaphosphollide**
$C_4H_{26}B_{18}FeP_2$
For complete entry see 62.39

71.C   **Cesium π - (1) - 2,3 - dicarbollyl rhenium tricarbonyl**
$C_5H_{11}B_9O_3Re^-$, Cs^+
For complete entry see 62.41

71.8   **Phenyl mercury(ii) bromide**
C_6H_5BrHg
V.I.Pakhomov   *Zh. Strukt. Khim.*, **4**, 594, 1963

71.C   **π - Allyl tricarbonyl iron iodide**
$C_6H_5FeIO_3$
For complete entry see 72.7

71.9   **Di - μ - chloro - dichlorodimethyl tetracarbonyl di - iridium(iii)**
$C_6H_6Cl_4Ir_2O_4$
N.A.Bailey, C.J.Jones, B.L.Shaw, E.Singleton
Chem. Communic., 1051, 1967

71.10   **cis - 1,2 - Difluorovinylpentacarbonyl manganese**
$C_7HF_2MnO_5$
F.W.B.Einstein, H.Luth, J.Trotter   *J. Chem. Soc. (A)*, 89, 1967

71.11   **α - (dd,ll) - 2 - Methoxycyclohexyl - 1 - mercuric chloride**
$C_7H_{13}ClHgO$
A.G.Brook, G.F.Wright   *Acta Cryst.*, **4**, 50, 1951
See also *Int. Distances,* M 214; *Structure Reports,* **15**, 453, 1951

71.12   **β - (dl,ld) - 2 - Methoxycyclohexyl - 1 - mercuric chloride**
$C_7H_{13}ClHgO$
A.G.Brook, G.F.Wright   *Acta Cryst.*, **4**, 50, 1951
See also *Int. Distances,* M 214; *Structure Reports,* **15**, 453, 1951

71.13   **bis(1,1,2,2 - Tetrafluoroethyl) iron(ii) tetracarbonyl**
$C_8H_2F_8FeO_4$
M.R.Churchill   *Inorg. Chem.*, **6**, 185, 1967

71.C **Iodocarbonyl - π - cyclopentadienyl - pentafluoroethyl rhodium**
$C_8H_5F_5IORh$
For complete entry see 73.7

71.14 **Pentacarbonyl methyl(methylamino)carbene chromium**
$C_8H_7CrNO_5$
P.E.Baikie, E.O.Fischer, O.S.Mills *Chem. Communic.*, 1199, 1967

71.C **Trimethyl tin - pentacarbonyl manganese**
$C_8H_9MnO_5Sn$
For complete entry see 69.14

71.15 **1,7 - Dioxa - 4,10 - dimercuracyclodecane**
Mercury diethylene oxide
$C_8H_{16}Hg_2O_2$
D.Grdenic *Acta Cryst.*, **5,** 367, 1952
See also *Int. Distances,* M 219; *Structure Reports,* **16,** 474, 1952

71.16 **Diethylmonobromogold dimer**
$C_8H_{20}Au_2Br_2$
A.Buraway, C.S.Gibson, G.C.Hampson, H.M.Powell
J. Chem. Soc., 1690, 1937
See also *Int. Distances,* M 220

71.17 **Methyl zinc methoxide**
$C_8H_{24}O_4Zn_4$
H.M.M.Shearer, C.B.Spencer *Chem. Communic.*, 194, 1966

71.18 **Dimethyl gold hydroxide tetramer**
$C_8H_{28}Au_4O_4$
G.E.Glass, J.Konnert, M.G.Miles, D.Britton, R.S.Tobias
J. Amer. Chem. Soc., **90,** 1131, 1968

71.C **Cyclopentadienyl dicarbonyl carboxymethyl iron**
$C_9H_8FeO_4$
For complete entry see 73.13

71.19 **α - Napthyl mercury iodide**
$C_{10}H_7HgI$
V.I.Pakhomov *Kristallografija,* **8,** 789, 1963
Also classified in 24

71.C **Cyclopentadienyl dicarbonyl carboxymethyl molybdenum**
$C_{10}H_8MoO_5$
For complete entry see 73.15

71.20 β - **Tetramethyl ferrocyanide**
$C_{10}H_{12}FeN_6$
R.Hulme, H.M.Powell   *J. Chem. Soc.*, 719, 1957
See also *Int. Distances*, M 160s; *Structure Reports*, **21**, 546, 1957

71.21   **cis - Dicyanotetra(methylisocyanide) iron(ii) chloroform solvate**
$C_{10}H_{12}FeN_6$, $4CHCl_3$
J.B.Wilford, N.O.Smith, H.M.Powell   *J. Chem. Soc. (A)*, 1544, 1968

71.C   **Potassium bis(acetylacetonato) chloroplatinate(ii)**
$C_{10}H_{14}ClO_4Pt^-$, K^+
For complete entry see 77.4

71.C   **Di - μ - iodo - tetramethyl - μ - (dimethylsulfide) - bis(dimethylsulfide) dirhodium**
$C_{10}H_{30}I_2Rh_2S_3$
For complete entry see 85.65

71.22   **Ethylidyne tri(cobalt tricarbonyl)**
$C_{11}H_3Co_3O_9$
P.W.Sutton, L.F.Dahl   *J. Amer. Chem. Soc.*, **89**, 261, 1967

71.C   **Tricarbonyl - π - cyclopentadienyl heptafluoropropyl molybdenum**
$C_{11}H_5F_7MoO_3$
For complete entry see 73.32

71.23   **Phenylethynyl (isopropylamine) gold(i)**
$C_{11}H_{14}AuN$
P.W.R.Corfield, H.M.M.Shearer   *Acta Cryst.*, **23**, 156, 1967
Also classified in 83

71.24   **Diethylamino(methyl)carbene - pentacarbonyl chromium**
$C_{11}H_{14}CrNO_5$
J.A.Connor, O.S.Mills   *J. Chem. Soc. (A)*, 334, 1969

71.C   **Methyl - di(neopentyl)sulfonium tri - iodo di(neopentyl) sulfonium - methylzincate**
$C_{11}H_{24}I_3Zn^-$, $C_{11}H_{25}S^+$
For complete entry see 11.45

71.25   **bis(Pentafluorophenyl) mercury**
$C_{12}F_{10}Hg$
N.R.Kunchur, M.Mathew   *Chem. Communic.*, 71, 1966

71.C π - **Cyclopentadienyl iron dicarbonyl 2,4 - cyclopentadiene**
$C_{12}H_{10}FeO_2$
For complete entry see 73.39

71.26 **Diphenyl mercury**
$C_{12}H_{10}Hg$
B.Ziolkovska, R.M.Myasnikova, A.I.Kitaigorodskij
Zh. Strukt. Khim., **5**, 737, 1964

71.27 π - **Cyclopentadienyl - dicarbonyl(4 - carboethoxy - 5 - hydroxy - 1,2,3 - molybdeno - diazacyclopenta - 3,5 - diene)**
$C_{12}H_{12}MoN_2O_5$
J.R.Knox, C.K.Prout *Acta Cryst. (B)*, **25**, 1952, 1969
Also classified in 73, 83

71.28 **Hexamethylisocyanido - ferrous chloride trihydrate**
$C_{12}H_{18}FeN_6{}^{2+}$, $2Cl^-$, $3H_2O$
H.M.Powell, G.W.R.Bartindale *J. Chem. Soc.*, 799, 1945
See also *Int. Distances*, M 235; *Structure Reports*, **10**, 214, 1945

71.29 **Trimethylplatinum chloride**
$C_{12}H_{36}Cl_4Pt_4$
R.E.Rundle, J.H.Sturdivant *J. Amer. Chem. Soc.*, **69**, 1561, 1947
See also *Int. Distances*, M 155; *Structure Reports*, **11**, 607, 1947

71.30 **Trimethyl platinum hydroxide tetramer**
$C_{12}H_{40}O_4Pt_4$
T.C.Spiro, D.H.Templeton, A.Zalkin *Inorg. Chem.*, **7**, 2165, 1968

71.31 **Trimethyl platinum hydroxide tetramer**
$C_{12}H_{40}O_4Pt_4$
H.S.Preston, J.C.Mills, C.H.L.Kennard
J. Organometal. Chem., **14**, 447, 1968

71.C **Sesquiethylenediamine trimethyl platinic iodide**
$C_{12}H_{42}N_6Pt_2{}^{2+}$, $2I^-$
For complete entry see 76.80

71.32 **Phenylmethoxycarbene - pentacarbonyl chromium**
$C_{13}H_8CrO_6$
O.S.Mills, A.D.Redhouse *J. Chem. Soc. (A)*, 642, 1968

71.C **Phenyltrimethylenemethane - iron tricarbonyl**
$C_{13}H_{10}FeO_3$
For complete entry see 72.41

71.33 **Di - iodocarbomethoxycarbonyl (2,2′ - bipyridyl) iridium**
$C_{13}H_{11}I_2IrN_2O_3$
V.G.Álbano, P.L.Bellon, M.Sansoni *Inorg. Chem.*, **8**, 298, 1969
Also classified in 83

71.C **Cyclopropane bis pyridine platinum(iv) dichloride**
$C_{13}H_{16}Cl_2N_2Pt$
For complete entry see 83.125

71.C **Cyclopropane bispyridine platinum(iv) tetrachloride carbon tetrachloride solvate**
$C_{13}H_{16}Cl_4N_2Pt$, CCl_4
For complete entry see 83.126

71.C **Dimethyl - π - cyclopentadienyl (methylcyclopentadiene) rhenium**
$C_{13}H_{19}Re$
For complete entry see 73.45

71.C **Cyclo - octenyl nickel(ii) acetylacetonate**
$C_{13}H_{20}NiO_2$
For complete entry see 75.28

71.34 **(δ - (β - Acetylvinyl) - π - cyclopentadienyl - iron - carbonyl) - μ - carbonyl - (iron tricarbonyl)**
$C_{14}H_{10}Fe_2O_6$
V.G.Andrianov, Yu.T.Struchkov *Zh. Strukt. Khim.*, **9**, 845, 1968
Also classified in 72, 73

71.C **(β - Acetylvinyl)dicarbonyl - (π - cyclopentadienyl) iron μ - carbonyl iron - tricarbonyl**
$C_{14}H_{10}Fe_2O_6$
For complete entry see 73.52

71.C **π - Cyclopentadienyl tricarbonyl phenyl tungsten**
$C_{14}H_{10}O_3W$
For complete entry see 73.53

71.35 **Di - p - tolyl mercury**
$C_{14}H_{14}Hg$
N.R.Kunchur, M.Mathew *Proc. Chem. Soc.*, 414, 1964

71.36 **Chloro - (2 - methoxycyclo - octa - 1,5 - dienyl) - pyridine platinum**
$C_{14}H_{18}ClNOPt$
C.Panattoni, G.Bombieri, E.Forsellini, B.Crociai
J. Chem. Soc. (D), 187, 1969
Also classified in 75, 83

71.37 π - **Cycloheptatrienyl dicarbonyl - pentafluorophenyl - molybdenum**
$C_{15}H_7F_5MoO_2$
M.D.Rausch, A.K.Ignatowicz, M.R.Churchill, J.AmeR.CheM.SoC.
Also classified in 75

71.38   **Ethoxy (phenylamino) carbene dichloro triethylphosphine platinum(ii)**
$C_{15}H_{25}Cl_2OPPt$
E.M.Badley, J.Chatt, R.L.Richards, G.A.Sim
J. Chem. Soc. (D), 1322, 1969
Also classified in 86

71.39   **Ethylenedioxy - O,O' - di(ω - o - tolyl) mercury**
$C_{16}H_{16}HgO_2$
F.W.Kupper, H.J.Lindner   *Z. Anorg. Allg. Chem.*, **359**, 41, 1968

71.40   **Triethylamine - (tricobalt enneacarbonyl carbon) - oxyborane**
$C_{16}H_{17}BCo_3NO_{10}$
F.Klanberg, W.B.Askew, L.J.Guggenberger   *Inorg. Chem.*, **7**, 2265, 1968
Also classified in 62

71.C   **Di(μ - allyl) - bis(platinum(ii) acetylacetonate)**
$C_{16}H_{24}O_4Pt_2$
For complete entry see 77.35

71.C   **2 - (Pentamethylbicyclo(2.2.0)hexa - 2,5 - diene methyl palladium acetylacetonate**
$C_{17}H_{24}O_2Pd$
For complete entry see 75.59

71.C   **trans - 1,4 - bis(Dicarbonyl - π - cyclopentadienyl iron) buta - 1,3 - diene**
$C_{18}H_{14}Fe_2O_4$
For complete entry see 73.76

71.C   **trans - 1,4 - bis(Dicarbonyl - π - cyclopentadienyl iron) buta - 1,3 - diene**
$C_{18}H_{14}Fe_2O_4$
For complete entry see 73.77

71.41   **N,N' - Dimethyl etylenediamine - N,N' - di(ω - o - tolyl) mercury**
$C_{18}H_{22}HgN_2$
F.W.Kupper, H.J.Lindner   *Z. Anorg. Allg. Chem.*, **359**, 41, 1968

71.C   **Trimethyl(acetylacetonyl) - 2,2' - bipyridyl platinum**
$C_{18}H_{24}N_2O_2Pt$
For complete entry see 83.164

71.C   **Ethyl - (trimethyl platinum) aceto - acetate dimer**
$C_{18}H_{36}O_6Pt_2$
For complete entry see 77.36

71.C   **μ - Ethylenediamine - bis(trimethyl(acetylacetonato) platinum(iv))**
$C_{18}H_{40}N_2O_4Pt_2$
For complete entry see 77.37

71.C   **Tricarbonyl iron N - p - tolylsulfonyl - 1,4 - dimethoxy - 4 - imino - i - but - 2 - enyl iron tricarbonyl**
$C_{19}H_{15}Fe_2No_{10}S$
For complete entry see 72.54

71.42   **Iridium(iii) benzylacetophenone dichloride bis(dimethylsulfoxide)**
$C_{19}H_{25}Cl_2IrO_3S_2$
M.McPartlin, R.Mason   *Chem. Communic.*, 545, 1967
Also classified in 85

71.C   **Dichloro tris(tetrahydrofuran) - p - tolyl chromium(iii)**
p - Tolylchromium(iii) dichloride tris(tetrahydrofuran)
$C_{19}H_{31}Cl_2CrO_3$
For complete entry see 84.46

71.43   **bis(Tricarbonyl phenylferroxycarbene)**
$C_{20}H_{10}Fe_2O_8$
E.O.Fischer, V.Kiener, D.Saint P.Bunbury, E.Frank, P.F.Lindley, O.S.Mills   *Chem. Communic.*, 1378, 1968

71.C   **Tetramethyl tritin tetra(iron tetracarbonyl)**
$C_{20}H_{12}Fe_4O_{16}Sn_3$
For complete entry see 69.35

71.C   **Benzaldehydato p - tolueneimine bis(iron carbonyl)**
$C_{20}H_{13}Fe_2NO_6$
For complete entry see 83.170

71.C   **Trimethyl(salicylaldehydato) platinum(iv)**
$C_{20}H_{22}O_4Pt$
For complete entry see 78.33

71.44   **Deca(methylisonitrile) dicobalt(ii) perchlorate**
$C_{20}H_{30}Co_2N_{10}^{4+}$, $4ClO_4^-$
F.A.Cotton, T.G.Dunne, J.S.Wood   *Inorg. Chem.*, **3**, 1495, 1964

71.45 **bis(Tricobalt enneacarbonyl) acetone**
$C_{21}Co_6O_{19}$
G.Allegra, S.Valle *Acta Cryst. (B)*, **25**, 107, 1969

71.46 **Octacarbonyl diphenylvinylidene di - iron**
$C_{22}H_{10}Fe_2O_8$
O.S.Mills, A.D.Redhouse *J. Chem. Soc. (A)*, 1282, 1968

71.C **Triphenylphosphine methallyl palladium chloride**
$C_{22}H_{22}ClPPd$
For complete entry see 72.59

71.47 **p - Chlorobenzoyl pentacarbonyl rhenium**
$C_{23}H_4O_6Re$
I.S.Astakhova, A.A.Johannsson, V.A.Semion, Yu.T.Struchkov,
K.N.Anisimov, N.E.Kolabova *J. Chem. Soc. (D)*, 488, 1969

71.C **Triphenyl tin pentacarbonyl manganese**
$C_{23}H_{15}MnO_5Sn$
For complete entry see 69.40

71.C **Tricarbonyl(triphenylphosphine) - σ - tetrafluoroethyl cobalt(i)**
$C_{23}H_{16}CoF_4O_3P$
For complete entry see 86.44

71.48 **Ruthenium carbonyl carbide - mesitylene**
$C_{24}H_{12}O_{14}Ru_6$
R.Mason, W.R.Robinson *Chem. Communic.*, 468, 1968
Also classified in 74

71.C **((o - Allylphenyl)(dimethylarsine)tribromo) - β - ((ethoxymethyl) - 2 - (dimethylarsino) phenylethyl) platinum(iv)**
$C_{24}H_{35}As_2Br_3OPt$
For complete entry see 86.46

71.49 **2 - (trans - bis(Triethylphosphine) - chloro palladium) azobenzene**
$C_{24}H_{39}ClN_2P_2Pd$
R.W.Siekman, D.L.Weaver *Chem. Communic.*, 1021, 1968
Also classified in 86

71.C **Trimethyl 4,6 dioxonyl platinum**
$C_{24}H_{48}O_4Pt_2$
For complete entry see 77.58

71.C **Methylmethoxycarbene triphenylphosphine tetracarbonyl chromium**
$C_{25}H_{21}CrO_5P$
For complete entry see 86.47

71.50 **Methyl zinc t - butyl sulfide**
$C_{25}H_{60}S_5Zn_5$
G.W.Adamson, H.M.M.Shearer, C.B.Spencer
Acta Cryst., **21**, A135, 1966

71.51 **trans - Dicarbonyl - π - cyclopentadienyl (triphenylphosphine) molybdenum acetyl**
$C_{27}H_{23}MoO_3P$
M.R.Churchill, J.P.Fennessey   *Inorg. Chem.*, **7**, 953, 1968
Also classified in 73, 86

71.C **trans - bis(Phenylethynyl) bis(triethylphosphine) nickel(ii)**
$C_{28}H_{40}NiP_2$
For complete entry see 86.51

71.C **trans - bis(Phenylethynyl) bis(triethylphosphine) nickel(ii)**
$C_{28}H_{40}NiP_2$
For complete entry see 86.52

71.C **π - Cyclopentadienyl - γ - pentafluorophenyl (triphenylphosphine) nickel**
$C_{29}H_{20}F_5NiP$
For complete entry see 73.112

71.C **π - Cyclopentadienyl (triphenylphosphine) - phenyl nickel**
$C_{29}H_{25}NiP$
For complete entry see 73.113

71.52 **Chlorobis(pentafluorophenyl) triphenylphosphine gold(iii)**
$C_{30}H_{15}AuClF_{10}$
R.W.Baker, P.Pauling   *J. Chem. Soc. (D)*, 745, 1969
Also classified in 86

71.53 **Iron π - cyclopentadienyl triphenyl phosphine monocarbonyl σ - benzoyl**
$C_{31}H_{25}FeO_2P$
Y.A.Chapovskii, V.A.Semion, V.G.Andrianov, Yu.T.Struchkov
Zh. Strukt. Khim., **9**, 1100, 1968
Also classified in 73, 86

71.54 **Methylzinc isopropyl sulfide octamer**
$C_{32}H_{80}S_8Zn_8$
G.W.Adamson, H.M.M.Shearer   *J. Chem. Soc. (D)*, 897, 1969
Also classified in 85

71.55   **trans - bis(Triphenylphosphine)methyl di - iodo rhodium(iii) benzene solvate**
$C_{37}H_{33}I_2P_2Rh$, C_6H_6
P.G.H.Troughton, A.C.Skapski   *Chem. Communic.*, 575, 1968
Residue 1 also classified in 86

71.C   **trans - Dimesityl bis(diethylphenylphosphine) cobalt(ii)**
$C_{38}H_{52}CoP_2$
For complete entry see 86.86

METAL PI-COMPLEXES (OPEN-CHAIN)

72.1   Potassium tribromoethylene platinate monohydrate
        $C_2H_4Br_3Pt^-$, K^+, H_2O
        G.B.Bokii, G.A.Kukina   *Zh. Strukt. Khim.*, **6,** 706, 1965

72.2   Potassium (ethylenetrichloroplatinum(ii)) monohydrate
        $C_2H_4Cl_3Pt^-$, K^+, H_2O
        M.Black, R.H.B.Mais, P.G.Owston   *Acta Cryst. (B),* **25,** 1753, 1969

72.3   cis Platinum(ii) ethylene ammine bromide
        $C_2H_7Br_2NPt$
        G.A.Kukina, G.B.Bokii, F.A.Brusentsev   *Zh. Strukt. Khim.*, **5,** 730, 1964

72.4   Ethylene palladium chloride
        trans - Di - μ - chloro bis(π - ethylene) dipalladium(ii)
        $C_4H_8Cl_4Pd_2$
        J.N.Dempsey, N.C.Baenziger   *J. Amer. Chem. Soc.*, **77,** 4984, 1955
        See also *Int. Distances,* M 169; *Structure Reports,* **19,** 538, 1955

72.5   trans - Ethylenedimethylamine dichloride platinum
        $C_4H_{11}Cl_2NPt$
        P.R.H.Alderman, P.G.Owston, J.M.Rowe   *Acta Cryst.*, **13,** 149, 1960

72.6 π - Allyl di(thiourea) nickel chloride
        $C_5H_{11}N_4NiS_2^+$, Cl^-
        A.Sirigu   *J. Chem. Soc. (D),* 256, 1969
        Residue 1 also classified in 79

72.7 π - Allyl tricarbonyl iron iodide
        $C_6H_5FeIO_3$
        M.Kh.Minasyants, Yu.T.Struchkov   *Zh. Strukt. Khim.*, **9,** 665, 1968
        Also classified in 71

72.8 **Allyl palladium chloride (three dimensional, at −140 °C)**
Di - μ - chlorobis(π - allyl) dipalladium(ii)
$C_6H_{10}Cl_2Pd_2$
A.E.Smith *Acta Cryst.*, **18,** 331, 1965

72.9 **Allyl palladium chloride**
Di - μ - chlorobis(π - allyl) dipalladium(ii)
$C_6H_{10}Cl_2Pd_2$
W.E.Oberhansli, L.F.Dahl *J. Organometal. Chem.*, **3,** 43, 1965

72.10 **Allyl palladium chloride (projections)**
Di - μ - chlorobis(π - allyl) dipalladium(ii)
$C_6H_{10}Cl_2Pd_2$
J.M.Rowe *Proc. Chem. Soc.*, 66, 1962

72.11 **Allyl palladium chloride (projections)**
Di - μ - chlorobis(π - allyl) dipalladium(ii)
$C_6H_{10}Cl_2Pd_2$
V.F.Levdik, M.A.Porai-Koshits *Zh. Strukt. Khim.*, **3,** 472, 1962

72.12 **bis(Allyl) chromium(ii)**
$C_6H_{10}Cr$
G.Albrecht, D.Stock *Z. Chem.*, **7,** 321, 1967

72.13 **Tetracarbonyl acrylonitrile iron**
$C_7H_3FeNO_4$
A.R.Luxmoore, M.R.Truter *Acta Cryst.*, **15,** 1117, 1962

72.14 **Butadiene iron tricarbonyl (at −40 °C)**
C_7H_6FeO
O.S.Mills, G.Robinson *Acta Cryst.*, **16,** 758, 1963

72.15 **Carbonyl piperidine - N - carbonitrile nickel(0)**
$C_7H_{10}N_2NiO$
K.Krogmann, R.Mattes *Angew. Chem.*, **78,** 1064, 1966

72.16 **Perfluorobutadiene iron tetracarbonyl**
$C_8F_6FeO_4$
P.B.Hitchcock, R.Mason *Chem. Communic.*, 242, 1967

72.17 **DL - Tetracarbonyl fumaric acid iron**
$C_8H_4FeO_8$
C.Pedone, A.Sirigu *Acta Cryst.*, **23,** 759, 1967
Also classified in 1

72.18   (−) - **Tetracarbonyl(fumaric acid) iron**
$C_8H_4FeO_8$
C.Pedone, A.Sirigu   *Inorg. Chem.*, **7**, 2614, 1968
Also classified in 81

72.19 π - **Allyl** - π - **cyclopentadienyl palladium**
$C_8H_{10}Pd$
M.Kh.Minasyants, Yu.T.Struchkov   *Zh. Strukt. Khim.*, **9**, 481, 1968
Also classified in 73

72.20   **bis(Butadiene) rhodium(i) chloride**
$C_8H_{12}Rh^+$, Cl^-
A.Immirzi, G.Allegra   *Acta Cryst. (B)*, **25**, 120, 1969

72.21 π - **2** - **Methallyl palladium chloride dimer**
Di - μ - chlorobis(methallyl) dipalladium(ii)
$C_8H_{14}Cl_2Pd_2$
R.Mason, A.G.Wheeler   *J. Chem. Soc. (A)*, 2549, 1968

72.22   **Methallyl nickel(ii)**
$C_8H_{14}Ni$
R.Uttech, H.Dietrich   *Z. Kristallogr.*, **122**, 60, 1965

72.23   **Rhodium chloride ethylene complex**
$C_8H_{16}Cl_2Rh_2$
K.A.Klanderman   *Dissert. Abstr.*, **25**, 6253, 1965

72.24   **Cyclopentadienyl rhodium diethylene complex**
$C_9H_{13}Rh$
K.A.Klanderman   *Dissert. Abstr.*, **25**, 6253, 1965
Also classified in 73

72.25   **pentakis(Methylisonitrile) cobalt(i) perchlorate**
$C_{10}H_{15}CoN_5^+$, ClO_4^-
F.A.Cotton, T.G.Dunne, J.S.Wood   *Inorg. Chem.*, **4**, 318, 1965

72.26   **Dichloro(2,7 - dimethyl - octa - 2,6 - diene - 1,8 - diyl) ruthenium(iv)**
$C_{10}H_{16}Cl_2Ru$
L.Porri, M.C.Gallazzi, A.Colombo, G.Allegra
Tetrahedron Letters, 4187, 1965

72.27   **2 - Carboxyethyl** - π - **allyl nickel bromide dimer**
$C_{10}H_{18}Br_2Ni_2O_4$
M.R.Churchill, T.A.O'Brien   *Inorg. Chem.*, **6**, 1386, 1967

72.28 π - **1,3** - **Dimethylallyl palladium chloride dimer**
Di - μ - chloro bis(π - 1,3 - dimethylallyl) dipalladium(ii)
$C_{10}H_{18}Cl_2Pd_2$
G.R.Davies, R.H.B.Mais, S.O'Brien, P.G.Owston
Chem. Communic., 1151, 1967

72.29 **Phenylethynyl(trimethylphosphine) silver(i)**
$C_{11}H_{14}AgP$
P.W.R.Corfield, H.M.M.Shearer *Acta Cryst.*, **20,** 502, 1966
Also classified in 86

72.30 **Phenylethynyl(trimethylphosphine) copper(i)**
$C_{11}H_{14}CuP$
P.W.R.Corfield, H.M.M.Shearer *Acta Cryst.*, **21,** 957, 1966
Also classified in 86

72.31 **trans** - **Butadiene iron carbonyl complex**
$C_{12}H_6FeO_8$
K.A.Klanderman *Dissert. Abstr.*, **25,** 6253, 1965

72.32 **1,5** - **Di(iron tricarbonyl)** - **3** - **methylene** - **penta** - **1,4** - **diene**
$C_{12}H_6Fe_2O_6$
P.Piret, J.Meunier-Piret, M.van Meerssche, G.S.D.King
Acta Cryst., **19,** 78, 1965

72.33 **Iron carbonyl hydride** - **but** - **2** - **yne complex**
$C_{12}H_8Fe_2O_8$
A.A.Hock, O.S.Mills *Acta Cryst.*, **14,** 139, 1961

72.34 **2,3** - **Dimethylbuta** - **1,3** - **diene** - **osmium carbonyl complex**
$C_{12}H_8O_6Os_2$
R.P.Dodge, O.S.Mills, V.Schomaker *Proc. Chem. Soc.*, 380, 1963

72.35 **Butadiene cobalt dicarbonyl**
$C_{12}H_{12}Co_2O_4$
R.O.Jones, E.N.Maslen *Z. Kristallogr.*, **123,** 330, 1966

72.C **Geijerene** - **silver nitrate complex**
3 - isoPropenyl - 4 - methyl - 4 - vinyl cyclohexene(i) - silver nitrate complex
$C_{12}H_{18}Ag_2{}^{2+}$, $2NO_3{}^-$
For complete entry see 75.25

72.36 **Di - μ - chloro - bis - (π - 1 - ethoxycarbonyl - 2 - hydroxyallyl) dipalladium(ii)**
$C_{12}H_{18}Cl_2O_3Pd_2$
K.Oda, N.Yasuoka, T.Ueki, N.Kasai, M.Kakudo, Y.Tezuka, T.Ogura, S.Kawaguchi *Chem. Communic.*, 989, 1968

72.37 **Dichloro(dodeca - 2,6,10 - triene - 1,12 - diyl) ruthenium(iv)**
$C_{12}H_{18}Cl_2Ru$
J.E.Lydon, M.R.Truter *J. Chem. Soc. (A),* 362, 1968

72.38 **trans - Dichloro cis - 2 - butene α - phenylethylamine platinum(ii)**
$C_{12}H_{19}Cl_2NPt$
P.Ganis, C.Pedone *Ric. Sci., 2, A,* **8,** 1462, 1965
Also classified in 83

72.39 **3 - Methylhepta - 1,6 - dienyl cobalt butadiene**
$C_{12}H_{19}Co$
G.Allegra, F.L.Giudice, G.Natta, U.Giannini, G.Fagherazzi, P.Pino *Chem. Communic.*, 1263, 1967

72.40 **bis - π - Allyl rhodium chloride**
$C_{12}H_{20}Cl_2Rh_2$
M.McPartlin, R.Mason *Chem. Communic.*, 16, 1967

72.41 **Phenyltrimethylenemethane - iron tricarbonyl**
$C_{13}H_{10}FeO_3$
M.R.Churchill, K.Gold *Inorg. Chem.*, **8,** 401, 1969
Also classified in 71

72.C **(δ - (β - Acetylvinyl) - π - cyclopentadienyl - iron - carbonyl) - μ - carbonyl - (iron tricarbonyl)**
$C_{14}H_{10}Fe_2O_6$
For complete entry see 71.34

72.C **(β - Acetylvinyl)dicarbonyl - (π - cyclopentadienyl) iron μ - carbonyl iron - tricarbonyl**
$C_{14}H_{10}Fe_2O_6$
For complete entry see 73.52

72.42 **π - 1,1,3,3 - Tetramethylallyl palladium chloride dimer**
$C_{14}H_{26}Cl_2Pd_2$
R.Mason, A.G.Wheeler *Inorg. Chem.*, **7,** 2543, 1968

72.43 μ - **Allene bis(acetylacetonato rhodium carbonyl)**
$C_{15}H_{18}O_6Rh_2$
P.Racanelli, G.Pantini, A.Immirzi, G.Allegra, L.Porri
J. Chem. Soc. (D), 361, 1969
Also classified in 77

72.44   **1,2 - Dimethylallyl - dicyclopentadienyl titanium(iii)**
$C_{15}H_{19}Ti$
R.B.Helmholdt, F.Jellinek, H.A.Martin, A.Vos
Rec. Trav. Chim. Pays-Bas, **86,** 1263, 1967
Also classified in 73

72.45   **Rhodium acetylacetonate bis(1,1 - dimethylallene)**
$C_{15}H_{23}O_2Rh$
P.Racanelli, G.Pantini, A.Immirzi, G.Allegra, L.Porri
J. Chem. Soc. (D), 361, 1969
Also classified in 77

72.C β - **Gorgonene** - **silver nitrate complex (absolute configuration)**
$C_{15}H_{24}AgNO_3$
For complete entry see 53.10

72.46   **Tetracobalt decacarbonyl (diethyl)acetylene**
$C_{16}H_{10}Co_4O_{10}$
L.F.Dahl, D.L.Smith   *J. Amer. Chem. Soc.,* **84,** 2450, 1962

72.47   **Styrene - palladium(ii) chloride dimer**
$C_{16}H_{16}Cl_4Pd_2$
N.C.Baenziger, J.R.Doyle, G.F.Richards, C.L.Carpenter
Adv. Chem. Coord. Compounds,
Proc. 6th Internat. Conf., Detroit, 131, 1961

72.48   **Dicyclopentadienyl nickel dimethylacetylene dicarboxylate**
$C_{16}H_{16}NiO_4$
L.F.Dahl, C.H.Wei   *Inorg. Chem.,* **2,** 713, 1963
Also classified in 73

72.C   **Di(μ - allyl) - bis(platinum(ii) acetylacetonate)**
$C_{16}H_{24}O_4Pt_2$
For complete entry see 77.35

72.49 Di - t - butylacetylene p - toluidine platinum(ii) chloride
$C_{17}H_{27}Cl_2NPt$
G.R.Davies, W.Hewertson, R.H.B.Mais, P.G.Owston
Chem. Communic., 423, 1967
Also classified in 16

72.50 Tricarbonyl - N - cinnamylideneaniline iron
$C_{18}H_{13}FeNO_3$
A.DeCian, R.Weiss *Chem. Communic.*, 348, 1968

72.51 μ - Butadiene bis(cyclopentadienyl manganese dicarbonyl)
$C_{18}H_{16}Mn_2O_4$
M.Ziegler *Z. Anorg. Allg. Chem.*, **355**, 12, 1967
Also classified in 73

72.52 Di(cobalt dicarbonyl) 1,6 - di(t - butyl) - 1,3,5 - hexatriene complex
$C_{18}H_{22}Co_2O_4$
O.S.Mills, G.Robinson *Proc. Chem. Soc.*, 187, 1964

72.53 1,5 - Hexadiene - silver perchlorate complex
$C_{18}H_{30}Ag_2{}^{2+}$, $2ClO_4{}^-$
I.W.Bassi, G.Fagherazzi *J. Organometal. Chem.*, **13**, 535, 1968

72.54 Tricarbonyl iron N - p - tolylsulfonyl - 1,4 - dimethoxy - 4 - imino - i - but - 2 - enyl iron tricarbonyl
$C_{19}H_{15}Fe_2No_{10}S$
L.Rodrique, M.van Meerssche, P.Piret *Acta Cryst. (B)*, **25**, 519, 1969
Also classified in 71, 83

72.55 bis(Tetramethylallene) rhodium(i) acetylacetonate
$C_{19}H_{31}O_2Rh$
T.G.Hewitt, K.Azenhofer, J.J.de Boer *J. Chem. Soc. (D)*, 312, 1969
Also classified in 77

72.56 Diphenylacetylene dicobalt hexacarbonyl
$C_{20}H_{10}Co_2O_6$
W.G.Sly *J. Amer. Chem. Soc.*, **81**, 18, 1959
See also *Int. Distances*, M 192s; *Structure Reports*, **23**, 696, 1959

72.57 Di - iron hexacarbonyl diphenylacetylene complex
$C_{20}H_{10}Fe_2O_6$
Y.Degreve, J.Piret, M.van Meerssche, P.Piret *Acta Cryst.*, **23**, 119, 1967

72.58 π - **Allyl(triphenylphosphine) palladium - trichlorotin acetone solvate**
$C_{21}H_{20}Cl_3PPdSn$, $0.5C_2H_6O$
R.Mason, G.B.Robertson, P.O.Whimp, D.A.White
Chem. Communic., 1655, 1968
Residue 1 also classified in 86

72.59   **Triphenylphosphine methallyl palladium chloride**
$C_{22}H_{22}ClPPd$
R.Mason, D.R.Russell   *Chem. Communic.*, 26, 1966
Also classified in 86, 71

72.60   **Vitamin A aldehyde iron tricarbonyl**
$C_{22}H_{26}FeO_4$
A.J.Birch, H.Fitton, R.Mason, G.B.Robertson, J.E.Stangroom
Chem. Communic., 613, 1966
Also classified in 54

72.61   **Diphenylacetylene - tri(iron tricarbonyl) complex**
$C_{23}H_{10}Fe_3O_9$
J.F.Blount, L.F.Dahl, C.Hoogzand, W.Hubel
J. Amer. Chem. Soc., **88,** 292, 1966

72.62 π - **Allyl triphenyl phosphine dicarbonyl iodo iron (monoclinic form)**
$C_{23}H_{20}FeIO_2P$
M.Kh.Minasyants, V.G.Andrianov, Yu.T.Struchkov
Zh. Strukt. Khim., **9,** 1055, 1968
Also classified in 86

72.63 π - **Allyl - triphenylphosphine dicarbonyl iodo iron (triclinic form)**
$C_{23}H_{20}FeIO_2P$
M.Kh.Minasyants, V.G.Andrianov, Yu.T.Struchkov
Zh. Strukt. Khim., **9,** 1055, 1968
Also classified in 86

72.64   **Diphenylacetylene bis(cyclopentadienyl nickel)**
$C_{24}H_{20}Ni_2$
O.S.Mills, B.W.Shaw   *J. Organometal. Chem.*, **11,** 595, 1968
Also classified in 73

72.65   **Di - iron hexacarbonyl tris(diethylacetylene)**
$C_{24}H_{30}Fe_2O_6$
J.A.M.Case   *Dissert. Abstr. (B)*, **28,** 2786, 1968

72.66   **Tetracarbonyl (diphenyl - 2 - (prop - cis - 1 - enyl)phenylphosphine) molybdenum(0)**
$C_{25}H_{17}MoO_4P$
H.Luth, M.R.Truter   *J. Chem. Soc. (A)*, 28, 1969
Also classified in 86

72.67   **But - 2 - enyl cobalt buta - 1,3 - diene triphenylphosphine benzene solvate**
$C_{26}H_{28}CoP$, $0.5C_6H_6$
L.Porri, G.Vitulli, M.Zocchi, G.Allegra   *J. Chem. Soc. (D)*, 276, 1969
Residue 1 also classified in 86

72.68   **Allene - di - iron hexacarbonyl - triphenylphosphine**
$C_{27}H_{19}Fe_2O_6P$
R.E.Davis   *Chem. Communic.*, 248, 1968
Also classified in 86

72.69   **Iron carbonyl phenylacetylene complex**
$C_{30}H_{18}Fe_2O_6$
G.S.D.King   *Acta Cryst.*, 15, 243, 1962

72.70   **Methallyl (bis - 1,2 - (diphenylphosphino)ethane) nickel bromide**
$C_{30}H_{31}BrNiP_2$
M.R.Churchill, T.A.O'Brien   *Chem. Communic.*, 246, 1968
Also classified in 86

72.71   **Tetraphenylbutatriene tetracarbonyl iron**
$C_{32}H_{22}FeO_4$
D.Bright, O.S.Mills   *Chem. Communic.*, 211, 1966

72.72   **Carbonyl bis(diphenylacetylene) - π - cyclopentadienyl niobium**
$C_{34}H_{25}NbO$
A.N.Nesmeyanov, A.I.Gusev, A.A.Pasynskii, K.N.Amisimov,
N.E.Kolobova, Yu.T.Struchkov   *J. Chem. Soc. (D)*, 277, 1969
Also classified in 73

72.73   **π - Cyclopentadienyl tantalum bis(diphenylacetylene) carbonyl**
π - Cyclopentadienyl tantalum carbonyl ditolan
$C_{34}H_{25}OTa$
G.G.Aleksandrov, A.I.Gusev, Yu.T.Struchkov
Zh. Strukt. Khim., 9, 333, 1968
Also classified in 73

72.74   **Iron carbonyl - diphenylacetylene complex (black form)**
$C_{36}H_{20}Fe_3O_8$
R.P.Dodge, V.Schomaker   *J. Organometal. Chem.*, 3, 274, 1965

72.75 **Iron carbonyl - diphenylacetylene complex (violet form)**
$C_{36}H_{20}Fe_3O_8$
R.P.Dodge, V.Schomaker *J. Organometal. Chem.*, **3**, 274, 1965

72.C **bis(Triphenylphosphine) tetrafluoroethylene - rhodium(i) chloride**
$C_{38}H_{30}ClF_4P_2Rh$
For complete entry see 86.84

72.76 **bis(Triphenylphosphine) ethylene nickel(0)**
$C_{38}H_{34}NiP$
C.D.Cook, C.H.Koo, S.C.Nyburg, M.T.Shiomi
Chem. Communic., 426, 1967
Also classified in 86

72.77 **bis(Triphenylphosphine) ethylene nickel(0)**
$C_{38}H_{34}NiP$
W.Dreissig, H.Dietrich *Acta Cryst. (B),* **24,** 108, 1968
Also classified in 86

72.C **bis(Triphenylphosphine) iridium carbonyl iodide tetrafluoroethylene**
$C_{39}H_{30}F_4IIrOP_2$
For complete entry see 86.87

72.78 **Iodobis(triphenylphosphine)allene - rhodium**
$C_{39}H_{34}IP_2Rh$
T.Kashiwaga, N.Yasuoka, N.Kasai, M.Kakudo
J. Chem. Soc. (D), 317, 1969
Also classified in 86

72.79 **Dicarbonyl bis(cyclopentadienyl) bis - μ - diphenylacetylene diniobium**
$C_{40}H_{30}Nb_2O_2$
A.N.Nesmeyanov, A.I.Gusev, A.A.Pasynskii, K.N.Anisimov,
N.E.Kolobova, Yu.T.Struchkov *Chem. Communic.*, 1365, 1968
Also classified in 73

72.80 **Dichloro - π - methylallyl bis(triphenylarsine) rhodium**
$C_{40}H_{37}As_2Cl_2Rh$
T.G.Hewitt, J.J.de Boer *Chem. Communic.*, 1413, 1968
Also classified in 86

72.81 **Tetracyanoethylene - platinum bis(triphenylphosphine)**
$C_{42}H_{30}N_4P_2Pt$
C.Panattoni, G.Bombieri, U.Belluco *J. Amer. Chem. Soc.*, **90**, 798, 1968
Also classified in 86

72.C   Iridium carbonyl bromide bis(triphenylphosphine) tetracyanoethylene
$C_{43}H_{30}BrIrN_4OP_2$
For complete entry see 86.98

72.82 π - Cyclopentadienyl - π - tetraphenylcyclobutadiene - π - diphenylacetylene
niobium carbonyl
$C_{48}H_{35}NbO$
A.N.Nesmeyanov, A.I.Gusev, A.A.Pasynskii, K.N.Anisimov,
N.E.Kolobova, Yu.T.Struchkov   J. Chem. Soc. (D), 739, 1969
Also classified in 73

72.83   bis(Triphenylphosphine)diphenylacetylene platinum
$C_{50}H_{40}P_2Pt$
J.O.Glanville, J.M.Stewart, S.O.Grim   J. Organometal. Chem., 7, P9, 1967
Also classified in 86

72.84   bis(π - Cyclopentadienyl niobium bis(diphenylacetylene) carbonyl)
bis(π - Cyclopentadienyl niobium carbonyl ditolan)
$C_{68}H_{50}Nb_2O_2$
G.G.Aleksandrov, A.I.Gusev, Yu.T.Struchkov
Zh. Strukt. Khim., 9, 333, 1968
Also classified in 73

METAL PI-COMPLEXES (CYCLOPENTADIENE)

73.1 Cyclopentadienyl dinitrosyl - isocyanato chromium
$C_6H_5CrN_3O_3$
M.A.Bush, G.A.Sim, G.R.Knox, M.Ahmad, C.G.Robertson
J. Chem. Soc. (D), 74, 1969

73.2 π - Cyclopentadienyl iron dicarbonyl - tin tribromide
$C_7H_5Br_3FeO_2Sn$
R.F.Bryan, P.T.Greene, G.A.Melson, P.F.Stokely, A.R.Manning
J. Chem. Soc. (D), 722, 1969

73.3 Dicarbonyl cyclopentadienyl cobalt - mecury(ii) chloride complex
$C_7H_5Cl_2CoHgO_2$
I.N.Nowell, D.R.Russell *Chem. Communic.*, 817, 1967

73.4 π - Cyclopentadienyl iron dicarbonyl - tin trichloride
$C_7H_5Cl_3FeO_2Sn$
R.F.Bryan, P.T.Greene, G.A.Melson, P.F.Stokely, A.R.Menning
J. Chem. Soc. (D), 722, 1969

73.5 π - Cyclopentadienyl - π - (1)2,3 - dicarbollyl iron(iii)
$C_7H_{16}B_9Fe$
A.Zalkin, D.H.Templeton, T.E.Hopkins
J. Amer. Chem. Soc., **87**, 3988, 1965
Also classified in 62

73.6 π - Cyclopentadienyl molybdenum tricarbonyl chloride
$C_8H_5ClMoO_3$
S.Chaiwasie, R.H.Fenn *Acta Cryst. (B)*, **24**, 525, 1968

73.7 Iodocarbonyl - π - cyclopentadienyl - pentafluoroethyl rhodium
$C_8H_5F_5IORh$
M.R.Churchill *Inorg. Chem.*, **4**, 1734, 1965
Also classified in 71

73.8   Cyclopentadienyl manganese tricarbonyl
$C_8H_5MnO_3$
A.F.Berndt, R.E.Marsh   *Acta Cryst.*, **16**, 118, 1963

73.C π - Allyl - π - cyclopentadienyl palladium
$C_8H_{10}Pd$
For complete entry see 72.19

73.9 π - Cyclopentadienyl cis - 1,2 - bis(trifluoromethyl) ethanedithione cobalt
$C_9H_5CoF_6S_2$
H.W.Baird, B.M.White   *J. Amer. Chem. Soc.*, **88**, 4744, 1966
Also classified in 85

73.10 π - Cyclopentadienyl(1,2 - dicyanoethene - 1,2 - dithiolato) cobalt
$C_9H_5CoN_2S_2$
M.R.Churchill, J.P.Fennessey   *Inorg. Chem.*, **7**, 1123, 1968
Also classified in 85

73.11   Cyclopentadienyl niobium carbonyl
$C_9H_5NbO_4$
W.Baird, L.F.Dahl   *Amer. Chem. Soc., Abstr. Papers*, **145**, 27N, 1963

73.12 π - Cyclopentadienyl vanadium tetracarbonyl
$C_9H_5O_4V$
J.B.Wilford, A.Whitla, H.M.Powell   *J. Organometal. Chem.*, **8**, 495, 1967

73.13   Cyclopentadienyl dicarbonyl carboxymethyl iron
$C_9H_8FeO_4$
M.L.H.Green, J.K.P.Ariyaratne, A.M.Bjerrum, M.Ishaq, C.K.Prout
Chem. Communic., 430, 1967
Also classified in 71

73.C   **Cyclopentadienyl rhodium diethylene complex**
$C_9H_{13}Rh$
For complete entry see 72.24

73.C   **(+) - α - Phenethylammonium (−) - tricarbonyl manganese - α - methyl - cyclopentadienyl - carboxylate (absolute config.)**
$C_{10}H_7MnO_5{}^-$, $C_8H_{12}N^+$
For complete entry see 3.57

73.14   **Ferrocene disulfonyl chloride**
$C_{10}H_8Cl_2FeO_4S_2$
O.V.Starovskii, Yu.T.Struchkov   *Zh. Strukt. Khim.*, **5**, 257, 1964

73.15   **Cyclopentadienyl dicarbonyl carboxymethyl molybdenum**
$C_{10}H_8MoO_5$
M.L.H.Green, J.K.P.Ariyaratne, A.M.Bjerrum, M.Ishaq, C.K.Prout
Chem. Communic., 430, 1967
Also classified in 71

73.16   **Cyclopentadienyl dinitrosyl chromium chloride**
$C_{10}H_{10}ClCrN_2O_2$
O.L.Carter, A.T.McPhail, G.A.Sim   *J. Chem. Soc. (A)*, 1095, 1966

73.17   **Dicyclopentadiene platinum chloride**
$C_{10}H_{10}Cl_2Pt$
N.C.Baenziger, J.R.Doyle, G.F.Richards, C.L.Carpenter
Adv. Chem. Coord. Compounds,
Proc. 6th Internat. Conf., Detroit, 131, 1961

73.18   **bis(Cyclopentadienyl titanium dichloride) oxide**
$C_{10}H_{10}Cl_4OTi_2$
P.Ganis, G.Allegra   *Atti Accad. Nazion. Lincei, R. C.,*
Cl. Sci. Fis. Mat. Nat., **33,** 303, 1962

73.19   **bis(Cyclopentadienyl) cobalt perchlorate**
$C_{10}H_{10}Co^+$, ClO_4^-
E.Frasson, G.Bombieri, C.Panattoni   *Acta Cryst.,* **16,** A68, 1963

73.20   **bis - Cyclopentadienyl chromium(ii)**
$C_{10}H_{10}Cr$
E.Weiss, E.O.Fischer   *Z. Anorg. Allg. Chem.,* **284,** 69, 1956
See also *Int. Distances,* M 158s; *Structure Reports,* **20,** 549, 1956

73.21   **bisCyclopentadienyl iron**
Ferrocene
$C_{10}H_{10}Fe$
P.F.Eiland, R.Pepinsky   *J. Amer. Chem. Soc.,* **74,** 4971, 1952
See also *Int. Distances,* M 225; *Structure Reports,* **16,** 505, 1952

73.22   **bisCyclopentadienyl iron**
Ferrocene
$C_{10}H_{10}Fe$
J.D.Dunitz, L.E.Orgel, A.Rich   *Acta Cryst.,* **9,** 373, 1956
See also *Int. Distances,* M 225; *Structure Reports,* **20,** 550, 1956

73.23   **Ferrocene (neutron study)**
$C_{10}H_{10}Fe$
B.T.M.Willis   *Acta Cryst.,* **13,** A114, 1960

73.C   **Ferrocene - tetracyanoethylene complex**
$C_{10}H_{10}Fe$, C_6N_4
For complete entry see 60.119

73.24   **Ferricinium tri - iodide**
$C_{10}H_{10}Fe^+$, I_3^-
T.Bernstein, F.H.Herbstein   *Acta Cryst. (B)*, **24**, 1640, 1968

73.25   **Ferricinium picrate**
$C_{10}H_{10}Fe^+$, $C_6H_2N_3O_7^-$
R.C.Pettersen   *Dissert. Abstr. (B)*, **27**, 3894, 1967
Residue 2 classified in 15

73.26   **Tricarbonyl - π - cyclopentadienyl ethyl molybdenum**
$C_{10}H_{10}MoO_3$
M.J.Bennett, R.Mason   *Proc. Chem. Soc.*, 273, 1963

73.27   **bis(Cyclopentadienyl oxomolybdenum sulfide)**
$C_{10}H_{10}Mo_2O_2S_2$
D.L.Stevenson, L.F.Dahl   *J. Amer. Chem. Soc.*, **89**, 3721, 1967

73.28   **Dicyclopentadienyl ruthenium**
Ruthenocene
$C_{10}H_{10}Ru$
G.L.Hardgrove, D.H.Templeton   *Acta Cryst.*, **12**, 28, 1959
See also *Int. Distances*, M 158s; *Structure Reports*, **23**, 692, 1959

73.29   **Dihydrido di - π - (cyclopentadienyl) molybdenum**
$C_{10}H_{12}Mo$
M.Gerloch, R.Mason   *J. Chem. Soc.*, 296, 1965

73.30   **Dihydrido di - (π - cyclopentadienyl) molybdenum (discussion)**
$C_{10}H_{12}Mo$
S.C.Abrahams, A.P.Ginsberg   *Inorg. Chem.*, **5**, 500, 1966

73.31   **Cyclopentadienyl dicarbonyl iron mercury tetracarbonyl cobalt**
$C_{11}H_5CoFeHgO_6$
R.F.Bryan, H.P.Weber   *Acta Cryst.*, **21**, A138, 1966

73.32   **Tricarbonyl - π - cyclopentadienyl heptafluoropropyl molybdenum**
$C_{11}H_5F_7MoO_3$
M.R.Churchill, J.P.Fennessey   *Inorg. Chem.*, **6**, 1213, 1967
Also classified in 71

73.33 Tricarbonyl trimethylene - 1,2 - cyclopentadienyl rhenium
$C_{11}H_9O_3Re$
K.K.Joshi, R.H.B.Mais, F.Nyman, P.G.Owston, A.M.Wood
J. Chem. Soc. (A), 318, 1968

73.34 Perfluorocyclopentadiene dicobalt heptacarbonyl
$C_{12}Co_2F_5O_7$
P.B.Hitchcock, R.Mason *Chem. Communic.*, 503, 1966

73.35 Tricarbonyl tetrakis(trifluoromethyl) cyclopentadienone iron
$C_{12}F_{12}FeO_4$
N.A.Bailey, R.Mason *Acta Cryst.*, **21**, 652, 1966

73.36 π - Cyclopentadienyl iron dicarbonyl manganese pentacarbonyl
$C_{12}H_5FeMnO_7$
P.J.Hansen, R.A.Jacobson *J. Organometal. Chem.*, **6**, 389, 1966

73.37 π - Cyclopentadienyl - tricarbonyl - molybdenum - manganese - pentacarbonyl
$C_{12}H_5MnMoO_7$
B.P.Biryukov, Yu.T.Struchkov *Zh. Strukt. Khim.*, **9**, 655, 1968

73.38 Di - iron hexacarbonyl - methylenecyclopentadiene complex
$C_{12}H_6Fe_2O_6$
J.Meunier-Piret, P.Piret, M.van Meerssche *Acta Cryst.*, **19**, 85, 1965

73.39 π - Cyclopentadienyl iron dicarbonyl 2,4 - cyclopentadiene
$C_{12}H_{10}FeO_2$
M.J.Bennett, F.A.Cotton, A.Davison, J.W.Faller, S.J.Lippard,
S.M.Morehouse *J. Amer. Chem. Soc.*, **88**, 4371, 1966
Also classified in 71

73.C π - Cyclopentadienyl - dicarbonyl(4 - carboethoxy - 5 - hydroxy - 1,2,3 - molybdeno - diazacyclopenta - 3,5 - diene)
$C_{12}H_{12}MoN_2O_5$
For complete entry see 71.27

73.C π - Cyclopentadienyl - π - cycloheptatrienyl vanadium
$C_{12}H_{12}V$
For complete entry see 75.24

73.40 bis - π - Cyclopentadienyl (2 - aminoethanethiolato) molybdenum iodide
$C_{12}H_{16}MoNS^+, I^-$
J.R.Knox, C.K.Prout *Acta Cryst. (B)*, **25**, 2482, 1969
Residue 1 also classified in 83, 85

73.41 π - **Cyclopentadienyl tricarbonyl molylbdenum - manganese pentacarbonyl**
$C_{13}H_5MnMoO_8$
B.P.Biryukov, Yu.T.Struchkov   *Zh. Strukt. Khim.*, **9**, 655, 1968

73.42 **Dicyclopentadienyl tricarbonyl rhodium**
$C_{13}H_{10}O_3Rh_2$
O.S.Mills, J.P.Nice   *J. Organometal. Chem.*, **10**, 337, 1967

73.43 α - **Keto - 1,1' - trimethyleneferrocene**
$C_{13}H_{12}FeO$
N.D.Jones, R.E.Marsh, J.H.Richards   *Acta Cryst.*, **19**, 330, 1965

73.44 **Quinidine (–) - 1,1' - dimethylferrocene - 3 - carboxylate monohydrate**
(absolute configuration)
$C_{13}H_{13}FeO_2^-$, $C_{20}H_{25}N_2O_2^+$, H_2O
O.L.Carter, A.T.McPhail, G.A.Sim   *J. Chem. Soc. (A)*, 365, 1967
Residue 1 also classified in 58

73.45 **Dimethyl -** π **- cyclopentadienyl (methylcyclopentadiene) rhenium**
$C_{13}H_{19}Re$
N.W.Alcock   *J. Chem. Soc. (A)*, 2001, 1967
Also classified in 71

73.46 **tetrakis(Trifluoromethyl) cyclopentadienone -** π **- cyclopentadiene cobalt**
$C_{14}H_5CoF_{12}O$
M.Gerloch, R.Mason   *Proc. R. Soc., A*, **279**, 170, 1964

73.47 **bis(Dicarbonyl -** π **- cyclopentadienyl iron) dichlorogermane**
$C_{14}H_{10}Cl_2Fe_2Ge$
M.A.Bush, P.Woodward   *J. Chem. Soc. (A)*, 1833, 1967
Also classified in 69

73.48 **Dichlorobis(dicarbonyl -** π **- cyclopentadienyl iron) tin(iv)**
$C_{14}H_{10}Cl_2Fe_2O_4Sn$
J.E.O'Connor, E.R.Corey   *Inorg. Chem.*, **6**, 968, 1967
Also classified in 69

73.49 **bis(Dicarbonyl -** π **- cyclopentadienyl iron) tin dinitrate**
$C_{14}H_{10}Fe_2N_2O_4Sn$
B.P.Biryukov, Yu.T.Struchkov   *Zh. Strukt. Khim.*, **9**, 488, 1968
Also classified in 69

73.50 **cis - bis(** π **- Cyclopentadienyl dicarbonyl iron) (at low temp.)**
$C_{14}H_{10}Fe_2O_4$
R.F.Bryan, P.T.Greene, D.S.Field, M.J.Newlands
J. Chem. Soc. (D), 1477, 1969

73.51 **trans - bis(π - Cyclopentadienyl dicarbonyl iron)**
$C_{14}H_{10}Fe_2O_4$
O.S.Mills *Acta Cryst.*, **11,** 620, 1958
See also *Int. Distances,* M 179s; *Structure Reports,* **22,** 680, 1958

73.52 **(β - Acetylvinyl)dicarbonyl - (π - cyclopentadienyl) iron μ - carbonyl iron - tricarbonyl**
$C_{14}H_{10}Fe_2O_6$
V.G.Andrianov, Yu.T.Struchkov *Chem. Communic.*, 1590, 1968
Also classified in 71, 72

73.C **(δ - (β - Acetylvinyl) - π - cyclopentadienyl - iron - carbonyl) - μ - carbonyl - (iron tricarbonyl)**
$C_{14}H_{10}Fe_2O_6$
For complete entry see 71.34

73.53 **π - Cyclopentadienyl tricarbonyl phenyl tungsten**
$C_{14}H_{10}O_3W$
V.A.Semion, Yu.T.Struchkov *Zh. Strukt. Khim.*, **9,** 1046, 1968
Also classified in 71

73.54 **bis(Cyclopentadienyl dicarbonyl ruthenium)**
$C_{14}H_{10}O_4Ru_2$
O.S.Mills, J.P.Nice *J. Organometal. Chem.*, **9,** 339, 1967

73.C **Dicarbonyl - π - cyclopentadienylbicyclo(2.2.1)hepta - 2π,5 - diene manganese(i)**
$C_{14}H_{13}MnO_2$
For complete entry see 75.34

73.55 **α - Keto - 1,5 - tetramethyleneferrocene**
$C_{14}H_{14}FeO$
E.B.Fleischer, S.W.Hawkinson *Acta Cryst.*, **22,** 376, 1967

73.56 **1,1' - Diacetylferrocene**
$C_{14}H_{14}FeO_2$
G.J.Palenik
Amer. Cryst. Assoc., Abstr. Papers (Winter Meeting), 62, 1967

73.57 **Diacetyl ruthenocene**
$C_{14}H_{14}O_2Ru_2$
J.Trotter *Acta Cryst.*, **16,** 571, 1963

73.58 **Cyclopentadiene - tetramethylcyclopentadienone cobalt(0)**
$C_{14}H_{17}CoO$
L.F.Dahl, D.L.Smith *J. Amer. Chem. Soc.*, **83,** 752, 1961

73.59 t - Butylferrocenyl nitroxide
$C_{14}H_{18}FeNO$
A.R.Forrester, S.P.Hepburn, R.S.Dunlop, H.H.Mills
J. Chem. Soc. (D), 698, 1969
Also classified in 12

73.60 cls - Di - μ - dimethylamido - bis(cyclopentadienyl - nitrosyl chromium)
$C_{14}H_{22}Cr_2N_4O_2$
M.A.Bush, G.A.Sim, G.R.Knox, M.Ahmad, C.G.Robertson
J. Chem. Soc. (D), 74, 1969
Also classified in 83

73.61 trans - Di - μ - dimethylamido - bis(cyclopentadienyl - nitrosyl chromium)
$C_{14}H_{22}Cr_2N_4O_2$
M.A.Bush, G.A.Sim, G.R.Knox, M.Ahmad, C.G.Robertson
Chem. Communic., 74, 1969
Also classified in 83

73.C p - Methyl - π - benzyl - π - cyclopentadienyl dicarbonyl molybdenum
$C_{15}H_{14}MoO_2$
For complete entry see 74.17

73.62 π - Tri - cyclopentadienyl uranium chloride
$C_{15}H_{15}ClU$
C.H.Wong, T.-M.Yen, T.Lee   *Acta Cryst.*, **18**, 340, 1965

73.63 Tri(cyclopentadienyl manganese) tetranitrosyl
$C_{15}H_{15}Mn_3N_4O_4$
R.C.Elder, F.A.Cotton, R.A.Schunn   *J. Amer. Chem. Soc.*, **89**, 3645, 1967

73.64 Tri(cyclopentadienyl nickel) - di - μ - sulfide
$C_{15}H_{15}N_{13}S_2$
H.Varenkamp, V.A.Uchtman, L.F.Dahl
J. Amer. Chem. Soc., **90**, 3272, 1968
Also classified in 85

73.65 tris(Cyclopentadienyl) samarium(iii)
$C_{15}H_{15}Sm$
C.H.Wong, T.Lee, Y.Lee   *Acta Cryst. (B)*, **25**, 2580, 1969

73.C 1,2 - Dimethylallyl - dicyclopentadienyl titanium(iii)
$C_{15}H_{19}Ti$
For complete entry see 72.44

73.66 **bisCyclopentadienyl molybdenum tricarbonyl**
$C_{16}H_{10}Mo_2O_6$
F.C.Wilson, D.P.Shoemaker *J. Chem. Phys.*, **27**, 809, 1957
See also *Int. Distances*, M 185s; *Structure Reports*, **21**, 572, 1957

73.67 π - **Cyclopentadienyl 1 - phenylcyclopentadiene cobalt**
$C_{16}H_{15}Co$
M.R.Churchill, R.Mason *Proc. R. Soc., A,* **279**, 191, 1964

73.68 **bis - (Dicarbonyl - π - cyclopentadienyl - iron) - dimethyl - lead**
$C_{16}H_{16}Fe_2O_4Pb$
B.P.Biryukov, Yu.T.Struchkov, K.N.Anisimov, N.E.Kolobova,
V.V.Skripkin *Zh. Strukt. Khim.*, **9**, 922, 1968
Also classified in 69

73.69 **bis(Dicarbonyl - π - cyclopentadienyl iron) dimethyl tin**
$C_{16}H_{16}Fe_2O_4Sn$
B.P.Biryukov, Yu.T.Struchkov *Zh. Strukt. Khim.*, **9**, 488, 1968
Also classified in 69

73.C **Dicyclopentadienyl nickel dimethylacetylene dicarboxylate**
$C_{16}H_{16}NiO_4$
For complete entry see 72.48

73.70 μ - **(Dimethylphosphido) hydrido di - π - cyclopentadienyl tetracarbonyl dimolybdenum**
$C_{16}H_{17}Mo_2O_4P$
R.J.Doedens, L.F.Dahl *J. Amer. Chem. Soc.*, **87**, 2576, 1965

73.71 **1,1' - 3,3' - bis(Trimethylene)ferrocene**
$C_{16}H_{18}Fe$
I.C.Paul *Chem. Communic.*, 377, 1966

73.72 **1,1' - Tetramethylethyleneferrocene**
$C_{16}H_{20}Fe$
M.B.Laing, K.N.Trueblood *Acta Cryst.*, **19**, 373, 1965

73.C π - **Cyclopentadienyl hexakis(trifluoromethyl) benzene rhodium**
$C_{17}H_5F_{18}Rh$
For complete entry see 74.19

73.73 π - **Cyclopentadienyl - (1 - benzoyl - cyclopentadienyl) cobalt**
$C_{17}H_{14}CoO$
M.R.Churchill *J. Organometal. Chem.*, **4**, 258, 1965

73.74 **Tricyclopentadienyl trinickel dicarbonyl**
$C_{17}H_{15}Ni_3O_2$
A.A.Hock, O.S.Mills *Adv. Chem. Coord. Compounds,*
Proc. 6th Internat. Conf., Detroit, 640, 1961

73.75 **bis(π - Cyclopentadienyl) (toluene - 3,4 - dithiolato) molybdenum**
$C_{17}H_{16}MoS_2$
J.R.Knox, C.K.Prout *Acta Cryst. (B),* **25,** 2013, 1969
Also classified in 85

73.76 **trans - 1,4 - bis(Dicarbonyl - π - cyclopentadienyl iron) buta - 1,3 - diene**
$C_{18}H_{14}Fe_2O_4$
R.E.Davis *Chem. Communic.,* 1218, 1968
Also classified in 71

73.77 **trans - 1,4 - bis(Dicarbonyl - π - cyclopentadienyl iron) buta - 1,3 - diene**
$C_{18}H_{14}Fe_2O_4$
M.R.Churchill, J.Worwald, W.P.Giering, G.F.Emerson
Chem. Communic., 1217, 1968
Also classified in 71

73.78 **bis - Indenyl ruthenium**
$C_{18}H_{14}Ru$
N.C.Webb, R.E.Marsh *Acta Cryst.,* **22,** 382, 1967

73.79 **tris(π - Cyclopentadienyl carbonyl rhodium)**
$C_{18}H_{15}O_3Rh_3$
O.S.Mills, E.F.Paulus *J. Organometal. Chem.,* **10,** 331, 1967

73.80 **π - Cyclopentadienyl rhodium carbonyl trimer**
$C_{18}H_{15}O_3Rh_3$
H.Muller, K.Dehnicke *J. Organometal. Chem.,* **10,** P3, 1967

73.81 **tris(Cyclopentadienyl - carbonyl - rhodium)**
$C_{18}H_{15}O_3Rh_3$
E.F.Paulus *Acta Cryst. (B),* **25,** 2206, 1969

73.C **μ - Butadiene bis(cyclopentadienyl manganese dicarbonyl)**
$C_{18}H_{16}Mn_2O_4$
For complete entry see 72.51

73.C **Cyclo - octatetraene di(cyclopentadienyl cobalt)**
$C_{18}H_{18}Co_2$
For complete entry see 75.61

73.C π - **Cyclopentadienyl - 1,2,3,4 - tetramethyl - 1 - exo - cyclopenta - 1',3' - dienecyclobutenyl nickel(ii)**
$C_{18}H_{22}Ni$
For complete entry see 75.62

73.82 **N - t - Butyl - tri(π - cyclopentadienyl nickel) amine**
$C_{19}H_{24}NNi$
S.Otsuka, A.Nakamura, T.Yoshida *Inorg. Chem.*, **7**, 261, 1968
Also classified in 83

73.83 π - **Cyclopentadienyl -** π - **2,6 - di - t - butyl - 1,4 - benzoquinone rhodium**
$C_{19}H_{25}O_2Rh$
G.G.Aleksandrov, A.I.Gusev, V.S.Khandarova, Yu.T.Struchkov,
S.P.Gubin *J. Chem. Soc. (D)*, 748, 1969
Also classified in 74

73.84 **bis(π - Cyclopentadienyl cobalt) bis(t - butylimine) carbonyl**
$C_{19}H_{28}Co_2N_2O$
Y.Matsu-ura, N.Yasuoka, T.Ueki, N.Kasai, M.Kakudo, T.Yoshida,
S.Otsuka *Chem. Communic.*, 1122, 1967
Also classified in 83

73.85 **bis(1 - (2' - Chloroferrocenyl))**
$C_{20}H_{16}Cl_2Fe_2$
Z.L.Kaluski, Yu.T.Struchkov
Bull. Acad. Polon. Sci., Ser. Sci. Chim., **14**, 719, 1966

73.86 **Biferrocenyl**
$C_{20}H_{18}Fe_2$
A.C.Macdonald, J.Trotter *Acta Cryst.*, **17**, 872, 1964

73.87 **Biferrocenyl**
$C_{20}H_{18}Fe_2$
Z.L.Kaluski, Yu.T.Struchkov, R.L.Avoyan
Zh. Strukt. Khim., **5**, 743, 1964

73.88 **Cyclotetra(μ - oxo - chloro -** π - **cyclopentadienyl titanium(iv))**
$C_{20}H_{20}Cl_4O_4Ti_4$
A.C.Skapski, P.G.H.Troughton, H.H.Sutherland
Chem. Communic., 1418, 1968

73.89 **Cyclopentadienyl iron sulfide (orthorhombic form)**
$C_{20}H_{20}Fe_4S_4$
R.A.Schunn, C.J.Fritchie, C.T.Prewitt *Inorg. Chem.*, **5**, 892, 1966

73.90   **Cyclopentadienyl iron sulfide (monoclinic form)**
$C_{20}H_{20}Fe_4S_4$
C.H.Wei, G.R.Wilkes, P.M.Treichel, L.F.Dahl   *Inorg. Chem.*, **5**, 900, 1966

73.91   **Cyclopentadienyl nickel sulfide**
$C_{20}H_{20}Ni_5S_4$
H.Vahrenkamp, L.F.Dahl   *Angew. Chem.*, **81**, 152, 1969
Also classified in 85

73.92   **Tetra(cyclopentadienyl) hydrido trirhodium**
$C_{20}H_{21}Rh_3$
O.S.Mills, E.F.Paulus   *J. Organometal. Chem.*, **11**, 587, 1968

73.93   **Di - π - (1,2 dicyclopentadienyl 1,2 bis (dimethyl amino)ethane)di - μ - carbonyl dicarbonyl di - iron**
$C_{20}H_{22}Fe_2N_2O_4$
P.McArdle, A.R.Manning, F.S.Stephens   *J. Chem. Soc. (D)*, 1310, 1969

73.94   **1,1' - bis(Pentamethyldisilanyl) - ferrocene**
$C_{20}H_{38}FeSi_4$
K.Hirotsu, T.Higuchi, A.Shimada   *Bull. Chem. Soc. Jap.*, **41**, 1557, 1968
Also classified in 63

73.95   **Diferrocenyl ketone**
$C_{21}H_{18}Fe_2O$
J.Trotter, A.C.Macdonald   *Acta Cryst.*, **21**, 359, 1966

73.96   **Chlorobis(π - cyclopentadienyl dicarbonyl iron) - (π - cyclopentadienyl tricarbonyl molybdenum) tin(iv)**
$C_{22}H_{15}ClFe_2MoO_7Sn$
J.E.O'Connor, E.R.Corey   *J. Amer. Chem. Soc.*, **89**, 3930, 1967

73.97   **4 - Biphenylylferrocene**
$C_{22}H_{18}Fe$
C.S.Williston   *Dissert. Abstr. (B)*, **28**, 2801, 1968

73.98   **2 - Biphenylylferrocene**
$C_{22}H_{18}Fe$
J.Trotter, C.S.Williston   *J. Chem. Soc. (A)*, 1379, 1967

73.99   **trans - Di - μ - phenylthio - dinitrosyl bis - (π - cyclopentadienyl) dichromium(i)**
$C_{22}H_{20}Cr_2N_2O_2S_2$
A.T.McPhail, G.A.Sim   *J. Chem. Soc. (A)*, 1858, 1968
Also classified in 85

73.100 Chloro - π - cyclopentadienyl bis(8 - quinolinolato) titanium(iv)
$C_{23}H_{17}ClN_2O_2Ti$
J.D.Matthews, A.G.Swallow *J. Chem. Soc. (D)*, 882, 1969
Also classified in 83, 84

73.101 Dibenzoylferrocene
$C_{24}H_{18}FeO_2$
Yu.T.Struchkov *Dokl. Akad. Nauk S. S. S. R.*, **110**, 67, 1956
See also *Int. Distances*, M 199s; *Structure Reports*, **20**, 551, 1956

73.102 Di - μ - phenylthio - bis(cyclopentadienyl carbonyl iron) (at −160 °C)
$C_{24}H_{20}Fe_2O_2S_2$
G.Ferguson, C.Hannaway, K.M.S.Islam *Chem. Communic.*, 1165, 1968
Also classified in 85

73.103 bis(Dicarbonyl - π - cyclopentadienyl iron) - dicyclopentadienyl tin
$C_{24}H_{20}Fe_2O_4Sn$
B.P.Biryukov, Yu.T.Struchkov, K.N.Anisimov, N.E.Kolobova,
V.V.Skripkin *Chem. Communic.*, 1193, 1968
Also classified in 69

73.C Diphenylacetylene bis(cyclopentadienyl nickel)
$C_{24}H_{20}Ni_2$
For complete entry see 72.64

73.104 Diacetyldiferrocenyl
$C_{24}H_{22}Fe_2O_4$
Z.L.Kaluski *Acta Cryst.*, **21**, A119, 1966

73.105 bis(1 - (1' - Carbomethoxy ferrocenyl)) (α form)
$C_{24}H_{22}Fe_2O_4$
Z.L.Kaluski, Yu.T.Struchkov
Bull. Acad. Polon. Sci., Ser. Sci. Chim., **16**, 557, 1968

73.106 1,12 - Dimethyl(1.1)ferrocenophane
$C_{24}H_{24}Fe_2$
J.S.McKechnie, B.H.Bersted, I.C.Paul, W.E.Watts
J. Organometal. Chem., **8**, 29, 1967

73.107 Diethylbiferrocene
$C_{24}H_{26}Fe_2$
Z.L.Kaluski, Yu.T.Struchkov *Zh. Strukt. Khim.*, **6**, 104, 1965

73.108 Di - π - cyclopentadienyl dicobalt tricyclopentadienyl
$C_{25}H_{24}Co_2$
O.V.Starovskii, Yu.T.Struchkov *Zh. Strukt. Khim.*, **6**, 248, 1965

73.109   Tricarbonyl - π - cyclopentadienyl tungstio (triphenylphosphine) gold
$C_{26}H_{20}AuO_3PW$
J.B.Wilford, H.M.Powell *J. Chem. Soc. (A)*, 8, 1969
Also classified in 86

73.110   bis(π - Cyclopentadienyl dicarbonyl iron) di(phenylsulfonyl) tin
$C_{26}H_{20}Fe_2O_8S_2Sn$
R.F.Bryan, A.R.Manning *Chem. Communic.*, 1220, 1968
Also classified in 69

73.C   trans - Dicarbonyl - π - cyclopentadienyl (triphenylphosphine) molybdenum
acetyl
$C_{27}H_{23}MoO_3P$
For complete entry see 71.51

73.111   Dicyclopentadienyl titanium diethyl aluminium dimer
$C_{28}H_{40}Al_2Ti_2$
P.Corradini, A.Sirigu *Inorg. Chem.*, **6,** 601, 1967
Also classified in 68

73.112 π - Cyclopentadienyl - γ - pentafluorophenyl (triphenylphosphine) nickel
$C_{29}H_{20}F_5NiP$
M.R.Churchill, T.A.O'Brien *J. Chem. Soc. (A)*, 2970, 1968
Also classified in 71, 86

73.113 π - Cyclopentadienyl (triphenylphosphine) - phenyl nickel
$C_{29}H_{25}NiP$
M.R.Churchill, T.A.O'Brien *J. Chem. Soc. (A)*, 266, 1969
Also classified in 71, 86

73.114 π - Cyclopentadienyl triphenylphosphine - monocarbonyl - σ - phenyl iron
$C_{30}H_{25}FeOP$
R.L.Avoyan, Yu.A.Chapovskii, Yu.T.Struchkov
Zh. Strukt. Khim., **7,** 900, 1966
Also classified in 86

73.115   Diferrocenylferrocene
$C_{30}H_{26}Fe_3$
Z.L.Kaluski *Acta Cryst.*, **21,** A119, 1966

73.C   Iron π - cyclopentadienyl triphenyl phosphine monocarbonyl σ - benzoyl
$C_{31}H_{25}FeO_2P$
For complete entry see 71.53

73.116 Dicarbonyl - 1 - carbomethoxy - 2 - phenyl - 2 - (π - 2′,4′ - dicarbomethoxy - 3′,5′ - diphenyl - 1′ - cyclopentadienyloxy) - σ - vinyl iron
$C_{33}H_{24}FeO_9$
L.F.Dahl, R.J.Doedens, W.Hubel, J.Nielsen
J. Amer. Chem. Soc., **88**, 446, 1966

73.C Carbonyl bis(diphenylacetylene) - π - cyclopentadienyl niobium
$C_{34}H_{25}NbO$
For complete entry see 72.72

73.C π - Cyclopentadienyl tantalum bis(diphenylacetylene) carbonyl
π - Cyclopentadienyl tantalum carbonyl ditolan
$C_{34}H_{25}OTa$
For complete entry see 72.73

73.117 Diphenylphosphino - cyclopentadienyl - cobalt dimer
$C_{34}H_{30}Co_2P_2$
J.M.Coleman, L.F.Dahl *J. Amer. Chem. Soc.*, **89**, 542, 1967
Also classified in 86

73.118 Diphenylphosphine - cyclopentadienyl - nickel dimer
$C_{34}H_{30}Ni_2P_2$
J.M.Coleman, L.F.Dahl *J. Amer. Chem. Soc.*, **89**, 542, 1967
Also classified in 86

73.C Dicarbonyl bis(cyclopentadienyl) bis - μ - diphenylacetylene diniobium
$C_{40}H_{30}Nb_2O_2$
For complete entry see 72.79

73.119 Triphenylphosphite diphenyl(o - π - cyclopentadienyl - phenyl)phosphite iron iodide
$C_{41}H_{33}FeIO_6P_2$
V.G.Andrianov, Yu.T.Struchkov *Zh. Strukt. Khim.*, **9**, 503, 1968
Also classified in 86

73.120 π - Cyclopentadienyl - bis(triphenylphosphite) iron iodide
$C_{41}H_{35}FeIO_6P_2$
V.G.Andrianov, Yu.T.Struchkov *Zh. Strukt. Khim.*, **9**, 240, 1968
Also classified in 64

73.C π - Cyclopentadienyl - π - tetraphenylcyclobutadiene - π - diphenylacetylene niobium carbonyl
$C_{48}H_{35}NbO$
For complete entry see 72.82

73.121   **Triphenylstannoxy tetraphenyl cyclopentadienyl tricarbonyl manganese**
$C_{50}H_{35}MnO_4Sn$
R.F.Bryan, H.P.Weber   *J. Chem. Soc. (A)*, 843, 1967

73.122   **bis(Diphenylphosphino) acetylene bis(π - cyclopentadienyl iron - di - μ - carbonyl (π - cyclopentadienyl - iron carbonyl))**
$C_{52}H_{40}Fe_4O_6P_2$
A.J.Carty, T.W.Ng, W.Carter, G.J.Palenik, T.Birchall
J. Chem. Soc. (D), 1101, 1969
Also classified in 86

73.C   **bis(π - Cyclopentadienyl niobium bis(diphenylacetylene) carbonyl)**
bis(π - Cyclopentadienyl niobium carbonyl ditolan)
$C_{68}H_{50}Nb_2O_2$
For complete entry see 72.84

METAL PI-COMPLEXES (ARENE)

74.1   **Silver perchlorate - benzene complex**
$(C_6H_6Ag^+)_n$, $(ClO_4^-)_n$
H.G.Smith, R.E.Rundle *J. Amer. Chem. Soc.*, **80,** 5075, 1958
See also *Int. Distances*, M 128s; *Structure Reports*, **22,** 698, 1958

74.2   **Silver(i) aluminium chloride - benzene complex**
$C_6H_6AgAlCl_4$
R.W.Turner, E.L.Amma *J. Amer. Chem. Soc.*, **88,** 3243, 1966

74.3   **Benzene - copper(i) aluminium tetrachloride complex**
$C_6H_6AlCl_4Cu$
R.W.Turner, E.L.Amma *J. Amer. Chem. Soc.*, **88,** 1877, 1966

74.4   **Benzene chromium tricarbonyl**
$C_9H_6CrO_3$
M.F.Bailey, L.F.Dahl *Inorg. Chem.*, **4,** 1314, 1965

74.5   **Benzene chromium tricarbonyl**
$C_9H_6CrO_3$
G.Allegra *Atti Accad. Nazion. Lincei, R. C.,*
Cl. Sci. Fis. Mat. Nat., **31,** 241, 1961

74.C   **Tricarbonyl chromium anisole - 1,3,5 - trinitrobenzene complex**
$C_{10}H_8CrO_4$, $C_6H_3N_3O_6$
For complete entry see 60.118

74.6   **Tricarbonyl chromium - o - toluidine**
$C_{10}H_9CrNO_3$
O.L.Carter, A.T.McPhail, G.A.Sim *J. Chem. Soc. (A)*, 228, 1967

74.C   **(−) - α - Phenethylammonium (−) - tricarbonyl chromium - m - toluate (absolute config.)**
$C_{11}H_7CrO_5^-$, $C_8H_{12}N^+$
For complete entry see 3.58

74.C $(-) - \alpha$ - **Phenethylammonium (+) - tricarbonyl chromium - o - toluate (absolute config.)**
$C_{11}H_7CrO_5{}^-$, $C_8H_{12}N^+$
For complete entry see 3.59

74.7   **Tricarbonyl chromium - methyl benzoate**
$C_{11}H_8CrO_5$
O.L.Carter, A.T.McPhail, G.A.Sim   *J. Chem. Soc. (A)*, 1619, 1967

74.8   **Palladium(ii) aluminium chloride benzene complex**
$C_{12}H_{12}Al_4Cl_{14}Pd_2$
G.Allegra, A.Immirzi, L.Porri   *J. Amer. Chem. Soc.*, **87**, 1394, 1965

74.9   **Dibenzene chromium(0)**
$C_{12}H_{12}Cr$
F.A.Cotton, W.A.Dollase, J.S.Wood   *J. Amer. Chem. Soc.*, **85**, 1543, 1963

74.10   **Dibenzene chromium(0) (refinement)**
$C_{12}H_{12}Cr$
J.A.Ibers   *J. Chem. Phys.*, **40**, 3129, 1964

74.11   **Dibenzene chromium(0) (at 100 ° K)**
$C_{12}H_{12}Cr$
E.Keulen, F.Jellinek   *J. Organometal. Chem.*, **5**, 490, 1966

74.12   **Dibenzene chromium**
$C_{12}H_{12}Cr$
F.Jellinek   *J. Organometal. Chem.*, **1**, 43, 1963

74.13   **Naphthalene chromium tricarbonyl (orthorhombic form)**
$C_{13}H_8CrO_3$
V.Kunz, W.Nowacki   *Helv. Chim. Acta*, **50**, 1052, 1967

74.14   **1 - Aminonaphthalene - tricarbonyl chromium**
$C_{13}H_9CrNO_3$
O.L.Carter, A.T.McPhail, G.A.Sim   *J. Chem. Soc. (A)*, 1866, 1968

74.15   **Ditoluene chromium iodide**
$C_{14}H_{16}Cr^+$, I^-
O.V.Starovskii, Yu.T.Struchkov   *Zh. Strukt. Khim.*, **2**, 162, 1961

74.16   **Ditoluene chromium 7,7,8,8 - tetracyanoquinodimethane**
$C_{14}H_{16}Cr^+$, $C_{12}H_4N_4{}^-$
R.P.Shibaeva, L.O.Atovmyan, L.P.Rozenberg
J. Chem. Soc. (D), 649, 1969
Residue 2 classified in 7, 12

74.C   **Ditoluenechromium - bis(7,7,8,8 - tetracyanoquinodimethane) complex**
$C_{14}H_{16}Cr^+$, $C_{12}H_4N_4^-$, $C_{12}H_4N_4$
For complete entry see 60.140

74.17   **p - Methyl - π - benzyl - π - cyclopentadienyl dicarbonyl molybdenum**
$C_{15}H_{14}MoO_2$
F.A.Cotton, M.D.LaPrade   *J. Amer. Chem. Soc.*, **90**, 5418, 1968
Also classified in 73

74.18   **Hexamethylbenzene chromium tricarbonyl**
$C_{15}H_{18}CrO_3$
M.F.Bailey, L.F.Dahl   *Inorg. Chem.*, **4**, 1298, 1965

74.19   **π - Cyclopentadienyl hexakis(trifluoromethyl) benzene rhodium**
$C_{17}H_5F_{18}Rh$
M.R.Churchill, R.Mason   *Proc. R. Soc., A*, **292**, 61, 1966
Also classified in 73

74.20   **Tricarbonylanthracene - chromium**
$C_{17}H_{10}CrO_3$
F.Hanic, O.S.Mills   *J. Organometal. Chem.*, **11**, 151, 1968

74.21   **Phenanthrene chromium tricarbonyl (monoclinic form)**
$C_{17}H_{10}CrO_3$
H.Deuschl, W.Hoppe   *Acta Cryst.*, **17**, 800, 1964

74.22   **Phenanthrene chromium tricarbonyl (orthorhombic form)**
$C_{17}H_{10}CrO_3$
K.W.Muir, G.Ferguson, G.A.Sim   *J. Chem. Soc. (B)*, 467, 1968

74.23   **9,10 - Dihydrophenanthrene chromium tricarbonyl**
$C_{17}H_{12}CrO_3$
K.W.Muir, G.Ferguson   *J. Chem. Soc. (B)*, 476, 1968

74.24   **bis(Tricarbonyl chromium) biphenyl (α form)**
$C_{18}H_{10}Cr_2O_6$
P.Corradini, G.Allegra   *J. Amer. Chem. Soc.*, **82**, 2075, 1960

74.25   **bis(Tricarbonyl chromium) biphenyl (β form)**
$C_{18}H_{10}Cr_2O_6$
G.Allegra   *Atti Accad. Nazion. Lincei, R. C.,*
Cl. Sci. Fis. Mat. Nat., **31**, 399, 1961

74.C   **π - Cyclopentadienyl - π - 2,6 - di - t - butyl - 1,4 - benzoquinone rhodium**
$C_{19}H_{25}O_2Rh$
For complete entry see 73.83

74.C **Ruthenium carbonyl carbide - mesitylene**
$C_{24}H_{12}O_{14}Ru_6$
For complete entry see 71.48

74.26 **bis(Cyclohexylbenzene) silver(i) perchlorate**
$C_{24}H_{32}Ag^+$, ClO_4^-
E.A.Hall, E.L.Amma   *Chem. Communic.*, 622, 1968

74.C **Tri(cyclo - octa - 1,5 - diene - platinum - di(tin trichloride)**
$C_{24}H_{36}Cl_6Pt_3Sn_2$
For complete entry see 69.43

74.C π - **Tetraphenylborato bis(trimethoxy phosphine) rhodium(i)**
$C_{30}H_{38}BO_6P_2Rh$
For complete entry see 86.57

METAL PI-COMPLEXES
(MISCELLANEOUS RING SYSTEMS)

75.1 Thiophene chromium tricarbonyl
$C_7H_4CrO_3S$
M.F.Bailey, L.F.Dahl *Inorg. Chem.*, **4**, 1306, 1965

75.2 Norbornadiene - silver nitrate complex
$C_7H_8Ag_2^{2+}$, $2NO_3^-$
N.C.Baenziger, H.L.Haight, R.Alexander, J.R.Doyle
Inorg. Chem., **5**, 1399, 1966

75.3 Norbornadiene dichloro palladium(ii)
C_7H_8ClPd
N.C.Baenziger, G.F.Richards, J.R.Doyle *Acta Cryst.*, **18**, 924, 1965

75.4 Norbornadiene dichloro palladium(ii) (liquid nitrogen temp.)
$C_7H_8Cl_2Pd$
N.C.Baenziger, J.R.Doyle, C.Carpenter *Acta Cryst.*, **14**, 303, 1961

75.5 Cyclo - octatetraene silver nitrate
$C_8H_8Ag^+$, NO_3^-
F.S.Mathews, W.N.Lipscomb *J. Phys. Chem.*, **63**, 845, 1959
See also *Int. Distances*, M 145s; *Structure Reports*, **23**, 639, 1959

75.6 Cyclo - octatetraene copper(i) chloride
$C_8H_8Cl_2Cu$
N.C.Baenziger, G.F.Richards, J.R.Doyle *Inorg. Chem.*, **3**, 1529, 1964

75.7 Cyclo - octatetraene palladium(ii) chloride (red form, at 100 ° K)
$C_8H_8Cl_2Pd$
C.V.Gorbel *Dissert. Abstr. (B)*, **28**, 625, 1967

75.8 Octafluorocyclohexa - 1,3 - diene iron tricarbonyl
$C_9H_8FeO_3$
M.R.Churchill, R.Mason *Proc. R. Soc., A*, **301**, 433, 1967

75.9   cis,cis,cis - 1,4,7 - Cyclononatriene silver nitrate complex
$C_9H_{12}Ag_3^{3+}$, $3NO_3^-$
R.B.Jackson, W.E.Streib   *J. Amer. Chem. Soc.*, **89**, 2539, 1967

75.10   Tropone iron tricarbonyl
$C_{10}H_6FeO_4$
R.P.Dodge   *J. Amer. Chem. Soc.*, **86**, 5429, 1964

75.11 π - Cycloheptatrienyl molybdenum tricarbonyl tetrafluoroborate
$C_{10}H_7MoO_3^+$, BF_4^-
G.R.Clark, G.J.Palenik   *J. Chem. Soc. (D)*, 667, 1969

75.12   Cycloheptatriene molybdenum tricarbonyl
$C_{10}H_8MoO_3$
J.D.Dunitz, P.Pauling   *Helv. Chim. Acta*, **43**, 2188, 1960

75.13   Cycloheptatrienyl vanadium(0) tricarbonyl
$C_{10}H_8O_3V$
G.Allegra, G.Perego   *Ric. Sci.*, 2, A, **1**, 362, 1961

75.14   N - Methoxycarbonyl - 1H - azepine - iron tricarbonyl
$C_{10}H_9FeNO_5$
I.C.Paul, S.M.Johnson, L.A.Paquette, J.H.Barrett, R.J.Haluska
J. Amer. Chem. Soc., **90**, 5023, 1968

75.15   Bullvalene - silver(i) tetrafluoroborate complex monohydrate
$C_{10}H_{10}Ag^+$, BF_4^- , H_2O
J.S.McKechnie, I.C.Paul   *Chem. Communic.*, 44, 1968

75.16   Bullvalene - aquo silver tetrafluoroborate complex
$C_{10}H_{12}AgO^+$, BF_4^-
J.S.McKechnie, I.C.Paul   *J. Chem. Soc. (B)*, 1445, 1968

75.17   Dichloro(1,3,5,7 - tetramethyl - 4,8,9 - trioxa - bicyclo(3.3.1)nona - 2,6 - diene) - platinum(ii)
$C_{10}H_{14}Cl_2O_3Pt$
R.Mason, G.B.Robertson   *J. Chem. Soc. (A)*, 492, 1969

75.18   Dipentene platinum(ii) chloride
$C_{10}H_{16}Cl_2Pt$
N.C.Baenziger, R.C.Medrud, J.R.Doyle   *Acta Cryst.*, **18**, 237, 1965

75.19   Cyclo - octatetraene iron tricarbonyl
$C_{11}H_8FeO_3$
B.Dickens, W.N.Lipscomb   *J. Chem. Phys.*, **37**, 2084, 1962

75.20 **Cyclo - octatetraene molybdenum tricarbonyl**
$C_{11}H_8MoO_3$
J.S.McKechnie, I.C.Paul   *J. Amer. Chem. Soc.*, **88,** 5927, 1966

75.21 **Bicyclo(3,2,1)octadienyl iron tricarbonyl tetrafluoroborate**
$C_{11}H_9FeO_3{}^+$, $BF_4{}^-$
T.N.Margulis, L.Schiff, M.Rosenblum
J. Amer. Chem. Soc., **87,** 3269, 1965

75.22 **Tricarbonyl - cyclo - octa - 1,3,5 - trienyl chromium**
$C_{11}H_{10}CrO_3$
V.S.Armstrong, C.K.Prout   *J. Chem. Soc.*, 3770, 1962

75.23 **3,4,5,5,6,6 - Hexafluorocyclo - 1 - yne - 3 - ene dicobalt**
$C_{12}Co_2F_6O_6$
N.A.Bailey, R.Mason   *J. Chem. Soc. (A)*, 1293, 1968

75.24 **π - Cyclopentadienyl - π - cycloheptatrienyl vanadium**
$C_{12}H_{12}V$
G.R.Engebretson, R.E.Rundle   *J. Amer. Chem. Soc.*, **85,** 481, 1963
Also classified in 73

75.25 **Geijerene - silver nitrate complex**
3 - isoPropenyl - 4 - methyl - 4 - vinyl cyclohexene(i) - silver nitrate complex
$C_{12}H_{18}Ag_2{}^{2+}$, $2NO_3{}^-$
D.J.Robinson, C.H.L.Kennard   *Chem. Communic.*, 914, 1968
Residue 1 also classified in 72

75.26 **Cyclo - octatetraene di - iron pentacarbonyl**
$C_{13}H_8Fe_2O_5$
E.B.Fleischer, A.L.Stone, R.B.K.Dewar, J.D.Wright, C.E.Keller, R.Pettit
J. Amer. Chem. Soc., **88,** 3158, 1966

75.27 **Acetylacetonate (cyclo - octa - 2,4 - dienyl) palladium**
$C_{13}H_{18}O_2Pd$
M.R.Churchill   *Inorg. Chem.*, **5,** 1608, 1966
Also classified in 77

75.28 **Cyclo - octenyl nickel(ii) acetylacetonate**
$C_{13}H_{20}NiO_2$
O.S.Mills, E.F.Paulus   *Chem. Communic.*, 738, 1966
Also classified in 77, 71

75.29 **Cyclo - octatetraene bis(tricarbonyl iron)**
$C_{14}H_8FeO_6$
B.Dickens, W.N.Lipscomb   *J. Chem. Phys.*, **37,** 2084, 1962

75.30 Cyclo - octatetraene diruthenium hexacarbonyl
$C_{14}H_8O_6Ru_2$
F.A.Cotton, W.T.Edwards *J. Amer. Chem. Soc.*, **90**, 5412, 1968

75.31 Tricarbonyl - (1,6 - methanocyclodecapentaene) - chromium
$C_{14}H_{10}CrO_3$
P.E.Baikie, O.S.Mills *J. Chem. Soc. (A)*, 328, 1969

75.32 (1,3,5 - Cyclo - octatriene) di - iron hexacarbonyl
$C_{14}H_{10}Fe_2O_6$
F.A.Cotton, W.T.Edwards *J. Amer. Chem. Soc.*, **91**, 843, 1969

75.33 1,2 - bis(Dimethylarsino) - 3,3,4,4 - tetrafluoro cyclobutene bis(iron tricarbonyl)
$C_{14}H_{12}As_2F_4Fe_2O_6$
F.W.B.Einstein, J.Trotter *J. Chem. Soc. (A)*, 824, 1967

75.34 Dicarbonyl - π - cyclopentadienylbicyclo(2.2.1)hepta - 2π,5 - diene manganese(i)
$C_{14}H_{13}MnO_2$
B.Granoff, R.A.Jacobson *Inorg. Chem.*, **7**, 2328, 1968
Also classified in 73

75.35 Norbornadiene dimer - silver(i) nitrate complex
$C_{14}H_{16}Ag^+$, NO_3^-
T.J.Katz, N.Acton, I.C.Paul *J. Amer. Chem. Soc.*, **91**, 206, 1969
Residue 1 also classified in 54

75.C Chloro - (2 - methoxycyclo - octa - 1,5 - dienyl) - pyridine platinum
$C_{14}H_{18}ClNOPt$
For complete entry see 71.36

75.36 Di - μ - bromo bis(π - cycloheptenyl) dipalladium
$C_{14}H_{24}Br_2Pd_2$
B.T.Kilbourn, R.H.B.Mais, P.G.Owston *Chem. Communic.*, 1438, 1968

75.37 trans - Dichloro(ethylenediamine)(cyclododeca - 1,5 - dienyl) rhodium(iii)
$C_{14}H_{27}Cl_2N_2Rh$
G.Paiaro, A.Musco, G.Diana *J. Organometal. Chem.*, **4**, 466, 1965
Also classified in 76

75.C π - Cycloheptatrienyl dicarbonyl - pentafluorophenyl - molybdenum
$C_{15}H_7F_5MoO_2$
For complete entry see 71.37

75.38   **Azulene di - iron pentacarbonyl**
$C_{15}H_8Fe_2O_5$
M.R.Churchill   *Inorg. Chem.*, **6,** 190, 1967

75.39   **1,3,5,7 - Tetramethyl cyclo - octatetraene chromium tricarbonyl**
$C_{15}H_{16}CrO_3$
M.J.Bennett, F.A.Cotton, J.Takats   *J. Amer. Chem. Soc.*, **90,** 903, 1968

75.40   **Hexamethylcyclohexadienyl rhenium tricarbonyl**
$C_{15}H_{19}O_3Re$
P.H.Bird, M.R.Churchill   *Chém. Communic.*, 777, 1967

75.41   **Germacratriene - silver nitrate complex**
$C_{15}H_{24}Ag^+$, NO_3^-
F.H.Allen, D.Rogers   *Chem. Communic.*, 588, 1967

75.42   **trans - Azulene dimanganese hexacarbonyl**
$C_{16}H_8Mn_2O_6$
M.R.Churchill, P.H.Bird   *Inorg. Chem.*, **7,** 1793, 1968

75.43   **Azulene dimolybdenum hexacarbonyl (triclinic form)**
$C_{16}H_8MoO_6$
M.R.Churchill, P.H.Bird   *Chem. Communic.*, 746, 1967

75.44   **Azulene dimolybdenum hexacarbonyl (disordered,apparent monoclinic cell)**
$C_{16}H_8MoO_6$
J.S.McKechnie, I.C.Paul   *Chem. Communic.*, 747, 1967

75.45   **Azulene dimolybdenum hexacarbonyl (disordered)**
$C_{16}H_8MoO_6$
A.W.Schlueter   *Dissert. Abstr. (B),* **27,** 3071, 1967

75.46   **Tricarbonyl - exo - 7 - phenylcyclohepta - 1,3,5 - triene chromium**
$C_{16}H_{12}CrO_3$
P.E.Baikie, O.S.Mills   *J. Chem. Soc. (A),* 2704, 1968

75.47   **Cyclo - octatetraene dimer - silver nitrate**
$C_{16}H_{16}Ag^+$, NO_3^-
S.C.Nyburg, J.Hilton   *Acta Cryst.*, **12,** 116, 1959
See also *Int. Distances*, M 185s; *Structure Reports,* **23,** 640, 1959

75.48   **bis(Cyclo - octatetraene) iron**
$C_{16}H_{16}Fe$
G.Allegra, A.Colombo, A.Immirzi, I.W.Bassi
J. Amer. Chem. Soc., **90,** 4455, 1968

75.49   Tetracarbonyl (hexamethylbicyclo(2.2.0)hexa - 2,5 - diene) chromium
$C_{16}H_{18}CrO_4$
G.Huttner, O.S.Mills   Chem. Communic., 344, 1968

75.50   Copper(i) chloride - cyclo - octa - 1,5 - diene complex
$C_{16}H_{24}Cl_2Cu_2$
J.H.van den Hende, W.C.Baird   J. Amer. Chem. Soc., 85, 1009, 1963

75.51   Rhodium(i) chloride 1,5 - cyclo - octadiene dimer
$C_{16}H_{24}Cl_2Rh_2$
J.A.Ibers, R.G.Snyder   Acta Cryst., 15, 923, 1962

75.52   bis(Cyclo - octa - 1,5 - diene) iridium trichlorotin
$C_{16}H_{24}Cl_3IrSn$
P.Porta, H.M.Powell, R.J.Mawby, L.M.Venanzi
J. Chem. Soc. (A), 455, 1967

75.53   bis(1,2,3,4 - Tetramethyl cyclobutadiene nickel(ii) chloride) benzene
$C_{16}H_{24}Cl_4Ni_2$, C_6H_6
J.D.Dunitz, H.C.Mez, O.S.Mills, H.M.M.Shearer
Helv. Chim. Acta, 45, 647, 1962

75.54   bis(Cyclo - octa - 1,5 - diene) nickel(0)
$C_{16}H_{24}Ni$
H.Dierks, H.Dietrich   Z. Kristallogr., 122, 1, 1965

75.55 π - Cyclo - octenyl - π - cyclo - octa - 1,5,diene cobalt
$C_{16}H_{25}Co$
S.Koda, A.Takenaka, T.Watanabe   J. Chem. Soc. (D), 1293, 1969

75.56   Acenaphthylene (pentacarbonyl) di - iron
$C_{17}H_8Fe_2O_5$
M.R.Churchill, J.Wormald   Chem. Communic., 1597, 1968

75.57   (1,3,5 - Trimethyl - 7 - methylene - 1,3,5 - cyclo - octatriene) di - iron
pentacarbonyl
$C_{17}H_{16}Fe_2O_5$
F.A.Cotton, J.Takats   J. Amer. Chem. Soc., 90, 2031, 1968

75.58   (1,3,5,7 - Tetramethyl cyclo - octatetraene) di - iron pentacarbonyl
$C_{17}H_{16}Fe_2O_5$
F.A.Cotton, M.D.LaPrade   J. Amer. Chem. Soc., 90, 2026, 1968

75.59 2 - (Pentamethylbicyclo(2.2.0)hexa - 2,5 - diene methyl palladium acetylacetonate
$C_{17}H_{24}O_2Pd$
J.F.Malone, W.S.McDonald, B.L.Shaw, G.Shaw
Chem. Communic., 869, 1968
Also classified in 77, 71

75.60 Di - indenyl iron
$C_{18}H_{14}Fe$
J.Trotter *Acta Cryst.*, **11**, 355, 1958
See also *Int. Distances*, M 188s; *Structure Reports*, **22**, 742, 1958

75.61 Cyclo - octatetraene di(cyclopentadienyl cobalt)
$C_{18}H_{18}Co_2$
E.Paulus, W.Hoppe, R.Huber *Naturwissenschaften*, **54**, 67, 1967
Also classified in 73

75.62 π - Cyclopentadienyl - 1,2,3,4 - tetramethyl - 1 - exo - cyclopenta - 1′,3′ - dienecyclobutenyl nickel(ii)
$C_{18}H_{22}Ni$
W.Oberhansli, L.F.Dahl *Inorg. Chem.*, **4**, 150, 1965
Also classified in 73

75.63 Cyclo - octadiene duroquinone nickel(0)
$C_{18}H_{24}NiO_2$
M.D.Glick, L.F.Dahl *J. Organometal. Chem.*, **3**, 200, 1965

75.64 π - Tetracyclo(8,6,0,0(2,9),0(3,8))hexadecapenta - 4,6,11,13,15 - ene - tricarbonyl iron
$C_{19}H_{16}FeO_3$
A.Robson, M.R.Truter *J. Chem. Soc. (A)*, 794, 1968

75.65 bis(Azulene) iron
$C_{20}H_{16}Fe$
M.R.Churchill, J.Wormald *Chem. Communic.*, 1033, 1968

75.66 bis(Cyclo - octatetraene) triruthenium tetracarbonyl
$C_{20}H_{16}O_4Ru_3$
M.J.Bennett, F.A.Cotton, P.Legzdins *J. Amer. Chem. Soc.*, **90**, 6335, 1968

75.67 Silver nitrate - bis(cyclodecene) complex
$C_{20}H_{36}Ag^+, NO_3^-$
P.Ganis, J.D.Dunitz *Helv. Chim. Acta*, **50**, 2379, 1967

75.68   Chromium tricarbonyl - 1 - ethoxy - 3 - oxo - 3a - phenyl - 3,3a - dihydroazulene
$C_{21}H_{16}CrO_5$
W.A.C.Brown, A.T.McPhail, G.A.Sim   *J. Chem. Soc. (B)*, 504, 1966

75.69   Guaiazulene dimolybdenum hexacarbonyl
$C_{21}H_{18}Mo_2O_6$
M.R.Churchill, P.H.Bird   *Inorg. Chem.*, **7**, 1545, 1968

75.70   4,6,8 - Trimethylazulene - tetraruthenium enneacarbonyl
$C_{22}H_{13}O_9Ru_4$
M.R.Churchill, P.H.Bird   *J. Amer. Chem. Soc.*, **90**, 800, 1968

75.71   Platinum chloride methoxydicyclopentadiene complex
$C_{22}H_{30}Cl_2O_2Pt_2$
W.A.Whitla, H.M.Powell, L.M.Venanzi   *Chem. Communic.*, 310, 1966

75.72   tris - Cyclo - octatetraene dititanium
$C_{24}H_{24}Ti_2$
H.Dierks, H.Dietrich   *Acta Cryst. (B)*, **24**, 58, 1968

75.73   2 - (Pentamethylbicyclo(2.2.0)hexa - 2,5 - diene) methyl palladium chloride dimer
$C_{24}H_{34}Cl_2Pt_2$
R.Mason, G.B.Robertson, P.O.Whimp, B.L.Shaw, G.Shaw
Chem. Communic., 868, 1968

75.74   Di - μ - chloro tris(trans - cyclo - octene) dicopper(i)
$C_{24}H_{42}Cl_2Cu_2$
P.Ganis, U.Lepore, G.Paiaro   *J. Chem. Soc. (D)*, 1054, 1969

75.75   2,4,6 - Triphenyltropone iron tricarbonyl
$C_{28}H_{18}FeO_4$
D.L.Smith, L.F.Dahl   *J. Amer. Chem. Soc.*, **84**, 1743, 1962

75.76   Di(azulene tricarbonyl methyl molybdenum)
$C_{28}H_{22}Mo_2O_6$
P.H.Bird, M.R.Churchill   *Inorg. Chem.*, **7**, 349, 1968

75.77   Norbornadiene copper(i) chloride
$C_{28}H_{32}Cl_4Cu_4$
N.C.Baenziger, H.L.Haight, J.R.Doyle   *Inorg. Chem.*, **3**, 1535, 1964

75.78 **Bullvalene - silver(i) fluoroborate complex**
$C_{30}H_{30}Ag^+$, BF_4^-
J.S.McKechnie, M.G.Newton, I.C.Paul
J. Amer. Chem. Soc., **89,** 4819, 1967

75.79 **Tetraphenylcyclobutadiene iron tricarbonyl**
$C_{31}H_{20}FeO_3$
R.P.Dodge, V.Schomaker *Acta Cryst.*, **18,** 614, 1965

75.80 **endo - 1 - Ethoxy - 1,2,3,4 - tetraphenylcyclobutenyl palladium chloride dimer**
$C_{60}H_{50}Cl_2O_2Pd_2$
L.F.Dahl, W.E.Oberhansli *Inorg. Chem.*, **4,** 629, 1965

75.81 **exo - 1 - Ethoxy - 1,2,3,4 - tetraphenylcyclobutenyl palladium chloride dimer**
$C_{60}H_{50}Cl_2O_2Pd_2$
L.F.Dahl, W.E.Oberhansli *Inorg. Chem.*, **4,** 629, 1965

METAL COMPLEXES (ETHYLENEDIAMINE)

76.1   **Diperoxoaquoethylenediamine chromium(iv) monohydrate**
$C_2H_{10}CrN_2O_5$, H_2O
R.Stomberg   *Ark. Kemi,* **24,** 47, 1965

76.2   **Trinitro (diethylenetriamine) cobalt(iii)**
$C_4H_{13}CoN_6O_6$
Y.Kushi, K.Watanabe, H.Kuroya   *Bull. Chem. Soc. Jap.,* **40,** 2985, 1967
Also classified in 83

76.3   **Trioxo(diethylenetriamine) molybdenum(vi)**
$C_4H_{13}MoN_3O_3$
F.A.Cotton, R.C.Elder   *Inorg. Chem.,* **3,** 397, 1964
Also classified in 83

76.4   **trans - Dibromo bis(ethylenediamine) cobalt(iii) bromide hydrobromide dihydrate**
$C_4H_{16}Br_2CoN_4{}^+$, H_5BrO_2 , Br^-
S.Ooi, Y.Komiyama, Y.Saito, H.Kuroya
Bull. Chem. Soc. Jap., **32,** 263, 1959
See also *Int. Distances,* M 118s; *Structure Reports,* **23,** 572, 1959

76.5   **trans - Chloronitrosyl bis(ethylenediamine) cobalt(iii) perchlorate**
$C_4H_{16}ClCoN_5O^+$, $ClO_4{}^-$
D.A.Snyder, D.L.Weaver   *J. Chem. Soc. (D),* 1425, 1969

76.6   **Nickel(ii) ethylenediamine chloride dimer**
$C_4H_{16}ClN_4Ni^+$, Cl^-
A.S.Ancyskina, M.A.Porai-Koshits
Dokl. Akad. Nauk S. S. S. R., **143,** 105, 1962

76.7   **trans - Dichloro - bis(ethylenediamine) cobalt(iii) chloride**
$C_4H_{16}Cl_2CoN_4{}^+$, Cl^-
K.A.Becker, G.Grosse, K.Plieth   *Z. Kristallogr.,* **112,** 375, 1959
See also *Int. Distances,* M 118s; *Structure Reports,* **23,** 572, 1959

76.8 trans - Dichloro - bis(ethylenediamine) cobalt(iii) chloride hydrochloride dihydrate
$C_4H_{16}Cl_2CoN_4{}^+$, Cl^- , HCl , $2H_2O$
A.Nakahara, Y.Saito, H.Kuroya *Bull. Chem. Soc. Jap.*, **25,** 331, 1952
See also *Int. Distances,* M 177; *Structure Reports,* **16,** 457, 1952

76.9 ($\pm$)cis - Dichloro - diethylenediamine cobalt(iii) chloride monohydrate
$C_4H_{16}Cl_2CoN_4{}^+$, Cl^- , H_2O
A.Hullen, K.Plieth, G.Ruban *Naturwissenschaften,* **52,** 618, 1965

76.10 ($\pm$)cis - Dichloro bis(ethylenediamine) cobalt(iii) chloride hydrate
$C_4H_{16}Cl_2CoN_4{}^+$, Cl^- , H_2O
K.Matsumoto, S.Ooi, H.Kuroya
J. Chem. Soc. Jap., Pure Chem. Sect., **89,** 167, 1968

76.11 trans - Dichloro - bis(ethylenediamine) cobalt(iii) nitrate
$C_4H_{16}Cl_2CoN_4{}^+$, $NO_3{}^-$
S.Ooi, H.Kuroya *Bull. Chem. Soc. Jap.,* **36,** 1083, 1963

76.12 trans - Dichloro bis(ethylenediamine) cobalt(iii) hexathionate monohydrate
$2C_4H_{16}Cl_2CoN_4{}^+$, $O_6S_6{}^{2-}$, H_2O
O.Foss, K.Maroy *Acta Chem. Scand.,* **19,** 2219, 1965

76.13 trans - Dichloro di(ethylenediamine) chromium(iii) chloride hydrochloride dihydrate
$C_4H_{16}Cl_2Cr^+$, $2Cl^-$, $H_5O_2{}^+$
S.Ooi, Y.Komiyama, H.Kuroya *Bull. Chem. Soc. Jap.,* **33,** 354, 1960

76.14 Palladium(ii) chloride bis(ethylenediamine) complex
$C_4H_{16}Cl_2N_4Pd$
J.R.Wiesner, E.C.Lingafelter *Inorg. Chem.,* **5,** 1770, 1966

76.15 bis(Ethylenediamine) copper(ii) fluoroborate
$C_4H_{16}CuN_4{}^{2+}$, $2BF_4{}^-$
D.S.Brown, J.D.Lee, B.G.A.Melsom *Acta Cryst. (B),* **24,** 730, 1968

76.16 bis(Ethylenediamine) copper(ii) perchlorate
$C_4H_{16}CuN_4{}^{2+}$, $2ClO_4{}^-$
A.Pajunen *Ann. Acad. Sci. Fenn., A2,* **138,** 3, 1967

76.17 Copper(ii) diethylenediamine nitrate
$C_4H_{16}CuN_4{}^{2+}$, $2NO_3{}^-$
Y.Komiyama, E.C.Lingafelter *Acta Cryst.,* **17,** 1145, 1964

76.18 bis(Ethylenediamine) copper(ii) thiocyanate
$C_4H_{16}CuN_4{}^{2+}$, $2CNS^-$
B.W.Brown, E.C.Lingafelter *Acta Cryst.*, **17**, 254, 1964

76.19 Mercury tetrathiocyanate - copper diethylenediamine
$C_4H_{16}CuN_4{}^{2+}$, $C_4HgN_4S_4{}^{2-}$
H.Scouloudi *Acta Cryst.*, **6**, 651, 1953
See also *Int. Distances*, M 177; *Structure Reports*, **17**, 673, 1953

76.20 bis(Ethylenediamine) nickel(ii) dibromoargentate
$nC_4H_{16}N_4Ni^{2+}$, $(AgBr_4{}^{2-})_n$
M.Kellerman *Dissert. Abstr. (B)*, **27**, 1855, 1966

76.21 Nickel(ii) di(ethylenediamine) nitrite tetrafluoroborate
$C_4H_{16}N_5NiO_2{}^+$, $BF_4{}^-$
M.G.B.Drew, D.M.L.Goodgame, M.A.Hitchman, D.Rogers
Chem. Communic., 477, 1965

76.22 Nitrito bis(ethylenediamine) nickel(ii) perchlorate
$C_4H_{16}N_5NiO_2{}^+$, $ClO_4{}^-$
F.J.Llewellyn, J.M.Waters *J. Chem. Soc.*, 3845, 1962

76.23 Dinitro bis(ethylenediamine) nickel
$C_4H_{16}N_6NiO_4$
L.Kh.Minaceva *Acta Cryst.*, **21**, A145, 1966

76.24 cis - Cobalt diazido bis(ethylanadiamine) nitrate
$C_4H_{16}N_{10}{}^+$, $NO_3{}^-$
V.M.Padanabhan, R.Balasubramanian, K.V.Muralidharan
Acta Cryst. (B), **24**, 1638, 1968

76.25 bis(Ethylenediamine)oxo - hydroxy rhenium perchlorate
$C_4H_{17}N_4O_2Re^{2+}$, $2ClO_4{}^-$
G.L.Betzner, W.Che'ng, N.C.Jayadavan, C.J.L.Lock, I.D.Brown
Acta Cryst., **21**, A136, 1966

76.26 Chloroaquo bis(ethylenediamine) copper(ii) chloride
$C_4H_{18}ClCuN_4O^+$, Cl^-
R.D.Ball, D.Hall, C.E.F.Rickard, T.N.Waters
J. Chem. Soc. (A), 1435, 1967

76.27 Isothiocyanato chloro bis(ethylenediamine) nickel(ii)
$C_5H_{16}ClN_5NiS$
A.E.Shvelashvili, M.A.Porai-Koshits, A.S.Antsyshkina
Zh. Strukt. Khim., **9**, 646, 1968

76.28 **trans - Sulfiteisothiocyanate - bis(ethylenediamine) cobalt(iii) dihydrate**
$C_5H_{16}CoN_5O_3S_2$, $2H_2O$
S.Baggio, L.N.Becka, *Acta Cryst.(B)*,**25**, 946, 1969

76.29 **Nitrito - thiocyanato bis(ethylenediamine) nickel**
$C_5H_{16}N_6NiO_2S$
M.A.Porai-Koshits *Acta Cryst.*, **21**, A149, 1966

76.30 **cis - (+)d - bis(Ethylenediamine) cobalt(ii) dicyanide chloride monohydrate**
$C_6H_{16}CoN_6{}^+$, Cl^-, H_2O
K.Matsumoto, Y.Kushi, S.Ooi, H.Kuroya
Bull. Chem. Soc. Jap., **40**, 2988, 1967

76.31 **Nickel(ii) N,N,N′,N′ - tetramethylethylenediamine nitrite**
$C_6H_{16}N_4NiO_4$
M.G.B.Drew, D.Rogers *Chem. Communic.*, 476, 1965

76.32 **trans - bis(Ethylenediamine) bis(isothiocyanato) nickel(ii)**
$C_6H_{16}N_6NiS_2$
B.W.Brown, E.C.Lingafelter *Acta Cryst.*, **16**, 753, 1963

76.33 **DL - β(Chloroaquo - triethylenetetramine) cobalt(iii) perchlorate**
$C_6H_{20}ClCoN_4O^{2+}$, $2ClO_4{}^-$
D.A.Buckingham, P.A.Marzilli, I.E.Maxwell, A.M.Sargeson, M.Fehlmann,
H.C.Freeman *Chem. Communic.*, 488, 1968
Residue 1 also classified in 83

76.34 **(+) - cis - Dinitro bis - (−) - propylenediamine cobalt(iii) chloride**
$C_6H_{20}ClCoN_6O_4$
G.A.Barclay, E.Goldschmied, N.C.Stephenson, A.M.Sargeson
Chem. Communic., 540, 1966

76.35 **trans - Dichloro - bis - l - propylenediamine cobalt(iii) chloride**
hydrochloride dihydrate
$C_6H_{20}Cl_2CoN_4{}^+$, $2Cl^-$, $H_5O_2{}^+$
Y.Saito, H.Iwasaki *Bull. Chem. Soc. Jap.*, **35**, 1131, 1962

76.36 **Sodium - hexa - aquo - D - tris(ethylenediamine) cobalt(iii) chloride**
$3C_6H_{24}CoN_6{}^+$, H_2NaO^+ , $4Cl^-$
Y.Saito, K.Nakatsu, M.Shiro, H.Kuroya
Bull. Chem. Soc. Jap., **30**, 795, 1957
See also *Int. Distances*, M 136s; *Structure Reports*, **21**, 547, 1957

76.37 **D - tris(Ethylenediamine) cobalt(iii) bromide monohydrate**
$C_6H_{24}CoN_6{}^{3+}$, $3Br^-$, H_2O
K.Nakatsu *Bull. Chem. Soc. Jap.*, **35**, 832, 1962

76.38 (+)D - tris(Ethylenediamine) cobalt(iii) chloride monohydrate
$C_6H_{24}CoN_6{}^{3+}$, $3Cl^-$, H_2O
M.Iwata, K.Nakatzu, Y.Saito *Acta Cryst. (B)*, **25**, 2562, 1969

76.39 tris(Ethylenediamine) chromium(iii) pentacyano nickelate(ii) sesquihydrate
$C_6H_{24}CrN_6{}^{3+}$, $C_5N_5Ni^{3-}$, $1.5H_2O$
K.N.Raymond, P.W.R.Corfield, J.A.Ibers *Inorg. Chem.*, **7**, 1362, 1968

76.40 tris(Ethylenediamine) chromium(iii) hexacyano - cobaltate(iii) hexahydrate
$C_6H_{24}CrN_6{}^{3+}$, $C_6CoN_6{}^{3-}$, $6H_2O$
K.N.Raymond, J.A.Ibers *Inorg. Chem.*, **7**, 2333, 1968

76.41 Potassium tris(ethylenediamine) nickel(ii) selenocyanate
$C_6H_{24}N_6Ni^{2+}$, K^+ , $3CNSe^-$
M.Kellerman *Dissert. Abstr. (B)*, **27**, 1855, 1966

76.42 Tri(ethylenediamine) nickel(ii) nitrate
$C_6H_{24}N_6Ni^{2+}$, $2NO_3{}^-$
L.N.Swink, M.Atoji *Acta Cryst.*, **13**, 639, 1960

76.43 L - Nickel(ii) tris(ethylenediamine) silicate hydrate
$3C_6H_{24}N_6Ni^{2+}$, $O_{15}Si_6{}^{6-}$, $26H_2O$
Yu.I.Smolin *J. Chem. Soc. (D)*, 395, 1969

76.44 cis - Diethylenetriamine chromium tricarbonyl
$C_7H_{13}CrN_3O_3$
F.A.Cotton, D.C.Richardson *Inorg. Chem.*, **5**, 1851, 1966
Also classified in 83

76.45 Zinc(ii) 2,2',2'' - triaminotriethylamine thiocyanate isothiocyanate
$C_7H_{15}N_5SZn^+$, CNS^-
P.C.Jain, E.C.Lingafelter, P.Paoletti *J. Amer. Chem. Soc.*, **90**, 519, 1968
Residue 1 also classified in 83

76.46 β,β',β'' - Triaminotriethylamine - isothiocyanato copper(ii) thiocyanate
$C_7H_{18}CuN_5S^+$, CNS^-
P.C.Jain, E.C.Lingafelter *J. Amer. Chem. Soc.*, **89**, 6131, 1967
Residue 1 also classified in 83

76.C (–)(589) - Sarcosinato bis(ethylenediamine) cobalt(iii) di - iodide dihydrate
$C_7H_{22}CoN_5O_2{}^{2+}$, $2I^-$, $2H_2O$
For complete entry see 82.24

76.47 **2,2′,2″ - Triaminotriethylamine nickel(ii) dithiocyanate**
$C_8H_{18}N_6NiS_2$
S.E.Rasmussen *Acta Cryst.*, **13**, A51, 1960
Also classified in 83

76.48 **DαβS(Chloro cobalt(iii) tetraethylene - penta - amine) perchlorate**
$C_8H_{23}ClCoN_5{}^{2+}$, $2ClO_4{}^-$
M.R.Snow, D.A.Buckingham, P.A.Marzilli, A.M.Sargeson
J. Chem. Soc. (D), 891, 1969
Residue 1 also classified in 83

76.49 **DαβR(Chloro cobalt(iii) tetraethylene - penta - amine) perchlorate**
$C_8H_{23}ClCoN_5{}^{2+}$, $2ClO_4{}^-$
M.R.Snow, D.A.Buckingham, P.A.Marzilli, A.M.Sargeson
J. Chem. Soc. (D), 891, 1969
Residue 1 also classified in 83

76.C **4 - (2 - Aminoethyl) - 1,4,7,10 - tetra - azadecane azido cobalt(iii) nitrate monohydrate**
$C_8H_{23}CoN_8{}^{2+}$, $2NO_3{}^-$, H_2O
For complete entry see 83.64

76.50 **Dinitrito bis(N,N - dimethylethylenediamine) nickel(ii)**
$C_8H_{24}N_6NiO_4$
M.G.B.Drew, D.M.L.Goodgame, M.A.Hitchman, D.Rogers
Proc. Chem. Soc., 363, 1964

76.51 **bis(Diethylenetriamine) copper(ii) nitrate**
$C_8H_{26}CuN_6{}^{2+}$, $2NO_3{}^-$
F.S.Stephens *J. Chem. Soc. (A)*, 883, 1969
Residue 1 also classified in 83

76.52 **bis(Di(2 - aminoethyl)amine) nickel(ii) chloride monohydrate**
$C_8H_{26}N_6Ni^{2+}$, $2Cl^-$, H_2O
P.Paoletti, S.Biagini, M.Cannas *J. Chem. Soc. (D)*, 513, 1969
Residue 1 also classified in 83

76.53 **DL - μ - Amido - μ - superoxo - bis(bis(ethylenediamine) cobalt(iii)) tetranitrate monohydrate**
$C_8H_{34}Co_2N_9O_2{}^{4+}$, $4NO_3{}^-$, $2H_2O$
U.Thewalt, R.Marsh *J. Amer. Chem. Soc.*, **89**, 6364, 1967

76.54 **DL - tetrakis(Ethylenediamine) - μ - amido - μ - nitro dicobalt(iii) nitrate**
$C_8H_{34}Co_2N_{10}O_2{}^{4+}$, $4NO_3{}^-$
C.E.Wilkes *Dissert. Abstr.*, **26**, 2509, 1965

76.55 **DL - tetrakis(Ethylenediamine) -** μ **- amido -** μ **- nitro dicobalt(iii) nitrate**
$C_8H_{34}Co_2N_{10}O_2^{4+}$, $4NO_3^-$
P.Goldstein *Dissert. Abstr.*, **24**, 2708, 1964

76.56 **DL -** μ **- Amido -** μ **- hydroperoxo - bis(bis(ethylenediamine) cobalt(iii)) tetranitrate dihydrate**
$C_8H_{35}Co_2N_9O_2^{4+}$, $4NO_3^-$, $2H_2O$
U.Thewalt, R.Marsh *J. Amer. Chem. Soc.*, **89**, 6364, 1967

76.57 **Cobalt(ii) chloride - bis(2 - dimethylaminoethyl) methylamine complex**
$C_9H_{23}Cl_2CoN_3$
M.di Vaira, P.L.Orioli *Chem. Communic.*, 590, 1965
Also classified in 83

76.C **D - (Cobalt bis - ethylenediamine - L - glutamate) perchlorate**
$C_9H_{24}CoN_6O_3^+$, ClO_4^-
For complete entry see 82.37

76.58 **Cobalt(iii) bromide tris(L - propylene - 1,2 - diamine)**
$C_9H_{30}CoN_6^{3+}$, $3Br^-$
H.Iwasaki, Y.Saito *Bull. Chem. Soc. Jap.*, **39**, 92, 1966

76.59 **Ammonium ethylenediaminetetra - acetato cobaltate(iii) dihydrate**
$C_{10}H_{12}CoN_2O_8^-$, H_4N^+ , H_2O
H.A.Weakliem, J.L.Hoard *J. Amer. Chem. Soc.*, **81**, 549, 1959
See also *Int. Distances*, M 159s; *Structure Reports*, **23**, 577, 1959

76.60 **Potassium copper(ii) ethylenediamine tetra - acetate trihydrate**
$C_{10}H_{12}CuN_2O_8^{2-}$, $2K^+$, $3H_2O$
N.V.Novozhilova, T.N.Polynova, M.A.Porai-koshits, L.I.Martynenko
Zh. Strukt. Khim., **8**, 553, 1967
Residue 1 also classified in 84

76.61 **Sodium hexa - oxo -** μ **- ethylenediaminetetra - acetato - dimolybdate(vi) octahydrate**
$C_{10}H_{12}Mo_2N_2O_{14}^{4-}$, $4Na^+$, $8H_2O$
J.J.Park, M.D.Glick, J.L.Hoard *J. Amer. Chem. Soc.*, **91**, 301, 1969
Residue 1 also classified in 81

76.62 **Sodium terbium(iii) ethylenediamine - tetra - acetate octahydrate**
$C_{10}H_{12}N_2O_8Tb^-$, Na^+
B.Lee *Dissert. Abstr. (B)*, **28**, 84, 1967
Residue 1 also classified in 84

76.63 **Lithium monaquo - ethylenediaminetetra - acetato ferrate(iii) dihydrate**
$C_{10}H_{14}FeN_2O_9{}^-$, Li^+ , $2H_2O$
M.J.Hamor, T.A.Hamor, J.L.Hoard *Inorg. Chem.*, **3**, 34, 1964
Residue 1 also classified in 84

76.64 **Rubidium monoaquo - ethylenediaminetetra - acetato ferrate(iii) monohydrate**
$C_{10}H_{14}FeN_2O_9{}^-$, Rb^+
M.D.Lind, J.L.Hoard *Inorg. Chem.*, **3**, 34, 1964
Residue 1 also classified in 84

76.65 **Manganese(ii) monohydrogen ethylenediamine tetra - acetate decahydrate**
bisHydrogen (ethylenedinitrilo) tetra - acetato trimanganese decahydrate
$2C_{10}H_{14}MnN_2O_9{}^-$, $H_8MnO_4{}^{2+}$, $4H_2O$
S.Richards, B.Pedersen, J.V.Silverton, J.L.Hoard *Inorg. Chem.*, **3**, 27, 1964
Residue 1 also classified in 84

76.66 **Hydrogen aquoethylenediaminetetra - acetato chromate(iii)**
$C_{10}H_{15}CrN_2O_9$
J.L.Hoard, C.H.L.Kennard, G.S.Smith *Inorg. Chem.*, **2**, 1316, 1963
Also classified in 84

76.67 **Hydrogen aquoethylenediaminetetra - acetato ferrate(iii)**
$C_{10}H_{15}FeN_2O_9$
J.L.Hoard, C.H.L.Kennard, G.S.Smith *Inorg. Chem.*, **2**, 1316, 1963
Also classified in 84

76.68 **Dichloro(tetrahydrogen ethylenediaminetetra - acetato) palladium(ii) pentahydrate**
$C_{10}H_{16}Cl_2N_2O_8Pd$, $5H_2O$
D.J.Robinson, C.H.L.Kennard *Chem. Communic.*, 1236, 1967
Residue 1 also classified in 83

76.69 **Dihydrogen ethylenediaminetetra - acetato - aquo nickelate(ii)**
$C_{10}H_{16}N_2NiO_9$
G.S.Smith, J.L.Hoard *J. Amer. Chem. Soc.*, **81**, 556, 1959
Also classified in 81, 82
See also *Int. Distances*, M 164s; *Structure Reports*, **23**, 576, 1959

76.70 **bis(Ethylenediamine) isothiocyanato nickel perchlorate**
$C_{10}H_{16}N_{10}Ni_2S_2{}^{2+}$, $2ClO_4{}^-$
A.E.Svelasvili *Acta Cryst.*, **21**, A153, 1966

76.71   Potassium lanthanum(iii) ethylenediaminetetra - acetate octahydrate
$C_{10}H_{18}KLaN_2O_{11}$, $5H_2O$
J.L.Hoard, B.Lee, M.D.Lind   *J. Amer. Chem. Soc.*, **87**, 1612, 1965
Residue 1 also classified in 84

76.72   Lanthanum(iii) hydrogen ethylenediaminetetra - acetate septahydrate
$C_{10}H_{21}LaN_2O_{12}$, $3H_2O$
M.D.Lind, B.Lee, J.L.Hoard   *J. Amer. Chem. Soc.*, **87**, 1611, 1965
Residue 1 also classified in 84

76.73   (+) - (N,N,N',N' - tetrakis(2' - Aminoethyl) - 1,2 - diaminoethane) cobalt(iii) hexacyanocobaltate(iii) dihydrate
$C_{10}H_{28}CoN_6^{3+}$, $C_6CoN_6^{3-}$, $2H_2O$
A.Muto, F.Marumo, Y.Saito   *Inorg. Nucl. Chem. Letters*, **5**, 85, 1969
Residue 1 also classified in 83

76.74   bis(Ethylenediamine) isothiocyanato nickel(ii) iodide
$C_{10}H_{32}N_{10}Ni_2S_2^{2+}$, $2I^-$
A.E.Svelasvili   *Acta Cryst.*, **21**, A153, 1966

76.C   D(+) - Triethylenetetramine - S - proline cobalt(iii) tetrachlorozincate
$C_{11}H_{26}CoN_5O_2^{2+}$, Cl_4Zn^{2-}
For complete entry see 82.43

76.C   L(−) - Triethylenetetramine - S - proline cobalt(iii) di - iodide dihydrate
$C_{11}H_{26}CoN_5O_2^{2+}$, $2I^-$, $2H_2O$
For complete entry see 82.44

76.75   1,1,7,7 - Tetraethyldiethylenetriamine cobalt(ii) chloride
$C_{12}H_{29}Cl_2CoN_3$
Z.Dori, R.Eisenberg, H.B.Gray   *Inorg. Chem.*, **6**, 483, 1967
Also classified in 83

76.C   tris(2 - Dimethylaminoethyl)amine copper(ii) bromide
$C_{12}H_{30}BrCuN_4^+$, Br^-
For complete entry see 83.119

76.76   tris - (2 - Dimethylaminoethyl)amine iron(ii) bromide
$C_{12}H_{30}BrFe^+$, Br^-
M.di Vaira, P.L.Orioli   *Acta Cryst. (B)*, **24**, 1269, 1968
Residue 1 also classified in 83

76.77   tris - (2 - Dimethylaminoethyl)amine manganese(ii) bromide
$C_{12}H_{30}BrMn^+$, Br^-
M.di Vaira, P.L.Orioli   *Acta Cryst. (B)*, **24**, 1269, 1968
Residue 1 also classified in 83

76.C   tris(2 - Dimethylaminoethyl)amine nickel(ii) bromide
$C_{12}H_{30}BrN_4Ni^+$, Br^-
For complete entry see 83.120

76.78   tris - (2 - Dimethylaminoethyl)amine zinc(ii) bromide
$C_{12}H_{30}BrN_4Zn^+$, Br^-
M.di Vaira, P.L.Orioli *Acta Cryst. (B)*, **24**, 1269, 1968

76.79   Tri(2 - dimethylaminoethyl)amine cobalt(ii) bromide
$C_{12}H_{30}Br_2CoN_4$
M.di Vaira, P.L.Orioli *Inorg. Chem.*, **6**, 955, 1967
Also classified in 83

76.C   DL - β - (Triethylenetetramine - O - ethylglycylglycine) cobalt(iii)
perchlorate monohydrate
$C_{12}H_{30}CoN_6O_3{}^{3+}$, $3ClO_4{}^-$, H_2O
For complete entry see 82.54

76.80   Sesquiethylenediamine trimethyl platinic iodide
$C_{12}H_{42}N_6Pt_2{}^{2+}$, $2I^-$
M.R.Truter, E.G.Cox *J. Chem. Soc.*, 948, 1956
Residue 1 also classified in 71
See also *Int. Distances*, M 173s; *Structure Reports*, **20**, 504, 1956

76.81   Hexa(ethylenediamine) hexa(μ - hydroxo) tetrachromium hexa - azide
tetrahydrate
$C_{12}H_{54}Cr_4N_{12}O_6{}^{6+}$, $6N_3{}^-$, $4H_2O$
M.T.Flood, R.E.Marsh, H.B.Gray *J. Amer. Chem. Soc.*, **91**, 193, 1969

76.C   trans - Dichloro(ethylenediamine)(cyclododeca - 1,5 - dienyl) rhodium(iii)
$C_{14}H_{27}Cl_2N_2Rh$
For complete entry see 75.37

76.C   N,N' - Disalicylidene ethylenediamine zinc(ii) monohydrate
$C_{16}H_{16}N_2O_3Zn$
For complete entry see 78.25

76.C μ - Ethylenediamine - bis(trimethyl(acetylacetonato) platinum(iv))
$C_{18}H_{40}N_2O_4Pt_2$
For complete entry see 77.37

76.C μ - Oxo - bis(iron bis(salicylidene - ethylenediamine)) pyridine solvate
$C_{32}H_{28}Fe_2N_4O_5$, $2C_5H_5N$
For complete entry see 78.60

METAL COMPLEXES (ACETYLACETONE)

77.1   bis(Aquoacetylacetonato) copper(ii) picrate
$C_5H_{11}CuO_4{}^+$, $C_6H_2N_2O_7{}^-$
R.D.Gillard, D.Rogers, R.D.Diamand, D.J.Williams
Acta Cryst., **16**, A67, 1963
Residue 2 classified in 15

77.2   Hexafluoro acetylacetonato - bis(carbonyl) rhodium(i)
$C_7HF_6O_4Rh$
N.A.Bailey, E.Coates, G.B.Robertson, F.Bonati, R.Ugo
Chem. Communic., 1041, 1967
Also classified in 84

77.3   Acetylacetonato - bis(carbonyl) rhodium(i)
$C_7H_7O_4Rh$
N.A.Bailey, E.Coates, G.B.Robertson, F.Bonati, R.Ugo
Chem. Communic., 1041, 1967
Also classified in 84

77.4   Potassium bis(acetylacetonato) chloroplatinate(ii)
$C_{10}H_{14}ClO_4Pt^-$, K^+
R.Mason, G.B.Robertson, P.J.Pauling   *J. Chem. Soc. (A)*, 485, 1969
Residue 1 also classified in 71

77.5   Rhenium(iv) bis(acetylacetonate) dichloride
Rhenium(iv) bis(pentane - 2,4 - dionate) dichloride
$C_{10}H_{14}Cl_2O_4Re$
C.J.L.Lock, C.Wan   *Chem. Communic.*, 1109, 1967

77.6   Vanadyl bisacetylacetonate
$C_{10}H_{14}O_5V$
R.P.Dodge, D.H.Templeton, A.Zalkin   *J. Chem. Phys.*, **35**, 55, 1961

77.7   Vanadyl bisacetylacetonate (refinement)
$C_{10}H_{14}O_5V$
P.Hon, R.L.Belford, C.E.Pfluger   *J. Chem. Phys.*, **43**, 3111, 1965

77.8 **Monoaquo bisacetylacetone zinc**
$C_{10}H_{16}O_5Zn$
H.Montgomery, E.C.Lingafelter *Acta Cryst.*, **16**, 748, 1963

77.9 **Monoaquo bisacetylacetone zinc**
$C_{10}H_{16}O_5Zn$
E.L.Lippert, M.R.Truter *J. Chem. Soc.*, 4996, 1960

77.10 **Uranyl acetylacetonate monohydrate**
$C_{10}H_{16}O_7U$
E.Frasson, G.Bombieri, C.Panattoni *Coordin. Chem. Rev.*, **1**, 145, 1966

77.11 **Cobalt diaquo bisacetylacetonate**
$C_{10}H_{18}CoO_6$
G.J.Bullen *Acta Cryst.*, **12**, 703, 1959
See also *Int. Distances*, M 165s; *Structure Reports*, **23**, 527, 1959

77.12 **Manganese(ii) diaquo bisacetylacetonate**
$C_{10}H_{18}MnO_6$
H.Montgomery, E.C.Lingafelter *Acta Cryst. (B)*, **24**, 1127, 1968

77.13 **Diaquo bisacetylacetonato nickel(ii)**
$C_{10}H_{18}NiO_6$
H.Montgomery, E.C.Lingafelter *Acta Cryst.*, **17**, 1481, 1964

77.14 **Copper(ii) 3 - methylpentane - 2,4 - dionate (at −93 ° C)**
$C_{12}H_{18}CuO_4$
I.Robertson, M.R.Truter *J. Chem. Soc. (A)*, 309, 1967
Also classified in 84

77.15 **Copper(ii) ethylacetoacetate**
$C_{12}H_{18}CuO_6$
G.A.Barclay, A.Cooper *J. Chem. Soc.*, 3746, 1965
Also classified in 84

77.16 **Copper(ii) ethylacetoacetate (discussion)**
$C_{12}H_{18}CuO_6$
D.Hall, A.J.McKinnon, T.N.Waters *J. Chem. Soc. (A)*, 615, 1966
Also classified in 84

77.17 **Aquo - N,N' - ethylene bis(acetylacetoneiminato) copper(ii)**
$C_{12}H_{20}CuN_2O_3$
G.R.Clark, D.Hall, T.N.Waters *J. Chem. Soc. (A)*, 823, 1969
Also classified in 83, 84

77.18 **Trimethyl(nonane - 4,6 - dionato) platinum(iv) (neutron diffraction)**
$C_{12}H_{24}O_2Pt$
R.N.Hargreaves, M.R.Truter *Chem. Communic.*, 473, 1968
Also classified in 84

77.C **Acetylacetonate (cyclo - octa - 2,4 - dienyl) palladium**
$C_{13}H_{18}O_2Pd$
For complete entry see 75.27

77.C **Cyclo - octenyl nickel(ii) acetylacetonate**
$C_{13}H_{20}NiO_2$
For complete entry see 75.28

77.C **μ - Allene bis(acetylacetonato rhodium carbonyl)**
$C_{15}H_{18}O_6Rh_2$
For complete entry see 72.43

77.19 **Trisilver dinitrate tris(acetylacetonato) nickelate(ii) monohydrate**
$C_{15}H_{21}Ag_3NiO_6{}^{2+}$, $2NO_3{}^-$, H_2O
W.H.Watson, C.-T.Lin *Inorg. Chem.*, **5**, 1074, 1966

77.20 **Cobalt(iii) acetylacetonate**
$C_{15}H_{21}CoO_6$
L.M.Shkol'nikova, E.A.Shugam *Zh. Strukt. Khim.*, **2**, 72, 1961

77.21 **Tetrabutylammonium tris(acetylacetonato) cobalt(ii)**
$C_{15}H_{21}CoO_6{}^-$, $C_{16}H_{36}N^+$
B.Granoff *Dissert. Abstr. (B)*, **27**, 4341, 1967
Residue 2 classified in 3

77.22 **Chromium acetylacetone**
$C_{15}H_{21}CrO_6$
B.Morosin *Acta Cryst.*, **19**, 131, 1965

77.23 **Iron(iii) acetylacetonate**
$C_{15}H_{21}FeO_6$
J.Iball, C.H.Morgan *Acta Cryst.*, **23**, 239, 1967

77.24 **trisAcetylacetonato manganese(iii)**
$C_{15}H_{21}MnO_6$
B.Morosin, J.R.Brathovde *Acta Cryst.*, **17**, 705, 1964

77.25 **tris - Acetylacetonato - rhodium(iii)**
$C_{15}H_{21}O_6Rh$
E.B.Parker Junior *Dissert. Abstr.*, **26**, 3654, 1966

77.26 **Vanadium(iii) acetylacetonate (α form)**
Tris⁺ 2,4 - pentanedionato) vanadium(iii)
$C_{15}H_{21}O_6V$
B.Morosin, H.Montgomery   *Acta Cryst. (B)*, **25**, 1354, 1969

77.27 **Vanadium(iii) acetylacetonate (β form)**
Tris(2,4 - pentanedionato) vanadium(iii)
$C_{15}H_{21}O_6V$
B.Morosin, H.Montgomery   *Acta Cryst. (B)*, **25**, 1354, 1969

77.28 **Ytterbium(iii) acetylacetonate**
$C_{15}H_{21}O_6Yb$
E.F.Korytnyi, L.A.Aslanov, M.A.Porai-Koshits, O.M.Petrukhin
Zh. Strukt. Khim., **9**, 540, 1968

77.C **Rhodium acetylacetonate bis(1,1 - dimethylallene)**
$C_{15}H_{23}O_2Rh$
For complete entry see 72.45

77.29 **tris(Acetylacetonato) aquo ytterbium(iii)**
$C_{15}H_{23}O_7Yb$
J.A.Cunningham, D.E.Sands, W.F.Wagner, M.F.Richardson
Inorg. Chem., **8**, 22, 1969

77.30 **tris(Acetylacetonato) aquo ytterbium(iii) benzene solvate**
$C_{15}H_{23}O_7Yb$, $0.5C_6H_6$
E.D.Watkins, J.A.Cunningham, T.Phillips, D.E.Sands, W.F.Wagner
Inorg. Chem., **8**, 29, 1969

77.31 **Holmium diaquo tris(acetylacetonate) hydrate**
$C_{15}H_{25}HoO_8$, H_2O
L.A.Aslanov, E.F.Korytnyi, M.A.Porai-Koshits, A.I.Byrke
Zh. Strukt. Khim., **9**, 331, 1968

77.32 **Diaquo tris(acetylacetonato) lanthanum(iii)**
$C_{15}H_{25}LaO_8$
T.Phillips, D.E.Sands, W.F.Wagner   *Inorg. Chem.*, **7**, 2295, 1968

77.33 **Neodymium diaquo - tris(acetylacetonate) monohydrate**
$C_{15}H_{25}NdO_8$, H_2O
L.A.Aslanov, M.O.Dekaprilevich, M.A.Porai-Koshits, V.I.Ivanov
Zh. Strukt. Khim., **8**, 1106, 1967

77.34 **Yttrium(iii) tris(acetylacetonate) diaquo monohydrate**
$C_{15}H_{25}O_8Y$, H_2O
J.A.Cunningham, D.E.Sands, W.F.Wagner   *Inorg. Chem.*, **6**, 499, 1967

77.35   Di(μ - allyl) - bis(platinum(ii) acetylacetonate)
$C_{16}H_{24}O_4Pt_2$
W.S.McDonald, B.E.Mann, G.Raper, B.L.Shaw, G.Shaw
J. Chem. Soc. (D), 1254, 1969
Also classified in 72, 71

77.C   2 - (Pentamethylbicyclo(2.2.0)hexa - 2,5 - diene methyl palladium
acetylacetonate
$C_{17}H_{24}O_2Pd$
For complete entry see 75.59

77.36   Ethyl - (trimethyl platinum) aceto - acetate dimer
$C_{18}H_{36}O_6Pt_2$
A.C.Hazell, M.R.Truter   *Proc. R. Soc., A*, **254**, 218, 1960
Also classified in 71

77.37 μ - Ethylenediamine - bis(trimethyl(acetylacetonato) platinum(iv))
$C_{18}H_{40}N_2O_4Pt_2$
A.Robson, M.R.Truter   *J. Chem. Soc.*, 630, 1965
Also classified in 71, 76

77.38   Copper(ii) acetylacetonate - quinoline complex
$C_{19}H_{21}CuNO_4$
S.Ooi, Q.Fernando   *Chem. Communic.*, 532, 1967
Also classified in 83

77.C   bis(Tetramethylallene) rhodium(i) acetylacetonate
$C_{19}H_{31}O_2Rh$
For complete entry see 72.55

77.39   Cesium tetrakis(hexafluoroacetylacetonato) yttrate(iii)
$C_{20}H_4F_{24}O_8Y^-$, Cs^+
M.J.Bennett, F.A.Cotton, P.Legzdins, S.J.Lippard
Inorg. Chem., **7**, 1770, 1968

77.40   Copper(ii) benzoylacetonate
$C_{20}H_{16}CuO_4$
P.Hon, C.E.Pfluger, R.L.Belford   *Inorg. Chem.*, **5**, 516, 1966
Also classified in 84

77.41   Palladium(ii) bis(1 - phenyl - 1,3 - butanedionate)
$C_{20}H_{18}O_4Pd$
P.Hon, C.E.Pfluger, R.L.Belford   *Inorg. Chem.*, **6**, 730, 1967
Also classified in 84

77.42 **Vanadyl 1 - phenyl - 1,3 - butanedionate**
$C_{20}H_{18}O_5V$
P.Hon, R.L.Belford, C.E.Pfluger *J. Chem. Phys.*, **43,** 1323, 1965
Also classified in 84

77.43 **Vanadyl 1 - phenyl - 1,3 - butanedionate (discussion)**
$C_{20}H_{18}O_5V$
P.Hon, R.L.Belford, C.E.Pfluger *J. Chem. Phys.*, **43,** 3111, 1965
Also classified in 84

77.44 **trans - bis(Acetylacetonato) dipyridine cobalt(ii)**
trans - bis(2,4 - Pentanedionato) dipyridine cobalt(ii)
$C_{20}H_{24}CoN_2O_4$
R.C.Elder *Inorg. Chem.*, **7,** 1117, 1968
Also classified in 83

77.45 **trans - bis(2,4 - Pentanedionato) dipyridine - nickel(ii)**
trans - bis(Acetylacetonato) dipyridine - nickel(ii)
$C_{20}H_{24}N_2NiO_4$
R.C.Elder *Inorg. Chem.*, **7,** 2316, 1968
Also classified in 83

77.46 **bis(Acetylacetonato) bis(pyridine N - oxide) nickel(ii)**
bis(2,4 - Pentanedionato) bis(pyridine N - oxide) nickel(ii)
$C_{20}H_{24}N_2NiO_6$
W.DeW.Horrocks Junior, D.H.Templeton, A.Zalkin
Inorg. Chem., **7,** 1552, 1968
Also classified in 84

77.47 **Cerium(iv) acetylacetone**
$C_{20}H_{28}CeO_8$
B.Matkovic, D.Grdenic *Acta Cryst.*, **16,** 456, 1963

77.48 **μ - Oxo - bis(chlorobis(2,4 - pentanedionato) titanium(iv)) - chloroform**
$C_{20}H_{28}Cl_2O_9Ti_2$, $CHCl_3$
K.Watenpaugh, C.N.Caughlan *Inorg. Chem.*, **6,** 963, 1967

77.49 **Thorium(iv) acetylacetonate**
$C_{20}H_{28}O_8Th$
D.Grdenic, B.Matkovic *Nature*, **182,** 465, 1958
See also *Int. Distances*, M 193s; *Structure Reports*, **22,** 596, 1958

77.50 **Zirconium(iv) acetylacetonate**
$C_{20}H_{28}O_8Zr$
J.V.Silverton, J.L.Hoard *Inorg. Chem.*, **2,** 243, 1963

77.51 **Dimeric bis(acetylacetonato) aquo cobalt(ii)**
$C_{20}H_{32}Co_2O_{10}$
F.A.Cotton, R.C.Elder *Inorg. Chem.*, **5**, 423, 1966

77.52 **N,N' - Ethylene - bis(salicylaldehydeiminato) acetylacetonate cobalt(iii) monohydrate**
$C_{21}H_{21}CoN_2O_4$, H_2O
M.Calligaris, G.Nardin, L.Randaccio *J. Chem. Soc. (D)*, 1248, 1969
Residue 1 also classified in 78

77.53 **tris - Acetylacetonato - oxo molybdenum triethanolate**
$C_{21}H_{36}Mo_3O_{13}$
S.Scavnicar, M.Herceg *Acta Cryst.*, **21**, A151, 1966

77.54 **4 - Methylpyridine bis(o - hydroxyacetophenonato) copper(ii)**
$C_{22}H_{21}CuNO_4$
V.F.Duckworth, N.C.Stephenson *Acta Cryst. (B)*, **25**, 2245, 1969
Also classified in 84, 83

77.55 **Copper(ii) 3 - phenyl - 2,4 - pentanedionate**
$C_{22}H_{22}CuO_4$
J.W.Carmichael, L.K.Steinrauf, R.L.Belford
J. Chem. Phys., **43**, 3959, 1965
Also classified in 84

77.56 **bis(Dipivaloylmethanido) nickel(ii)**
$C_{22}H_{38}NiO_4$
F.A.Cotton, J.J.Wise *Inorg. Chem.*, **5**, 1200, 1966
Also classified in 84

77.57 **bis(Dipivaloylmethanido) zinc(ii)**
$C_{22}H_{38}O_4Zn$
F.A.Cotton, J.S.Wood *Inorg. Chem.*, **3**, 245, 1964
Also classified in 84

77.58 **Trimethyl 4,6 dioxonyl platinum**
$C_{24}H_{48}O_4Pt_2$
A.G.Swallow, M.R.Truter *Proc. R. Soc., A*, **252**, 205, 1960
Also classified in 71, 84

77.59 **Copper(ii) 1,3 - diphenyl - 1,3 - propanedionate**
$C_{30}H_{22}CuO_4$
M.Blackstone, J.van Thuijl, C.Romers
Rec. Trav. Chim. Pays-Bas, **85**, 557, 1966
Also classified in 84

77.60 **Palladium(ii) dibenzoylmethanate**
$C_{30}H_{22}O_4Pd$
E.A.Shugam, L.M.Shkol'nikova, A.N.Knyazeva
Zh. Strukt. Khim., **9**, 222, 1968
Also classified in 84

77.61 **tris(1 - Phenyl - 1,3 - butanedionato) aquo yttrium(iii)**
$C_{30}H_{29}O_7Y$
F.A.Cotton, P.Legzdins *Inorg. Chem.*, **7**, 1777, 1968
Also classified in 84

77.62 **bis(Acetylacetonato) nickel(ii)**
$C_{30}H_{42}Ni_3O_{12}$
G.J.Bullen, R.Mason, P.Pauling *Inorg. Chem.*, **4**, 456, 1965

77.63 **Zinc(ii) acetylacetonate trimer**
$C_{30}H_{42}O_{12}Zn_3$
M.J.Bennett, F.A.Cotton, R.Eiss *Acta Cryst. (B)*, **24**, 904, 1968

77.64 **Hexa(acetylacetonato) aquo tricobalt(ii)**
$C_{30}H_{44}Co_3O_{13}$
F.A.Cotton, R.Eiss *J. Amer. Chem. Soc.*, **90**, 38, 1968

77.65 **Ammonium tetrakis(4,4,4 - trifluoro - 1 - (2 - thienyl) - 1,3 - butanedione) praseodymate(iii) monohydrate**
$C_{32}H_{16}F_{12}O_8PrS_4{}^-, H_4N^+, H_2O$
R.A.Lalancette, M.Cefola, W.C.Hamilton, S.J.La Placa
Inorg. Chem., **6**, 2127, 1967
Residue 1 also classified in 84

77.66 **Piperidinium gadolinium tetrakis(benzoylacetonate)**
$C_{40}H_{36}GdO_8{}^-, C_5H_{12}N^+$
L.A.Butman, L.A.Aslanov, M.A.Porai-Koshits
Zh. Strukt. Khim., **9**, 274, 1968
Residue 2 classified in 84, 3

77.67 **Cobalt(ii) acetylacetonate**
$C_{40}H_{56}Co_4O_{16}$
F.A.Cotton, R.C.Elder *Inorg. Chem.*, **4**, 1145, 1965

77.68 **bis(Triphenylphosphine) acetylacetonato rhenium chloride**
$C_{41}H_{37}Cl_2O_2P_2Re$
G.L.Betzner, W.Che'ng, N.C.Jayadavan, C.J.L.Lock, I.D.Brown
Acta Cryst., **21**, A136, 1966
Also classified in 86

77.69   **Di(μ - diphenylphosphinatoacetylacetonato chromium(iii))**
$C_{44}H_{48}Cr_2O_{12}P_2$
C.E.Wilkes, R.A.Jacobson   *Inorg. Chem.*, **4,** 99, 1965
Also classified in 84

METAL COMPLEXES
(SALICYLIC DERIVATIVES)

78.1 N - Salicylideneglycinato aquo copper(ii) hemihydrate
$C_9H_9CuNO_4$, $0.5H_2O$
T.Ueki, T.Ashida, Y.Sasada, M.Kakudo *Acta Cryst.*, **22,** 870, 1967

78.2 N - Salicylideneglycinato aquo copper(ii) tetrahydrate
$C_9H_9CuNO_4$, $4H_2O$
T.Veki, T.Ashida, Y.Sasada, M.Kakudo *Acta Cryst. (B),* **25,** 328, 1969
Residue 1 also classified in 82

78.3 bis - (5 - Chlorosalicylaldoximato) copper(ii)
$C_{14}H_{10}Cl_2CuN_2O_4$
P.L.Orioli, E.C.Lingafelter, B.W.Brown *Acta Cryst.*, **17,** 1113, 1964

78.4 Copper(ii) bis(salicylaldehyde) (form A)
$C_{14}H_{10}CuO_4$
A.J.McKinnon, T.N.Waters, D.Hall *J. Chem. Soc.*, 3290, 1964

78.5 Copper(ii) bis salicylaldehyde) (form B)
$C_{14}H_{10}CuO_4$
D.Hall, A.J.McKinnon, T.N.Waters *J. Chem. Soc.*, 425, 1965

78.6 Copper(ii) bis salicylaldehyde) (form A)
$C_{14}H_{10}CuO_4$
J.A.Bevan, D.P.Graddon, J.F.McConnell *Nature,* **199,** 373, 1963

78.7 Copper(ii) bis(salicylaldiminate)
$C_{14}H_{12}CuN_2O_2$
E.N.Baker, D.Hall, T.N.Waters *J. Chem. Soc. (A),* 680, 1966

78.8 bis(Salicylaldoximato) copper(ii)
$C_{14}H_{12}CuN_2O_4$
M.A.Jarski, E.C.Lingafelter *Acta Cryst.*, **17,** 1109, 1964

78.9 **bis - Salicylideneiminato nickel(ii)**
$C_{14}H_{12}N_2NiO_2$
J.M.Stewart, E.C.Lingafelter *Acta Cryst.*, **12**, 842, 1959
See also *Int. Distances*, M 179s; *Structure Reports*, **23**, 655, 1959

78.10 **Nickel(ii) salicylaldoxime**
$C_{14}H_{12}N_2NiO_4$
R.C.Srivastava, E.C.Lingafelter, P.C.Jain *Acta Cryst.*, **22**, 922, 1967

78.11 **Diaquo - bis(salicylaldehydato) nickel (trans form)**
$C_{14}H_{14}NiO_6$
J.M.Stewart, E.C.Lingafelter, J.D.Breazeale *Acta Cryst.*, **14**, 888, 1961

78.12 **Zinc - diaquo - salicylate**
$C_{14}H_{14}O_8Zn$
H.P.Klug, L.E.Alexander, G.G.Sumner *Acta Cryst.*, **11**, 41, 1958
See also *Int. Distances*, M 137s; *Structure Reports*, **22**, 711, 1958

78.13 **N,N' - bis(Salicylidene)ethylenediamine iron(iii) chloride**
$C_{16}H_{14}ClFeN_2O_2$
M.Gerloch, F.E.Mabbs *J. Chem. Soc. (A)*, 1900, 1967

78.14 **N,N' - bis(Salicylideneiminato) - iron(iii) chloride nitromethane solvate**
$C_{16}H_{14}ClFeN_2O_2$, CH_3NO_2
M.Gerloch, F.E.Mabbs *J. Chem. Soc. (A)*, 1598, 1967

78.15 **bis(Salicylaldehyde) ethylenedi - imine cobalt(ii) monochloroformate**
$C_{16}H_{14}CoN_2O_2$, $CHCl_3$
W.P.Schaefer, R.E.Marsh *Acta Cryst. (B)*, **25**, 1675, 1969

78.16 **N,N' - Disalicylidene ethylenediamine copper**
$C_{16}H_{14}CuN_2O_2$
D.Hall, T.N.Waters *J. Chem. Soc.*, 2644, 1960

78.17 **N,N' - Ethylene bis(salicylaldiminato) copper(ii) chloroform complex**
$C_{16}H_{14}CuN_2O_2$, $CHCl_3$
E.N.Baker, D.Hall, A.J.McKinnon, T.N.Waters
Chem. Communic., 134, 1967

78.18 **N,N' - Ethylene bis(salicylaldiminato) copper(ii) p - nitrophenol complex**
$C_{16}H_{14}CuN_2O_2$, $C_6H_5NO_3$
E.N.Baker, D.Hall, A.J.McKinnon, T.N.Waters
Chem. Communic., 134, 1967

78.C N,N - Ethylenebis(salicylaldiminato) copper(ii) - sodium perchlorate complex p - xylene solvate
$2C_{16}H_{14}CuN_2O_2$, Na^+ , ClO_4^- , C_8H_{10}
For complete entry see 60.145

78.19 bis(N - Methylsalicylaldiminato) copper(ii) (α form)
$C_{16}H_{16}CuN_2O_2$
E.C.Lingafelter, G.L.Simmons, B.Morosin, C.Scheringer, C.Freiburg
Acta Cryst., **14**, 1222, 1961

78.20 bis(N - Methylsalicylaldiminato) copper(ii) (α form)
$C_{16}H_{16}CuN_2O_2$
B.Methuen, M.V.Stackelberg *Z. Anorg. Allg. Chem.*, **305**, 279, 1960

78.21 bis(N - Methylsalicylaldiminato) copper(ii) (γ form)
$C_{16}H_{16}CuN_2O_2$
D.Hall, S.V.Sheat, T.N.Waters *J. Chem. Soc. (A)*, 460, 1968

78.22 bis - (N - Methylsalicylideneiminato) nickel(ii) (monoclinic form)
$C_{16}H_{16}N_2NiO_2$
E.Frasson, C.Panattoni, L.Sacconi *J. Phys. Chem.*, **63**, 1908, 1959
See also *Int. Distances*, M 185s; *Structure Reports*, **23**, 656, 1959

78.23 bis - (N - Methylsalicylideneiminato) nickel(ii) (orthorhombic form)
$C_{16}H_{16}N_2NiO_2$
M.R.Fox, E.C.Lingafelter *Acta Cryst.*, **22**, 943, 1967

78.24 Zinc N - methylsalicylaldiminate
$C_{16}H_{16}N_2O_2Zn$
P.L.Orioli, M.di Vaira, L.Sacconi *Inorg. Chem.*, **5**, 400, 1966

78.25 N,N' - Disalicylidene ethylenediamine zinc(ii) monohydrate
$C_{16}H_{16}N_2O_3Zn$
D.Hall, F.H.Moore *J. Chem. Soc. (A)*, 1822, 1966
Also classified in 76

78.26 N,N' - Disalicylidene - propane - 1,2 - diamine copper monohydrate
$C_{17}H_{16}CuN_2O_2$, H_2O
F.J.Llewellyn, T.N.Waters *J. Chem. Soc.*, 2639, 1960

78.27 Copper(ii) N - ethylsalicylaldiminate
$C_{18}H_{20}CuN_2O_2$
E.N.Baker, G.R.Clark, D.Hall, T.N.Waters *J. Chem. Soc. (A)*, 251, 1967

78.28   Copper(ii) N - ethylsalicylaldiminate
$C_{18}H_{20}CuN_2O_2$
C.Panattoni, G.Bombieri, R.Graziani   *Acta Cryst.*, **23**, 537, 1967

78.29   bis - (N - 2 - Hydroxyethylsalicylaldiminato) copper(ii)
$C_{18}H_{20}CuN_2O_4$
E.R.Boyko, D.Hall, M.E.Kinloch, T.N.Waters
Acta Cryst., **21**, 614, 1966

78.30   bis - (N - Ethylsalicylaldiminato) - nickel
$C_{18}H_{20}N_2NiO_2$
L.M.Shkol'nikova, A.N.Knazeva   *Zh. Strukt. Khim.*, **8**, 94, 1967

78.31   bis - N - Ethylsalicylaldimine palladium
$C_{18}H_{20}N_2O_2Pd$
E.Frasson, C.Panattoni, L.Sacconi   *Acta Cryst.*, **17**, 85, 1964

78.32   Nickel(ii) - N - β - diethyl aminoethyl - salicylaldimine catecholate
$C_{19}H_{24}N_2NiO_3$
L.Sacconi, P.L.Orioli, M.di Vaira   *Chem. Communic.*, 849, 1967
Also classified in 83

78.33   Trimethyl(salicylaldehydato) platinum(iv)
$C_{20}H_{22}O_4Pt$
M.R.Truter, R.C.Watling   *J. Chem. Soc. (A)*, 1955, 1967
Also classified in 71

78.34   bis(N - iso - Propylsalicylaldiminato) copper(ii)
$C_{20}H_{24}CuN_2O_2$
P.L.Orioli, L.Sacconi   *J. Amer. Chem. Soc.*, **88**, 277, 1966

78.35   bis(N - n - Propylsalicylaldiminato) copper(ii)
$C_{20}H_{24}CuN_2O_2$
G.Bombieri, C.Panattoni, E.Forsellini, R.Graziani
Acta Cryst. (B), **25**, 1208, 1969

78.36   bis(N - iso - Propylsalicylaldiminato) nickel(ii)
$C_{20}H_{24}N_2NiO_2$
M.R.Fox, P.L.Orioli, E.C.Lingafelter, L.Sacconi
Acta Cryst., **17**, 1159, 1964

78.C   N,N' - Ethylene - bis(salicylaldehydeiminato) acetylacetonate cobalt(iii) monohydrate
$C_{21}H_{21}CoN_2O_4$, H_2O
For complete entry see 77.52

78.37 **bis((Salicylidene - γ - iminopropyl)methylamine)) nickel(ii)**
$C_{21}H_{25}N_3NiO_2$
P.L.Orioli, M.di Vaira, L.Sacconi *Chem. Communic.*, 300, 1966
Also classified in 83

78.38 **bis(N - Isopropyl - 3 - methylsalicylaldiminato) palladium (monoclinic form)**
$C_{22}H_{26}N_2O_2Pd$
P.C.Jain, E.C.Lingafelter *Acta Cryst.*, **23**, 127, 1967

78.39 **bis(N - Isopropyl - 3 - methylsalicylaldiminato) palladium (tetragonal form)**
$C_{22}H_{26}N_2O_2Pd$
P.C.Jain, E.C.Lingafelter *Acta Cryst.*, **23**, 127, 1967

78.40 **bis(N - t - Butylsalicylaldiminato) copper(ii)**
$C_{22}H_{28}CuN_2O_2$
T.P.Cheeseman, D.Hall, T.N.Waters *J. Chem. Soc. (A)*, 685, 1966

78.41 **bis(N - n - Butylsalicylaldiminato) copper(ii)**
$C_{22}H_{28}CuN_2O_2$
D.Hall, R.H.Sumner, T.N.Waters *J. Chem. Soc. (A)*, 420, 1969

78.42 **bis - (N - Isobutylsalicylaldiminato) nickel(ii)**
$C_{22}H_{28}N_2NiO_2$
M.D.Fox *Dissert. Abstr.*, **27**, 127B, 1966

78.43 **bis(N - isoPropyl - 3 - methylsalicylaldiminato) nickel(ii)**
$C_{22}H_{28}N_2NiO_2$
R.L.Braun, E.C.Lingafelter *Acta Cryst.*, **21**, 546, 1966

78.44 **bis - N - n - Butylsalicylaldimine palladium(ii)**
$C_{22}H_{28}N_2O_2Pd$
E.Frasson, C.Panattoni, L.Sacconi *Acta Cryst.*, **17**, 477, 1964

78.45 **bis(N - t - Butylsalicylaldiminato) palladium(ii)**
$C_{22}H_{28}N_2O_2Pd$
V.W.Day, M.D.Glick, J.L.Hoard *J. Amer. Chen. Soc.*, **90**, 4803, 1968

78.46 **bis(N - Methyl - 2 - hydroxy - 1 - naphthaldiminato) copper(ii)**
$C_{24}H_{20}CuN_2O_2$
D.Hall, A.J.McKinnon, J.M.Waters, T.N.Waters *Nature,* **201**, 607, 1964

78.47 **Nickel(ii) N - isopropyl - 3 - ethylsalicylaldiminate**
$C_{24}H_{32}N_2NiO_2$
R.L.Braun, E.C.Lingafelter *Acta Cryst.*, **22**, 780, 1967

78.48 **Palladium(ii) N - isopropyl - 3 - ethylsalicylaldiminate**
$C_{24}H_{32}N_2O_2Pd$
R.L.Braun, E.C.Lingafelter *Acta Cryst.*, **22**, 787, 1967

78.49 **Nickel(ii) N - γ - dimethylaminopropyl salicylaldiminate**
$C_{24}H_{34}N_4NiO_2$
M.di Vaira, P.L.Orioli *Inorg. Chem.*, **6**, 490, 1967
Also classified in 83

78.50 **N,N' - (2,2' - Biphenyl) - bis(salicylaldiminato) copper(ii)**
$C_{26}H_{18}CuN_2O_2$
T.P.Cheeseman, D.Hall, T.N.Waters *J. Chem. Soc. (A)*, 1396, 1966

78.51 **bis(N - Phenylsalicylaldiminato) copper(ii)**
$C_{26}H_{20}CuN_2O_2$
L.Wei, R.M.Stogsdill, E.C.Lingafelter *Acta Cryst.*, **17**, 1058, 1964

78.52 **bis - (N - Phenylsalicylaldiminato) nickel**
$C_{26}H_{20}N_2NiO_2$
R.V.Chastain *Dissert. Abstr.*, **27**, 124B, 1966

78.53 **Nickel(ii) 5 - chloro - N - (2' - diethylaminoethyl) - salicylaldiminate**
$C_{26}H_{36}Cl_2N_4NiO_2$
P.L.Orioli, M.di Vaira, L.Sacconi *J. Amer. Chem. Soc.*, **88**, 4383, 1966
Also classified in 83

78.54 **2,2' - bis(Salicylaldiminato) - (+) - (R) - 6,6' - dimethyl - biphenyl cobalt(ii)**
$C_{28}H_{22}CoN_2O_2$
L.H.Pignolet, R.P.Taylor, W.DeW.Horrocks Junior
Chem. Communic., 1443, 1968

78.55 **bis(N - o - Methoxyphenyl - salicylaldiminato) zinc**
$C_{28}H_{24}N_2O_4Zn$
L.M.Shkol'nikova, V.A.Kogan, L.G.Makarevich, O.A.Osipov,
A.E.Obodovskaya *Zh. Strukt. Khim.*, **9**, 543, 1968
Also classified in 84

78.56 **bis - D - N - Phenylethyl salicylaldiminato copper(ii) (orthorhombic form)**
$C_{30}H_{28}CuN_2O_2$
Z.A.Starikova *Acta Cryst.*, **21**, A154, 1966

78.57 **bis - DL - N - Phenylethyl salicylaldiminato copper(ii)**
$C_{30}H_{28}CuN_2O_2$
Z.A.Starikova *Acta Cryst.*, **21**, A154, 1966

78.58   **bis(N - Phenylsalicylaldiminato) copper(ii) pyridine**
$C_{31}H_{45}CuN_3O_2$
D.Hall, S.V.Sheat-Rumball, T.N.Waters   *J. Chem. Soc. (A),* 2721, 1968
Also classified in 83

78.59   **N,N' - Ethylene bis(salicylaldehydeiminato) cobalt(ii) (inactive form)**
$C_{32}H_{28}Co_2N_4O_4$
S.Bruckner, M.Calligaris, G.Nardin, L.Randaccio
Acta Cryst. (B), **25,** 1671, 1969
Also classified in 83, 84

78.60   **μ - Oxo - bis(iron bis(salicylidene - ethylenediamine)) pyridine solvate**
$C_{32}H_{28}Fe_2N_4O_5$, $2C_5H_5N$
M.Gerloch, E.D.McKenzie, A.D.C.Towl   *Nature,* **220,** 906, 1968
Residue 1 also classified in 83, 76

78.61   **tetrakis - N - Ethylsalicylaldiminato - titanium(iv)**
$C_{36}H_{40}N_4O_4Ti$
D.C.Bradley, M.B.Hursthouse, I.F.Rendall   *J. Chem. Soc. (D),* 672, 1969

78.62   **μ - Dioxygen - di(N,N' - ethylene bis(salicylaldehydeiminato) cobalt(ii) dimethylformamide)**
$C_{38}H_{42}Co_2N_6O_8$
M.Calligaris, G.Nardin, L.Randaccio   *J. Chem. Soc. (D),* 763, 1969
Also classified in 84

METAL COMPLEXES (THIOUREA)

79.1 **Copper thiourea nitrate**
$(CH_4CuN_2S^+)_n$, nNO_3^-
R.G.Vranka, E.L.Amma   *J. Amer. Chem. Soc.*, **88,** 4270, 1966

79.2 **Cadmium diaquo sulfate monothiourea complex**
$(CH_8CdN_2O_6S)_n$
L.Cavalca, P.Domiano, G.F.Gasparri, P.Boldrini
Acta Cryst., **22,** 878, 1967

79.3 **bis(Thiourea) silver(i) chloride**
$C_2H_8AgClN_4S_2$
E.A.Vizzini, I.F.Taylor, E.L.Amma   *Inorg. Chem.*, **7,** 1351, 1968

79.4 **Silver(i) chloride - thiourea complex**
$C_2H_8AgN_4S_2^+$, Cl^-
E.A.Vizzini, E.L.Amma   *J. Amer. Chem. Soc.*, **88,** 2872, 1966

79.5 **Dichloro bisthiourea cadmium**
$C_2H_8CdCl_2N_4S_2$
M.Nardelli, L.Cavalca, A.Braibanti   *Gazz. Chim. Ital.*, **87,** 137, 1957
See also *Int. Distances,* M 89s; *Structure Reports,* **21,** 528, 1957

79.6 **Cadmium nitrate bis(thiourea)**
$C_2H_8CdN_4O_6S_2$
S.Swaminathan, S.Natarajan   *Curr. Sci.*, **36,** 513, 1967

79.7 **bis(Thiourea) copper(i) chloride**
$(C_2H_8ClCuN_4S_2)_n$
W.A.Spofford, E.L.Amma   *Chem. Communic.*, 405, 1968

79.8 **bis(Thiourea) mercury(ii) chloride**
$C_2H_8ClHgN_4S_2^+$, Cl^-
K.K.Cheung, R.S.McEwen, G.A.Sim   *Nature,* **205,** 383, 1965

79.9   **Dichloro bisthiourea lead(ii)**
$C_2H_8Cl_2N_4PbS_2$
M.Nardelli, G.Fava   *Acta Cryst.*, **12**, 727, 1959
See also *Int. Distances*, M 90s; *Structure Reports*, **23**, 586, 1959

79.10   **Dichloro bisthiourea zinc**
$C_2H_8Cl_2N_4S_2Zn$
N.R.Kunchur, M.R.Truter   *J. Chem. Soc.*, 3478, 1958
See also *Int. Distances*, M 90s; *Structure Reports*, **22**, 605, 1958

79.11   **Mercury(ii) iodide - bis(thiourea) complex**
$C_2H_8HgI_2N_4S_2$
A.Korczynski   *Rocz. Chem.*, **42**, 393, 1968

79.12   **tris(Thiourea) mercury(ii) chloride**
$C_3H_{12}Cl_2HgN_6S_3$
A.Korczynski   *Rocz. Chem.*, **42**, 1207, 1968

79.13   **Copper(i) tris(thiourea) chloride**
$(C_3H_{12}CuN_6S_3{}^+)_n$, nCl^-
Y.Okaya, C.Knobler   *Acta Cryst.*, **17**, 928, 1964

79.14   **tris(Thiourea) zinc thiosulfate monohydrate**
$C_3H_{12}N_6O_3S_5Zn$, H_2O
G.D.Andreetti, L.Cavalca, P.Domiano, A.Musatti
Ric. Sci., **38**, 1100, 1968

79.15   **tris(Thiourea) zinc(ii) sulfate**
$C_3H_{12}N_6O_4S_4Zn$
G.D.Andreetti, L.Cavalca, A.Musatti   *Acta Cryst. (B)*, **24**, 683, 1968

79.16   **Dibromo tris(semicarbazide) disilver(i) polymer**
$(C_3H_{15}Ag_2Br_2N_9S_3)_n$
L.C.Capacchi, G.F.Gasparri, M.Ferrari, M.Nardelli
Ric. Sci., **38**, 974, 1968

79.17   **Mercury(ii) thiocyanate - thiourea complex**
$C_4H_8HgN_6S_4$
A.Korczynski   *Rocz. Chem.*, **40**, 547, 1966

79.18   **bis(Thiourea) nickel(ii) thiocyanate**
$C_4H_8N_6NiS_4$
M.Nardelli, G.F.Gasparri, G.G.Battistini, P.Domiano
Acta Cryst., **20**, 349, 1966

79.19 **bis(Dithiobiureto) palladium(ii)**
$C_4H_8N_6PdS_4$
R.L.Girling, E.L.Amma *Chem. Communic.*, 1487, 1968

79.20 **bis(Thiourea) cadmium formate**
$C_4H_{10}CdN_4O_4S_2$
M.Nardelli, G.F.Gasparri, P.Boldrini *Acta Cryst.*, **18,** 618, 1965
Also classified in 81

79.21 **Cadmium tetra - thiourea zinc tetrathiocyanate**
$C_4H_{16}CdN_8S_4{}^{2+}$, $C_4N_4S_4Zn^{2-}$
A.Korczynski *Rocz. Chem.*, **41,** 1197, 1967

79.22 **trans - Dichloro tetrakis(thiourea) cobalt(ii)**
$C_4H_{16}Cl_2CoN_8S_4$
J.E.O'Connor, E.L.Amma *Chem. Communic.*, 892, 1968

79.23 **Dichloro tetrakisthiourea nickel**
$C_4H_{16}Cl_2N_8NiS_4$
A.Lopez-Castro, M.R.Truter *J. Chem. Soc.*, 1309, 1963

79.24 **Mercury(ii) tetrathiourea cobalt(ii) tetrathiocyanate**
$C_4H_{16}HgN_8S_4{}^{2+}$, $C_4CoN_4S_4{}^{2-}$
A.Korczynski, M.A.Porai-Koshits *Rocz. Chem.*, **39,** 1567, 1965

79.25 **Tetrathiourea palladium(ii) chloride**
$C_4H_{16}N_8PdS_4{}^{2+}$, $2Cl^-$
S.Ooi, T.Kawase, K.Nakatsu, H.Kuroya
Bull. Chem. Soc. Jap., **33,** 861, 1960

79.26 **Tetra(thiourea) nickel(ii) thiosulphate monohydrate**
$C_4H_{26}N_8NiO_3S_6{}^{2+}$, $O_3S_2{}^{2-}$, H_2O
G.F.Gasparri, A.Mangia, A.Musatti, M.Nardelli
Acta Cryst. (B), **25,** 203, 1969

79.C **π - Allyl di(thiourea) nickel chloride**
$C_5H_{11}N_4NiS_2{}^+$, Cl^-
For complete entry see 72.6

79.27 **bis - Thiourea zinc acetate**
$C_6H_{14}N_4O_4S_2Zn$
L.Cavalca, G.F.Gasparri, G.D.Andreetti, P.Domiano
Acta Cryst., **22,** 90, 1967

79.28 **hexakis(Thiourea) nickel(ii) bromide**
$C_6H_{24}N_{12}NiS_6{}^{2+}$, $2Br^-$
M.S.Weiniger, J.E.O'Connor, E.L.Amma   *Inorg. Chem.*, **8**, 424, 1969

79.29 **Titanium(iii) iodide - hexa - urea complex**
$C_6H_{24}N_{12}O_6Ti^{3+}$, $3I^-$
A.Linek, J.Siskova, L.Jensovsky
Collect. Czechosl. Chem. Communic., **31**, 4453, 1966

79.30 **Hydroxo tris(carbamido) uranyl polyiodide**
$C_6H_{26}N_{12}O_{12}U_2{}^{2+}$, $I_4{}^{2-}$
Yu.Mikhailov, V.G.Kuznetsov, E.S.Kovaleva
Zh. Strukt. Khim., **9**, 710, 1968

79.31 **bis(Ethylenethiourea) nickel thiocyanate**
$(C_8H_{12}N_6NiS_4)_n$
M.Nardelli, G.F.Gasparri, A.Musatti, A.Manfredotti
Acta Cryst., **21**, 910, 1966
Also classified in 85

79.32 **Copper(i) nitrate - thiourea complex**
$C_9H_{36}Cu_4N_{18}S_9{}^{4+}$, $4NO_3^-$
R.G.Vranka   *Dissert. Abstr. (B)*, **27**, 3423, 1967

79.33 **bis - Ethylenethiourea cadmium thiocyanate**
$C_{10}H_{12}CdN_6S_4$
L.Cavalca, M.Nardelli, G.Fava   *Acta Cryst.*, **13**, 125, 1960
Also classified in 85

79.34 **Nickel(ii) chloride - tetra(ethylenethiourea) complex (triclinic form)**
$C_{12}H_{24}Cl_2N_8NiS_4$
W.T.Robinson, S.L.Holt Junior, G.B.Carpenter   *Inorg. Chem.*, **6**, 605, 1967
Also classified in 85

79.35 **Nickel(ii) chloride - tetra(ethylenethiourea) complex (monoclinic form)**
$C_{12}H_{24}Cl_2N_8NiS_4$
W.T.Robinson, S.L.Holt Junior, G.B.Carpenter   *Inorg. Chem.*, **6**, 605, 1967
Also classified in 85

79.36 **Dichloro tetrakis(trimethylenethiourea) nickel(ii)**
$C_{16}H_{32}Cl_2N_8NiS_4$
H.Luth, M.R.Truter   *J. Chem. Soc. (A)*, 1879, 1968
Also classified in 85

METAL COMPLEXES
(THIOCARBAMATE OR XANTHATE)

80.1   **Nickel(ii) dithiocarbamate**
$C_2H_4N_2NiS_4$
G.F.Gasparri, M.Nadelli, A.Villa   *Acta Cryst.*, **23**, 384, 1967

80.2   **Gold diethyldithiocarbamate**
$(C_5H_{10}AuNS_2)_n$
R.Hesse   *Acta Cryst.*, **13**, A51, 1960

80.3   **Nickel xanthate**
$C_6H_{10}NiO_2S_4$
M.Franzini   *Z. Kristallogr.*, **118**, 393, 1963

80.4   **Zinc ethylxanthate**
$C_6H_{10}O_2S_4Zn$
T.Ikeda, H.Hagihara   *Acta Cryst.*, **21**, 919, 1966

80.5   **Nitroso - (dimethyldithiocarbamato) cobalt**
$C_6H_{12}CoN_3OS_4$
P.R.H.Alderman, P.G.Owston, J.M.Rowe   *J. Chem. Soc.*, 668, 1962

80.6   **Nitrosyl iron bis - (N,N - dimethyldithiocarbamate)**
$C_6H_{12}FeN_3OS_4$
G.R.Davies, R.H.B.Mais, P.G.Owston   *Chem. Communic.*, 81, 1968

80.7   **Cobalt(iii) tris(o - ethylxanthate)**
$C_9H_{15}CoO_3S_6$
S.Merlino   *Acta Cryst. (B)*, **25**, 2270, 1969

80.8   **Dibromo - N,N - di - n - butyldithiocarbamato gold(iii)**
$C_9H_{18}AuBr_2NS_2$
P.T.Beurskens, J.A.Cras, J.J.Steggerda   *Inorg. Chem.*, **7**, 810, 1968

80.9   **Dibromo - N,N - di - n - butyldithiocarbamato copper(iii)**
$C_9H_{18}Br_2CuNS_2$
P.T.Beurskens, J.A.Cras, J.J.Steggerda   *Inorg. Chem.*, **7**, 810, 1968

80.10 **Cadmium n - butyl xanthate**
$C_{10}H_{18}CdO_2S_4$
H.M.Rietveld, E.N.Maslen *Acta Cryst.*, **18,** 429, 1965

80.11 **Cadmium diethyldithiocarbamate**
$C_{10}H_{20}CdN_2S_4$
E.A.Shugam, V.M.Agre *Kristallografija*, **13,** 253, 1968

80.12 **Iron(iii) bis(diethyldithiocarbamato) chloride**
$C_{10}H_{20}ClFeN_2S_4$
B.F.Hoskins, R.L.Martin, A.H.White *Nature,* **211,** 627, 1966

80.13 **Copper(ii) diethyldithiocarbamate**
$C_{10}H_{20}CuN_2S_4$
M.Bonamico, G.Dessy, A.Mugnoli, A.Vaciago, L.Zambonelli
Acta Cryst., **19,** 886, 1965

80.14 **Copper(ii) diethyldithiocarbamate**
$C_{10}H_{20}CuN_2S_4$
B.H.O'Connor, E.N.Maslen *Acta Cryst.,* **21,** 828, 1966

80.15 **Copper(ii) diethyldithiocarbamate**
$C_{10}H_{20}CuN_2S_4$
R.Bally *These Doct. Sci. Phys.- Paris,* 1966

80.16 **Nitrosyl iron bis(N,N - diethyldithiocarbamate)**
$C_{10}H_{20}FeN_3OS_4$
M.Colapietro, A.Domenicano, L.Scaramuzza, A.Vaciago, L.Zambonelli
Chem. Communic., 583, 1967

80.17 **Nickel(ii) diethyldithiocarbamate**
$C_{10}H_{20}N_2NiS_4$
M.Bonamico, G.Dessy, C.Mariani, A.Vaciago, L.Zambonelli
Acta Cryst., **19,** 619, 1965

80.18 **Nickel(ii) diethyldiselenocarbamate**
$C_{10}H_{20}N_2NiSe_4$
M.Bonamico, G.Dessy *Chem. Communic.,* 1114, 1967
Also classified in 85

80.19 **Platinum(ii) diethyldithiocarbamate**
$C_{10}H_{20}N_2PtS_4$
A.Z.Amanov, G.A.Kukina, M.A.Porai-Koshits
Zh. Strukt. Khim., **8,** 174, 1967

80.20 **Zinc diethyldithiocarbamate**
$C_{10}H_{20}N_2S_4Zn$
M.Bonamico, G.Mazzone, A.Vaciago, L.Zambonelli
Acta Cryst., **19,** 898, 1965

80.21 **Zinc diethyldithiocarbamate**
$C_{10}H_{20}N_2S_4Zn$
Z.V.Zvonkova, A.N.Khvatkina, N.S.Ivanova
Kristallografija, **12,** 1065, 1967

80.22 **bis(N,N - Dimethyldithiocarbamato) pyridine zinc benzene solvate**
$2C_{11}H_{17}N_3S_4Zn$, C_6H_6
K.A.Fraser, M.M.Harding *Acta Cryst.,* **22,** 75, 1967
Residue 1 also classified in 83

80.23 **Oxo molybdenum(v) - ethyl xanthate complex**
$C_{12}H_{20}Mo_2O_7S_8$
A.B.Blake, F.A.Cotton, J.S.Wood *J. Amer. Chem. Soc.,* **86,** 3024, 1964

80.24 **Dimethyldithiocarbamate zinc**
$C_{12}H_{24}N_4S_8Zn_2$
H.P.Klug *Acta Cryst.,* **21,** 536, 1966

80.25 **Copper(ii) hexamethylenedithiocarbamate**
$C_{14}H_{24}CuN_2S_4$
Z.V.Zvonkova, V.I.Yakovenko *Kristallografija,* **13,** 169, 1968

80.26 **bis - (N,N - Di - n - propyldithiocarbamato) copper(ii)**
$C_{14}H_{28}CuN_2S_4$
A.Pignedoli, G.Peyronel *Gazz. Chim. Ital.,* **92,** 745, 1962

80.27 **Nickel(ii) N,N - di - n - propyldithiocarbamate (discussion)**
$C_{14}H_{28}N_2NiS_4$
A.Pignedoli, G.Peyronel *Acta Cryst. (B),* **24,** 433, 1968

80.28 **Nickel(ii) N,N - di - n - propyldithiocarbamate**
$C_{14}H_{28}N_2NiS_4$
G.Peyronel, A.Pignedoli *Acta Cryst.,* **23,** 398, 1967

80.29 **tris(Diethyldithiocarbamato) cobalt(iii)**
$C_{15}H_{30}CoN_3S_6$
T.Brennan, I.Bernal *J. Phys. Chem.,* **73,** 443, 1969

80.30 **Cobalt(iii) tris(N,N - diethyldithiocarbamate**
$C_{15}H_{30}CoN_3S_6$
S.Merlino *Acta Cryst. (B),* **24,** 1441, 1968

80.31 **Tetramethylammonium uranyl tris(diethyldithiocarbamate)**
$C_{15}H_{30}N_3O_2S_6U^-$, $C_4H_{12}N^+$
K.Bowman, Z.Dori *Chem. Communic.*, 636, 1968
Residue 2 classified in 3

80.32 **Ruthenium tris(N,N - diethyldithiocarbamate**
$C_{15}H_{30}N_3RuS_6$
A.Domenicano, A.Vaciago, L.Zambonelli, P.L.Loader, L.M.Venanzi
Chem. Communic., 476, 1966

80.33 **Nitrosyl ruthenium tris(N,N - diethyldithiocarbamate)**
$C_{15}H_{30}N_4ORuS_6$
A.Domenicano, A.Vaciago, L.Zambonelli, P.L.Loader, L.M.Venanzi
Chem. Communic., 476, 1966

80.34 **bis(N,N - Di - n - butyldithiocarbamato) gold(iii) dibromoaurate(i)**
$C_{18}H_{36}AuN_2S_4^+$, $AuBr_2^-$
P.T.Beurskens, H.J.A.Blaauw, J.A.Cras, J.J.Steggerda
Inorg. Chem., **7**, 805, 1968

80.35 **Cadmium diethyldithiocarbamate**
$C_{20}H_{40}Cd_2N_4S_8$
A.Domenicano, L.Torelli, A.Vaciago, L.Zambonelli
J. Chem. Soc. (A), 1351, 1968

80.36 **Cadmium diethyldithiocarbamate**
$C_{20}H_{40}Cd_2N_4S_8$
E.A.Shugam, V.M.Agre *Kristallografija*, **13**, 253, 1968

80.37 **Copper(i) diethyldithiocarbamate**
$C_{20}H_{40}Cu_4N_4S_8$
R.Hesse *Ark. Kemi*, **20**, 481, 1963

80.38 **Tetraethylammonium neptunium tetrakis (N,N - diethyldithiocarbamate)**
$C_{20}H_{40}N_4NpS_4^-$, $C_8H_{20}N^+$
D.Brown, D.G.Holah, C.E.F.Rickard *J. Chem. Soc. (D)*, 280, 1969
Residue 2 classified in 3

80.39 **Thorium tetrakis(N,N - diethyldithiocarbamate)**
$C_{20}H_{40}N_4S_8Th$
D.Brown, D.G.Holah, C.E.F.Rickard *J. Chem. Soc. (D)*, 280, 1969

80.40 **tris(N,N - Di - n - butyldithiocarbamato) iron(iii)**
$C_{27}H_{54}FeN_3S_6$
B.F.Hoskins, B.P.Kelley *Chem. Communic.*, 1517, 1968
Also classified in 85

80.41 **Silver diethyldithiocarbamate**
$C_{30}H_{60}Ag_6N_6S_{12}$
R.Hesse *Acta Cryst.*, **13,** A51, 1960

METAL COMPLEXES (CARBOXYLIC ACID)

81.1 Sodium ammonium oxy molybdenum oxalate dihydrate
$C_2MoO_7{}^{2-}$, H_4N^+, Na^+, $2H_2O$
L.O.Atovmyan, G.B.Bokii *Zh. Strukt. Khim.*, **4**, 576, 1963

81.2 Silver oxalate
$C_2O_4{}^{2-}$, $2Ag^+$
R.L.Griffith *J. Chem. Phys.*, **11**, 499, 1943
See also *Int. Distances*, M 142; *Structure Reports*, **9**, 319, 1943

81.3 Copper(ii) formate tetrahydrate (neutron study)
$(C_2H_2CuO_2)_n$, $4nH_2O$
K.Okada, M.I.Kay, D.T.Cromer, I.Almodovar
J. Chem. Phys., **44**, 1648, 1966

81.4 Copper(ii) formate tetrahydrate (refinement of x - ray data)
$(C_2H_2CuO_4)_n$, $4nH_2O$
K.Okada, M.I.Kay, D.T.Cromer, I.Almodovar
J. Chem. Phys., **44**, 1648, 1966

81.5 Cobalt formate dihydrate
$(C_2H_6CoO_6)_n$
A.S.Antsyshkina, M.K.Gusejnova, M.A.Porai-Koshits
Zh. Strukt. Khim., **8**, 365, 1967

81.6 Copper(ii) oxalate diammine (α form)
$C_2H_6CuN_2O_4$
J.Garaj *Chem. Communic.*, 904, 1968

81.7 Copper(ii) diaquo formate
$(C_2H_6CuO_6)_n$
M.Bukowska-Strzyzewska *Acta Cryst.*, **19**, 357, 1965

81.8 Copper(ii) diaquo formate
$(C_2H_6CuO_6)_n$
M.I.Kay, I.Almodovar, S.F.Kaplan *Acta Cryst. (B)*, **24**, 1312, 1968

81.9 Manganese (ii) diaquo formate
$(C_2H_6MnO_6)_n$
K.Osaki, Y.Nakai, T.Watanabe *J. Phys. Soc. Jap.*, **19**, 717, 1964

81.10 Manganese(ii) diaquo formate (neutron study)
$(C_2H_6MnO_6)_n$
M.I.Kay, I.Almodovar, S.F.Kaplan *Acta Cryst. (B)*, **24**, 1312, 1968

81.11 Nickel(ii) diaquo formate
$(C_2H_6NiO_6)_n$
K.Krogmann, R.Mattes *Z. Kristallogr.*, **118**, 291, 1963

81.12 Cobalt(iii) penta - ammine acetate chloride perchlorate
$C_2H_8CoN_5O_2{}^{2+}$, Cl^-, $ClO_4{}^-$
M.Bukowska-Strzyzewska *J. Amer. Chem. Soc.*, **87**, 3998, 1965
Residue 1 also classified in 83

81.13 Gadolinium formate
$C_3H_3GdO_3$
A.Pabst *J. Chem. Phys.*, **11**, 145, 1943
See also *Int. Distances,* M 107; *Structure Reports,* **9**, 313, 1943

81.14 Scandium formate
$(C_3H_3O_6Sc)_n$
M.K.Guseinova, A.S.Antsyshkina, M.A.Porai-Koshits
Zh. Strukt. Khim., **9**, 1040, 1968

81.15 Copper ammonium oxalate dihydrate
$C_4CuO_8{}^{2-}$, $2H_4N^+$, $2H_2O$
M.A.Viswamitra *J. Chem. Phys.*, **37**, 1408, 1962

81.16 Ammonium diperoxodioxalato niobate monohydrate
$C_4NbO_{12}{}^{3-}$, $3H_4N^+$, H_2O
G.Mathern, R.Weiss, R.Rohmer *J. Chem. Soc. (D)*, 70, 1969
Residue 1 also classified in 84

81.17 Potassium oxalato palladate(ii) tetrahydrate
$C_4O_8Pd^{2-}$, $2K^+$, $4H_2O$
K.Krogmann *Z. Anorg. Allg. Chem.*, **346**, 188, 1966

81.18 Potassium oxalato platinate dihydrate
$C_4O_8Pt^{2-}$, $2K^+$, $2H_2O$
R.Mattes, K.Krogmann *Z. Anorg. Allg. Chem.*, **332**, 247, 1964

81.19 **Magnesium dioxalato platinate hydrate**
$C_4O_8Pt^{2-}$, $0.82Mg^{2+}$, $5.3H_2O$
K.Krogmann *Z. Anorg. Allg. Chem.*, **358,** 97, 1968

81.20 **Ammonium yttrium(iii) aquo oxalate**
$C_4H_2O_9Y^-$, H_4N^+
T.R.R.McDonald, J.M.Spink *Acta Cryst.*, **23,** 944, 1967

81.21 **trans - Potassium dioxalato diaquo chromate**
$C_4H_4CrO_{10}^-$, K^+
J.N.van Nickerk, F.R.L.Schoening *Acta Cryst.*, **4,** 35, 1951
See also *Int. Distances*, M 161; *Structure Reports*, **15,** 424, 1951

81.22 **Copper(ii) formate dioxan complex**
$C_4H_4Cu_2O_8$, $C_4H_8O_2$
M.Bukowska-Strzyzewska *Rocz. Chem.*, **40,** 567, 1966
Residue 2 classified in 38

81.23 **Copper(ii) formate**
$(C_4H_4Cu_2O_8)_n$
G.A.Barclay, C.H.L.Kennard *J. Chem. Soc.*, 3289, 1961

81.24 **Ammonium dimolybdomalate hydrate**
$C_4H_4Mo_2O_{11}^{2-}$, $2H_4N^+$, $2.5H_2O$
L.A.Aslanov, G.V.Sokolova, M.A.Porai-Koshits, T.N.Polynova,
G.A.Gordeeva *Zh. Strukt. Khim.*, **7,** 812, 1966

81.25 **Barium bis - dioxo molybdenum(v) oxalate pentahydrate**
$C_4H_4Mo_2O_{14}^{2-}$, H_2BaO^{2+}, $2H_2O$
F.A.Cotton, S.M.Morehouse *Inorg. Chem.*, **4,** 1377, 1965

81.26 **Rhodium formate hemihydrate**
$C_4H_4O_8Rh_2$, H_2O
A.S.Ancyskina *Acta Cryst.*, **21,** A135, 1966

81.27 **bis(Glycollato) copper(ii)**
$C_4H_6CuO_6$
C.K.Prout, R.A.Armstrong, J.R.Carruthers, J.G.Forrest, P.Murray-Rust,
F.J.C.Rossotti *J. Chem. Soc. (A)*, 2791, 1968

81.28 **Thorium diaquo tetraformate monohydrate**
$(C_4H_8O_{10}Th)_n$, nH_2O
E.G.Arutnyan, A.S.Antsyshkina, E.Ya.Balta
Zh. Strukt. Khim., **7,** 471, 1966

81.C   **bis(Thiourea) cadmium formate**
$C_4H_{10}CdN_4O_4S_2$
For complete entry see 79.20

81.29   **Zinc diaquo acetate**
$C_4H_{10}O_6Zn$
J.N.van Niekerk, F.R.L.Schoening, J.H.Talbot
Acta Cryst., **6**, 720, 1953
See also *Int. Distances*, M 174; *Structure Reports*, **17**, 623, 1953

81.30   **Diaquo bis(glycollato) zinc(ii)**
$C_4H_{10}O_8Zn$
A.J.Fischinger, L.E.Webb   *J. Chem. Soc. (D)*, 407, 1969

81.31   **Copper(ii) diammine acetate**
$C_4H_{12}CuN_2O_4$
M.Bukowska-Strzyzewska   *Rocz. Chem.*, **37**, 1335, 1963
Also classified in 83

81.32   **Copper(ii) diammine acetate**
$C_4H_{12}CuN_2O_4$
Yu.A.Simonov, A.V.Ablov, T.I.Malinovskii   *Kristallografija*, **8**, 270, 1963
Also classified in 83

81.33   **Copper(ii) aquo succinate monohydrate**
$C_4H_{12}Cu_2O_{10}$, $2H_2O$
B.H.O'Connor, E.N.Maslen   *Acta Cryst.*, **20**, 824, 1966

81.34   **Cobalt tetra - aquo acetate**
$C_4H_{14}CoO_8$
J.N.van Niekerk, F.R.L.Schoening   *Acta Cryst.*, **6**, 609, 1953
See also *Int. Distances*, M 177; *Structure Reports*, **17**, 627, 1953

81.35   **Zinc dihydrazine diacetate**
$C_4H_{14}N_4O_4Zn$
A.Ferrari, A.Braibanti, G.Bigliardi, A.M.Lanfredi
Acta Cryst., **19**, 548, 1965
Also classified in 83

81.36   **Nickel tetra - aquo acetate**
$C_4H_{14}NiO_8$
J.N.van Niekerk, F.R.L.Schoening   *Acta Cryst.*, **6**, 609, 1953
See also *Int. Distances*, M 177; *Structure Reports*, **17**, 627, 1953

81.37 **Ammonium trioxalato chromate(iii)**
$C_6CrO_{12}{}^{3-}$, $3H_4N^+$
J.N.van Niekerk, F.R.L.Schoening *Acta Cryst.*, **5**, 499, 1952
See also *Int. Distances*, M 188; *Structure Reports*, **16**, 466, 1952

81.38 **Ammonium oxotrioxalato niobate monohydrate**
$C_6NbO_{13}{}^{3-}$, $3H_4N^+$, H_2O
G.Mathern, R.Weiss, R.Rohmer *J. Chem. Soc. (D)*, 70, 1969

81.39 **Potassium tris - oxalato vanadate(ii) trihydrate**
$C_6O_{12}V^{3-}$, $3K^+$, $3H_2O$
R.H.Fenn, A.J.Graham, R.D.Gillard *Nature*, **213**, 1012, 1967

81.40 **Ammonium diuranyl trioxalate**
$C_6O_{16}U_2{}^{2-}$, $2H_4N^+$
N.W.Alcock *Chem. Communic.*, 1327, 1968

81.C **Tetramethylammonium copper(ii) thiocyanate formate**
$C_6H_4Cu_2N_2O_8S_2{}^{2-}$, $2C_4H_{12}N^+$
For complete entry see 3.38

81.41 **Hexa - aquo tris - oxalato dineodymium tetrahydrate**
$C_6H_6Nd_2O_{15}$, $4H_2O$
E.Hansson, J.Albertsson *Acta Chem. Scand.*, **22**, 1682, 1968

81.42 **Sodium uranyl acetate**
$C_6H_9O_8U^-$, Na^+
W.H.Zachariasen, H.A.Plettinger *Acta Cryst.*, **12**, 526, 1959
See also *Int. Distances*, M 131s; *Structure Reports*, **23**, 532, 1959

81.43 **Neodymium - diaquo - nitrilotriacetate monohydrate (form ii)**
$C_6H_{10}NNdO_8$, H_2O
K.F.Belyaeva, M.A.Porai-Koshits, N.D.Mitrofanova, L.I.Martynenko
Zh. Strukt. Khim., **9**, 541, 1968
Residue 1 also classified in 82

81.44 **Cerium triacetate hydrate**
$(C_6H_{11}CeO_7)_n$
G.G.Sadikov, G.A.Kukina, M.A.Porai-Koshits
Zh. Strukt. Khim., **8**, 551, 1967

81.45 **Aquo bis(DL - lactato) copper(ii) hemihydrate**
$C_6H_{12}CuO_7$, $0.5H_2O$
C.K.Prout, R.A.Armstrong, J.R.Carruthers, J.G.Forrest, P.Murray-Rust,
F.J.C.Rossotti *J. Chem. Soc. (A)*, 2791, 1968

81.46 μ - **Oxalato bis(oxalato) hexa - aquo titanium(iii) tetrahydrate**
$C_6H_{12}O_{18}Ti_2$, $4H_2O$
M.G.B.Drew, G.W.A.Fowles, D.F.Lewis   *J. Chem. Soc. (D)*, 876, 1969

81.47 **Tetra - aquo - dibarium copper formate**
$(C_6H_{14}Ba_2CuO_{16})_n$
R.V.G.S.Rao, K.Sundarama, G.S.Rao   *Z. Kristallogr.*, **110,** 231, 1958
See also *Int. Distances*, M 129s; *Structure Reports*, **22,** 587, 1958

81.48 **Diaquo bis(methoxyacetato) copper(ii)**
$C_6H_{14}CuO_8$
C.K.Prout, R.A.Armstrong, J.R.Carruthers, J.G.Forrest, P.Murray-Rust,
F.J.C.Rossotti   *J. Chem. Soc. (A)*, 2791, 1968

81.49 **Vanadyl(iv) diaquopyridine - 2,6 - dicarboxylate dihydrate**
$C_7H_7NO_7V$, $2H_2O$
B.H.Bersted, R.L.Belford, I.C.Paul   *Inorg. Chem.*, **7,** 1557, 1968
Residue 1 also classified in 83

81.50 **Tetraphenylarsonium tetrakis(trifluoroacetato) cobaltate(ii)**
$C_8CoF_{12}O_8{}^{2-}$, $2C_{24}H_{20}As^+$
J.G.Bergman, F.A.Cotton   *Inorg. Chem.*, **5,** 1420, 1966
Residue 2 classified in 86

81.51 **Potassium thorium tetroxalate tetrahydrate**
$(C_8O_{16}Th^{4-})_n$, $4nK^+$, $4nH_2O$
M.N.Akhtar, A.J.Smith   *J. Chem. Soc. (D)*, 705, 1969

81.52 **Sodium tetrakisoxalato zirconate(iv) trihydrate**
$C_8O_{16}Zr^{4-}$, $4H_4N^+$, $3H_2O$
G.L.Glen, J.V.Silverton, J.L.Hoard   *Inorg. Chem.*, **2,** 250, 1963

81.C **(−) - Tetracarbonyl(fumaric acid) iron**
$C_8H_4FeO_8$
For complete entry see 72.18

81.53 **Mercury(i) o - phthalate**
$(C_8H_4Hg_2O_4)_n$
B.Lindh   *Acta Chem. Scand.*, **21,** 2743, 1967

81.54 **Tetrasodium divanadyl(iv) D - tartrate L - tartrate dodecahydrate**
$C_8H_4O_{14}V_2{}^{4-}$, $4Na^+$, $12H_2O$
R.E.Tapscott, R.L.Belford, I.C.Paul   *Inorg. Chem.*, **7,** 356, 1968

81.55 **Ammonium vanadyl (+) - tartrate monohydrate**
$C_8H_4O_{14}V_2^{4-}$, $4H_4N^+$, $2H_2O$
J.G.Forrest, C.K.Prout *J. Chem. Soc. (A)*, 1312, 1967

81.56 **bis(Diacetatobromo rhodium(iii)) - bisguanidine**
$C_8H_{12}Br_2O_8Rh_2$, $2CH_5N_3$
L.M.Dikareva *Acta Cryst.*, **21,** A140, 1966
Residue 1 also classified in 60; residue 2 classified in 8, 60

81.57 **Diammonium bis(diacetatobromo rhodate(iv)) - bis(ammonium bromide)**
$C_8H_{12}Br_2O_8Rh_2^{2-}$, $2Br^-$, $4H_4N^+$
L.M.Dikareva *Acta Cryst.*, **21,** A140, 1966

81.58 **bis(Diacetatochloro rhodium(iii)) - bisguanidine**
$C_8H_{12}Cl_2O_8Rh_2$, $2CH_5N_3$
L.M.Dikareva *Acta Cryst.*, **21,** A140, 1966
Residue 1 also classified in 60; residue 2 classified in 8, 60

81.59 **Molybdenum(ii) acetate**
$C_8H_{12}Mo_2O_8$
D.Lawton, R.Mason *J. Amer. Chem. Soc.*, **87,** 921, 1965

81.60 **Uranium(iv) acetate**
$(C_8H_{12}O_8U)_n$
I.Jelenic, D.Grdenic, A.Bezjak *Acta Cryst.*, **17,** 758, 1964

81.61 **Calcium cadmium acetate hexahydrate**
$(C_8H_{16}CaCdO_{10})_n$, $4nH_2O$
D.A.Langs, C.R.Hare *Chem. Communic.*, 890, 1967

81.62 **Calcium copper acetate hexahydrate**
$(C_8H_{16}CaCuO_{10})_n$, $4nH_2O$
D.A.Langs, C.R.Hare *Chem. Communic.*, 890, 1967

81.63 **Chromous diaquo acetate**
$C_8H_{16}Cr_2O_{10}$
J.N.van Niekerk, F.R.L.Schoening, J.F.de Wet
Acta Cryst., **6,** 501, 1953
See also *Int. Distances*, M 174; *Structure Reports*, **17,** 624, 1953

81.64 **Copper diaquoacetate**
tetrakis (μ - Aceto) diaquo dicopper(ii)
$C_8H_{16}Cu_2O_{10}$
J.N.van Niekerk, F.R.L.Schoening *Acta Cryst.*, **6,** 227, 1953
See also *Int. Distances*, M 177; *Structure Reports*, **17,** 625, 1953

81.65 **Copper diaquoacetate (neutron study)**
tetrakis (μ - Aceto) diaquo dicopper(ii)
$C_8H_{16}Cu_2O_{10}$
G.M.Brown, R.Chidambaram
Amer. Cryst. Assoc., Abstr. Papers (Winter Meeting), 21, 1967

81.66 **Diaquo bis(2 - hydroxy - 2 - methylpropionato) copper(ii)**
$C_8H_{18}CuO_8$
C.K.Prout, R.A.Armstrong, J.R.Carruthers, J.G.Forrest, P.Murray-Rust,
F.J.C.Rossotti *J. Chem. Soc. (A)*, 2791, 1968

81.C **Tetramethylammonium copper(ii) thiocyante acetate**
$C_{10}H_{12}Cu_2N_2O_8S_2{}^{2-}$, $2C_4H_{12}N^+$
For complete entry see 3.39

81.C **Sodium hexa - oxo - u - ethylenediaminetetra - acetato - dimolybdate(vi) octahydrate**
$C_{10}H_{12}Mo_2N_2O_{14}{}^{4-}$, $4Na^+$, $8H_2O$
For complete entry see 76.61

81.67 **Potassium penta - acetato cerate(iii) monohydrate**
$C_{10}H_{15}CeO_2{}^{2-}$, $2K^+$, H_2O
G.G.Sadikov, G.A.Kukina *Zh. Strukt. Khim.*, **9**, 145, 1968

81.68 **Zinc kainate dihydrate**
$C_{10}H_{15}NO_5Zn$, H_2O
H.Watase, I.Nitta *Bull. Chem. Soc. Jap.*, **30**, 889, 1957
See also *Int. Distances,* M 161s; *Structure Reports,* **21**, 613, 1957

81.C **Dihydrogen ethylenediaminetetra - acetato - aquo nickelate(ii)**
$C_{10}H_{16}N_2NiO_9$
For complete entry see 76.69

81.C **Aquo (1,5 - diazacyclo - octane - N,N' - diacetato) nickel(ii) dihydrate**
$C_{10}H_{18}N_2NiO_5$, $2H_2O$
For complete entry see 82.40

81.69 **Sodium tris(oxydiacetato) gadolinate bis(sodium perchlorate) hexahydrate**
$C_{12}H_{12}GdO_{15}{}^{3-}$, $2ClO_4{}^-$, $5Na^+$, $6H_2O$
J.Albertsson *Acta Chem. Scand.*, **22**, 1563, 1968

81.70 **Potassium bis(nitrilotriacetato) zirconate(iv) monohydrate**
$C_{12}H_{12}N_2O_{12}Zr^{2-}$, $2K^+$, H_2O
J.L.Hoard, E.W.Silverton, J.V.Silverton
J. Amer. Chem. Soc., **90**, 2300, 1968
Residue 1 also classified in 83, 84

81.71 **Sodium neodymium(iii) diglycolate hexahydrate**
$C_{12}H_{12}NaO_{15}{}^{3-}$, $3Na^{+}$, $6H_2O$
N.-G.Vannerberg, J.Albertsson *Acta Chem. Scand.*, **19**, 1760, 1965

81.72 **Sodium tris(oxydiacetato) neodymate bis(sodium perchlorate) hexahydrate**
$C_{12}H_{12}NdO_{15}{}^{3-}$, $2ClO_4{}^{-}$, $5Na^{+}$, $6H_2O$
J.Albertsson *Acta Chem. Scand.*, **22**, 1563, 1968

81.73 **Sodium tris(oxydiacetato) ytterbate bis(sodium perchlorate) hexahydrate**
$C_{12}H_{12}O_{15}Yb^{3-}$, $2ClO_4{}^{-}$, $5Na^{+}$, $6H_2O$
J.Albertsson *Acta Chem. Scand.*, **22**, 1563, 1968

81.74 **Zinc oxyacetate**
$C_{12}H_{18}O_{13}Zn_4$
H.Kovama, Y.Saito *Bull. Chem. Soc. Jap.*, **27**, 112, 1954
See also *Int. Distances*, M 235; *Structure Reports*, **18**, 640, 1954

81.75 **Chromium(iii) acetate oxychloride pentahydrate**
$C_{12}H_{24}Cr_3O_{16}{}^{+}$, Cl^{-}, $2H_2O$
B.N.Figgis, G.B.Robertson *Nature*, **205**, 694, 1965

81.76 **Copper propionate**
$C_{12}H_{24}Cu_2O_8$
A.V.Ablov, T.N.Tarkhova, Yu.A.Simonov *Acta Cryst.*, **21**, A134, 1966

81.C **Neodymium tetra - aquo iminodiacetate hydrochloride trihydrate**
$C_{12}H_{25}N_3Nd_2O_{16}{}^{2+}$, $2Cl^{-}$, $3H_2O$
For complete entry see 82.52

81.77 **Copper(ii) benzoate trihydrate**
$C_{14}H_{10}CuO_4$, $3H_2O$
H.Koizumi, K.Osaki, T.Watanabe *J. Phys. Soc. Jap.*, **18**, 117, 1963

81.78 **Copper(ii) diaquo salicylate dihydrate**
$C_{14}H_{14}CuO_8$, $2H_2O$
F.Hanic, J.Michalov *Acta Cryst.*, **13**, 299, 1960

81.79 **Copper(ii) acetate tris - pyridine complex**
$C_{14}H_{16}CuN_2O_4$, C_5H_5N
K.Anzenhofer, L.N.A.Sten Rouwelaar
Rec. Trav. Chim. Pays-Bas, **86**, 801, 1967
Residue 1 also classified in 83

81.80   bis(Hydrogen o - phthalato) diaquo copper(ii)
$C_{16}H_{14}CuO_{10}$
M.B.Cingi, C.Guastini, A.Musatti, M.Nardelli
Acta Cryst. (B), **25,** 1833, 1969

81.81   Diaquo bis(phenoxyacetato) copper(ii)
$C_{16}H_{18}CuO_8$
C.K.Prout, R.A.Armstrong, J.R.Carruthers, J.G.Forrest, P.Murray-Rust,
F.J.C.Rossotti   *J. Chem. Soc. (A)*, 2791, 1968

81.82   Copper butyrate
$C_{16}H_{28}Cu_2O_8$
A.V.Ablov, T.N.Tarkhova, Yu.A.Simonov   *Acta Cryst.*, **21,** A134, 1966

81.83   Tetra - n - butyrato diruthenium chloride
$C_{16}H_{28}O_8Ru_2{}^+$, Cl^-
M.J.Bennett, K.G.Caulton, F.A.Cotton   *Inorg. Chem.*, **8,** 1, 1969

81.84   Copper(ii) acetylsalicylate
$C_{18}H_{14}CuO_8$
L.Manojlovic-Muir   *Chem. Communic.*, 1057, 1967

81.85   Monopyridine copper(ii) acetate (monoclinic form)
$C_{18}H_{22}Cu_2N_2O_8$
G.A.Barclay, C.H.L.Kennard   *J. Chem. Soc.*, 5244, 1961
Also classified in 83

81.86   Monopyridine copper (ii) acetate (orthorhombic form)
$C_{18}H_{22}Cu_2N_2O_8$
F.Hanic, D.Stempelova, K.Hanicova   *Acta Cryst.*, **17,** 633, 1964
Also classified in 83

81.87   Copper(ii) acetate - quinoline complex
$C_{26}H_{26}Cu_2N_2O_8$
T.N.Tarkhova, A.V.Ablov   *Kristallografija*, **13,** 611, 1968
Also classified in 83

81.88   Dichlorotetra(benzoato) dirhenium(iii) chloroform solvate
$C_{28}H_{20}Cl_2O_8Re_2$, $2CHCl_3$
M.J.Bennett, W.K.Bratton, F.A.Cotton, W.R.Robinson
Inorg. Chem., **7,** 1570, 1968

81.89   Oxopentachloropropionato bis(triphenyl phosphine) dirhenium(iv)
$C_{39}H_{35}Cl_5O_3P_2Re_2$
F.A.Cotton, B.M.Foxman   *Inorg. Chem.*, **7,** 1784, 1968
Also classified in 86

81.90   **bis(Uranyl acetate triphenylphosphine oxide)**
$C_{44}H_{42}O_{14}P_2U_2$
C.Panattoni, G.Bandoli, R.Graziani, U.Croatto
Chem. Communic., 278, 1968
Also classified in 64

81.C   **Hydrido acetato tris(triphenylphosphine) ruthenium(ii)**
$C_{57}H_{49}O_2P_3Ru$
For complete entry see 86.116

METAL COMPLEXES (AMINO-ACID)

82.1   **Glycine silver(i) nitrate**
       $C_2H_5AgN_2O_2{}^+$, $NO_3{}^-$
       J.K.M.Rao   *Indian J. Pure Appl. Phys.*, **6,** 51, 1968

82.2   **(+) - Sarcosinato tetra - ammine cobalt(iii) nitrate**
       $C_3H_{18}CoN_5O_2{}^{2+}$, $2NO_3{}^-$
       S.Larsen, K.J.Watson, A.M.Sargeson, K.R.Turnbull
       Chem. Communic., 847, 1968

82.3   **Copper(ii) monoglycylglycine trihydrate**
       $C_4H_6CuN_2O_3$, $3H_2O$
       B.Strandberg, I.Lindqvist, R.Rosenstein   *Z. Kristallogr.*, **116,** 266, 1961

82.4   **cis - Copper(ii) glycine monohydrate**
       $C_4H_8CuN_2O_4$, H_2O
       K.Tomita, I.Nitta   *Bull. Chem. Soc. Jap.*, **34,** 286, 1961

82.5   **cis - Copper(ii) glycine monohydrate**
       $C_4H_8CuN_2O_4$, H_2O
       H.C.Freeman, M.R.Snow, I.Nitta, K.Tomita
       Acta Cryst., **17,** 1463, 1964

82.6   **Nickel(ii) glycinate dihydrate**
       $C_4H_8N_2NiO_4$, $2H_2O$
       H.C.Freeman, J.M.Guss   *Acta Cryst. (B),* **24,** 1133, 1968

82.7   **trans - bis(Glycinato) platinum(ii)**
       $C_4H_8N_2O_4Pt$
       H.C.Freeman, M.L.Golomb   *Acta Cryst. (B),* **25,** 1203, 1969

82.8   **syn - Dichloro - bis - glycino - biaquo manganese(ii)**
       $C_4H_{14}Cl_2MnN_2O_6$
       M.Lee, Y.Okaya, R.Pepinsky   *Bull. Amer. Phys. Soc.*, **7,** 177, 1962

82.9   **Iron(ii) sulfate pentahydrate glycine**
$C_4H_{18}FeN_2O_8{}^{2+}$, $H_{12}FeO_6{}^{2+}$, $2O_4S^{2-}$
I.Lindqvist, R.Rosenstein *Acta Chem. Scand.*, **14**, 1228, 1960

82.10   **Copper aquo glutamate hydrate**
$C_5H_9CuNO_5$, H_2O
C.M.Gramaccioli, R.E.Marsh *Acta Cryst.*, **21**, 594, 1966

82.11   **Zinc aquo glutamate hydrate**
$C_5H_9NO_5Zn$, H_2O
C.M.Gramaccioli *Acta Cryst.*, **21**, 600, 1966

82.12   **($\pm$) - Methionine - palladium chloride complex**
$C_5H_{11}Cl_2NO_2PdS$
N.C.Stephenson, J.F.McConnell, R.Warren
Inorg. Nucl. Chem. Letters, **3**, 553, 1967

82.13   **Pyruvidene - β - alaninato aquo copper(ii) dihydrate**
$C_6H_9CuNO_5$, $2H_2O$
T.Ueki, T.Ashida, Y.Sasada, M.Kakudo
Acta Cryst. (B), **24**, 1361, 1968

82.14   **Disodium di - μ - oxo bis(oxocysteinato molybdenum(v)) pentahydrate**
$C_6H_{10}Mo_2N_2O_8S_2{}^{2-}$, $2Na^+$, $5H_2O$
J.R.Knox, C.K.Prout *Acta Cryst. (B)*, **25**, 1857, 1969
Residue 1 also classified in 85

82.C   **Neodymium - diaquo - nitrilotriacetate monohydrate (form ii)**
$C_6H_{10}NNdO_8$, H_2O
For complete entry see 81.43

82.15   **Glycylglycylglycino aquo copper(ii) chloride hemihydrate**
$(C_6H_{12}ClCuN_3O_5)_{2n}$, nH_2O
H.C.Freeman, G.Robinson, J.C.Schoone *Acta Cryst.*, **17**, 719, 1964

82.16   **cis - bis - (D - α - Alaninato) copper(ii)**
$C_6H_{12}CuN_2O_4$
R.D.Gillard, R.Mason, N.C.Payne, G.B.Robertson
Chem. Communic., 155, 1966

82.17   **trans - bis - (L - α - Alaninato) copper(ii)**
$C_6H_{12}CuN_2O_4$
A.Dijkstra *Acta Cryst.*, **20**, 588, 1966

82.18   bis(L - Serinato) copper(ii)
        $C_6H_{12}CuN_2O_6$
        D.van der Helm, W.A.Franks   *Acta Cryst. (B)*, **25**, 451, 1969

82.19   Nickel(ii) β - alanine dihydrate
        $C_6H_{12}N_2NiO_4$, $2H_2O$
        P.Jose, L.M.Pant, A.B.Biswas   *Acta Cryst.*, **17**, 24, 1964

82.20   bis(L - Serinato) zinc
        $C_6H_{12}N_2O_6Zn$
        W.A.Franks, A.F.Nicholas, C.G.Fisher, D.van derHelm
        Amer. Cryst. Assoc., Abstr. Papers (Winter Meeting), 32, 1967

82.21   Copper(ii) diaquo - β - alanine tetrahydrate
        $C_6H_{16}CuN_2O_6$, $4H_2O$
        K.Tomita   *Bull. Chem. Soc. Jap.*, **34**, 297, 1961

82.22   Diaquo bis(L - serinato) nickel(ii)
        $C_6H_{16}N_2NiO_8$
        D.van der Helm, M.B.Hossain   *Acta Cryst. (B)*, **25**, 457, 1969

82.23   Glycylglycinato - aquo - imidazole - copper(ii) hydrate
        $C_7H_{12}CuN_4O_4$, $1.5H_2O$
        J.D.Bell, H.C.Freeman, A.M.Wood, R.Driver, W.R.Walker
        J. Chem. Soc. (D), 1441, 1969
        Residue 1 also classified in 83

82.24   (−)(589) - Sarcosinato bis(ethylenediamine) cobalt(iii) di - iodide dihydrate
        $C_7H_{22}CoN_5O_2{}^{2+}$, $2I^-$, $2H_2O$
        J.F.Blount, H.C.Freeman, A.M.Sargeson, K.R.Turnbull
        Chem. Communic., 324, 1967
        Residue 1 also classified in 76

82.25   (Glycyl - L - histidinato) copper(ii) sesquihydrate
        $C_8H_{10}CuN_4O_3$, $1.5H_2O$
        J.F.Blount, K.A.Fraser, H.C.Freeman, J.T.Szymanski, C.-H.Wang
        Acta Cryst., **22**, 396, 1967

82.26   Disodium glycylglycylglycylglycino cuprate(ii) decahydrate
        $C_8H_{10}CuN_4O_5{}^{2-}$, $2Na^+$, $10H_2O$
        H.C.Freeman, M.R.Taylor   *Acta Cryst.*, **18**, 939, 1965

82.27   Disodium glycylglycylglycylglycinato nickelate(ii) octahydrate
        $C_8H_{10}N_4NiO_5{}^{2-}$, $2Na^+$, $8H_2O$
        H.C.Freeman, J.M.Guss, R.L.Sinclair   *Chem. Communic.*, 485, 1968
        Residue 1 also classified in 83

82.28 **Ammonium bis - glycylglycine cobalt(iii) dihydrate**
$C_8H_{12}CoN_4O_6^-$, H_4N^+, $2H_2O$
R.D.Gillard, E.D.McKenzie, R.Mason, G.B.Robertson
Nature, **209**, 1347, 1966

82.29 **Dipotassium bis(glycylglycinato) cuprate(ii) hexahydrate**
$C_8H_{12}CuN_4O_6^{2-}$, $2K^+$, $6H_2O$
A.Sugihara, T.Ashida, Y.Sasada, M.Kakudo
Acta Cryst. (B), **24**, 203, 1968

82.30 **Disodium bis(glycylglycinato) nickelate(ii) octahydrate**
$C_8H_{12}N_4NiO_6^{2-}$, $2Na^+$, $8H_2O$
H.C.Freeman, J.M.Guss, R.L.Sinclair *Chem. Communic.*, 485, 1968

82.31 **Disodium bis(glycylglycinato) nickelate(ii) nonahydrate**
$C_8H_{12}N_4NiO_6^{2-}$, $2Na^+$, $9H_2O$
H.C.Freeman, J.M.Guss, R.L.Sinclair *Chem. Communic.*, 485, 1968

82.32 **Zinc triaquo aspartate (absolute configuration)**
$C_8H_{18}N_2O_{11}Zn$
T.H.Doyne, R.Pepinsky, T.Watanabe *Acta Cryst.*, **10**, 438, 1957
See also *Int. Distances*, M 110s; *Structure Reports*, **21**, 537, 1957

82.33 **Nickel(ii) di - α - amino isobutyrate tetrahydrate**
$C_8H_{20}N_2NiO_6$, $2H_2O$
T.Neguchi *Bull. Chem. Soc. Jap.*, **35**, 99, 1962

82.C **N - Salicylideneglycinato aquo copper(ii) tetrahydrate**
$C_9H_9CuNO_4$, $4H_2O$
For complete entry see 78.2

82.34 **α - (+) - (Cobalt(iii) alaninate)**
$C_9H_{12}CoN_3O_6$
M.G.B.Drew, J.H.Dunlop, R.D.Gillard, D.Rogers
Chem. Communic., 42, 1966

82.35 **Monoaquo (β - alanyl - L - histidinato) copper(ii) monohydrate**
Copper(ii) carnosine monohydrate
$C_9H_{14}CuN_4O_4$, H_2O
H.C.Freeman, J.T.Szymanski *Acta Cryst.*, **22**, 406, 1967

82.36 **Diglycylglycinato - aquo - imidazole - copper(ii) monohydrate**
$C_9H_{15}CuN_5O_5$, H_2O
J.D.Bell, H.C.Freeman, A.M.Wood, R.Driver, W.R.Walker
J. Chem. Soc. (D), 1441, 1969
Residue 1 also classified in 83

82.37 **D - (Cobalt bis - ethylenediamine - L - glutamate) perchlorate**
$C_9H_{24}CoN_6O_3{}^+$, $ClO_4{}^-$
J.H.Dunlop, R.D.Gillard, N.C.Payne, G.B.Robertson
Chem. Communic., 877, 1966
Residue 1 also classified in 76

82.38 **Glycylglycinato bis(imidazole) copper(ii) perchlorate**
$C_{10}H_{15}CuN_6O_3{}^+$, $ClO_4{}^-$
J.D.Bell, H.C.Freeman, A.M.Wood, R.Driver, W.R.Walker
J. Chem. Soc. (D), 1441, 1969
Residue 1 also classified in 83

82.C **Dihydrogen ethylenediaminetetra - acetato - aquo nickelate(ii)**
$C_{10}H_{16}N_2NiO_9$
For complete entry see 76.69

82.39 **L - Histidinato - L - threoninato aquo copper(ii) hydrate**
$C_{10}H_{18}CuN_4O_6$, H_2O
H.C.Freeman, J.M.Guss, M.J.Healey, R.-P.Martin
J. Chem. Soc. (D), 225, 1969

82.40 **Aquo (1,5 - diazacyclo - octane - N,N' - diacetato) nickel(ii) dihydrate**
$C_{10}H_{18}N_2NiO_5$, $2H_2O$
J.J.Legg, D.O.Nielson, D.L.Smith, N.L.Larson
J. Amer. Chem. Soc., **90**, 5030, 1968
Residue 1 also classified in 81

82.41 **bis(Methioninato) copper(ii)**
$C_{10}H_{20}CuN_2O_4S_2$
M.V.Veldis, G.J.Palenik *J. Chem. Soc. (D)*, 1277, 1969

82.42 **Copper diaquo proline**
$C_{10}H_{20}CuN_2O_6$
A.M.McL.Mathieson, H.K.Welsh *Acta Cryst.*, **5**, 599, 1952
See also *Int. Distances,* M 183; *Structure Reports,* **16**, 447, 1952

82.43 **D(+) - Triethylenetetramine - S - proline cobalt(iii) tetrachlorozincate**
$C_{11}H_{26}CoN_5O_2{}^{2+}$, Cl_4Zn^{2-}
D.A.Buckingham, L.G.Marzilli, I.E.Maxwell, A.M.Sargeson, H.C.Freeman
J. Chem. Soc. (D), 583, 1969
Residue 1 also classified in 76

82.44 **L(−) - Triethylenetetramine - S - proline cobalt(iii) di - iodide dihydrate**
$C_{11}H_{26}CoN_5O_2{}^{2+}$, $2I^-$, $2H_2O$
D.A.Buckingham, L.G.Marzilli, I.E.Maxwell, A.M.Sargeson, H.C.Freeman
J. Chem. Soc. (D), 583, 1969
Residue 1 also classified in 76

82.45 **Cadmium(ii) L - histidinate dihydrate**
$C_{12}H_{16}CdN_6O_4$, H_2O
R.Candlin, M.M.Harding *J. Chem. Soc. (A)*, 421, 1967

82.46 **bis - (L - Histidino) cobalt(ii) monohydrate**
$C_{12}H_{16}CoN_6O_4$, H_2O
M.M.Harding, H.A.Long *J. Chem. Soc. (A)*, 2554, 1968

82.47 **Sodium glycylglycylglycino cuprate(ii) monohydrate**
$C_{12}H_{16}Cu_2N_6O_8{}^{2-}$, $2Na^+$, $2H_2O$
H.C.Freeman, J.C.Schoone, J.G.Sime *Acta Cryst.*, **18**, 381, 1965

82.48 **bis(Histidino) nickel(ii) monohydrate**
$C_{12}H_{16}N_6NiO_4$, H_2O
K.A.Fraser, M.M.Harding *J. Chem. Soc. (A)*, 415, 1967

82.49 **Di - (L - histidino) - zinc(ii) dihydrate**
$C_{12}H_{16}N_6O_4Zn$, $2H_2O$
R.H.Kretsinger, F.A.Cotton, R.F.Bryan *Acta Cryst.*, **16**, 651, 1963

82.50 **Di - (histidino) zinc pentahydrate**
$C_{12}H_{16}N_6O_4Zn$, $5H_2O$
M.M.Harding, S.J.Cole *Acta Cryst.*, **16**, 643, 1963

82.51 **bis(L - Histidine) copper(ii) nitrate dihydrate**
$C_{12}H_{22}CuN_6O_6{}^{2+}$, $2NO_3{}^-$
B.Evertsson *Acta Cryst. (B)*, **25**, 30, 1969

82.52 **Neodymium tetra - aquo iminodiacetate hydrochloride trihydrate**
$C_{12}H_{25}N_3Nd_2O_{16}{}^{2+}$, $2Cl^-$, $3H_2O$
J.Albertsson, A.Oskarsson *Acta Chem. Scand.*, **22**, 1700, 1968
Residue 1 also classified in 81

82.53 **Aquo bis(L - leucinato) copper(ii)**
$C_{12}H_{26}CuN_2O_5$
C.M.Weeks, A.Cooper, D.A.Norton *Acta Cryst. (B)*, **25**, 443, 1969

82.54 **DL - β - (Triethylenetetramine - O - ethylglycylglycine) cobalt(iii) perchlorate monohydrate**
$C_{12}H_{30}CoN_6O_3{}^{3+}$, $3ClO_4{}^-$, H_2O
D.A.Buckingham, P.A.Marzilli, I.E.Maxwell, A.M.Sargeson, M.Fehlmann, H.C.Freeman *Chem. Communic.*, 488, 1968
Residue 1 also classified in 76

82.55 **Copper(ii) pyridoxylidene ($\pm$) - valine complex**
$C_{13}H_{16}CuN_2O_4$
J.F.Cutfield, D.Hall, T.N.Waters *Chem. Communic.*, 785, 1967
Also classified in 83

82.56 **Manganese(ii) pyridoxylidene valine tetrahydrate**
$C_{13}H_{16}MnN_2O_4$, $4H_2O$
E.Willstadter, T.A.Hamor, J.L.Hoard *J. Amer. Chem. Soc.*, **85**, 1205, 1963
Residue 1 also classified in 83

82.57 **($\pm$) - Phenylalanine - (pyridoxylidene - 5 - phosphate) aquo copper(ii)**
$C_{17}H_{18}CuN_2O_8P$
G.A.Bentley, J.M.Waters, T.N.Waters *Chem. Communic.*, 988, 1968
Also classified in 46, 84

82.58 **Copper(ii) glycyl - L - leucyl - L - tyrosinate tetrahydrate diethylether solvate**
$C_{17}H_{23}CuN_3O_5$, $0.5C_4H_{10}O$, $4H_2O$
D.van der Helm, W.A.Franks *J. Amer. Chem. Soc.*, **90**, 5627, 1968

82.59 **Ferrichrome - A tetrahydrate**
$C_{41}H_{58}FeN_9O_{20}$, $4H_2O$
A.Zalkin, J.D.Forrester, D.H.Templeton
J. Amer. Chem. Soc., **88**, 1810, 1966
Residue 1 also classified in 84

METAL COMPLEXES (NITROGEN LIGAND)

83.C Nickel(ii) S - methylaminodithionitride S - aminodithionitride
$CH_4N_4NiS_4$
For complete entry see 85.1

83.C Thiosemicarbazide zinc chloride
$CH_5Cl_2N_3SZn$
For complete entry see 85.3

83.1 Tetraphenylarsonium oxo - tetrabromo - acetonitrile rhenate(v)
$C_2H_3Br_4NORe^-$, $C_{24}H_{20}As^+$
F.A.Cotton, S.J.Lippard *Inorg. Chem.*, **5,** 416, 1966
Residue 1 also classified in 84; residue 2 classified in 65

83.2 Copper(i) chloride - acetonitrile complex
$(C_2H_3ClCuN)_n$
M.Massaux, M.-J.Bernard, M.-T.LeBihan
Bull. Soc. Fr. Mineral. Cristallogr., **92,** 118, 1969

83.3 Copper(ii) chloride 1,2,4 - triazole
$C_2H_3Cl_2CuN_3$
J.A.J.Jarvis *Acta Cryst.*, **15,** 964, 1962

83.4 bis(Hydrazinecarboxylato - N',O) cadmium
$C_2H_6CdN_4O_4$
A.Braibanti, A.Tiripicchio, A.M.M.Lanfredi, F.Bigoli
Z. Kristallogr., **126,** 307, 1968
Also classified in 84

83.5 bis(Hydrazinecarboxylato - N',O) cadmium monohydrate
$C_2H_6CdN_4O_4$, H_2O
A.Braibanti, A.M.M.Lanfredi, A.Tiripicchio, F.Bigoli
Acta Cryst. (B), **25,** 100, 1969
Residue 1 also classified in 84

83.6   **Copper(i) chloride azomethane complex**
$C_2H_6Cl_2Cu_2N_2$
I.D.Brown, J.D.Dunitz   *Acta Cryst.*, **13**, 28, 1960

83.7   **bis(Hydrazinecarboxylato - N′,O) - manganese(ii) dihydrate**
$C_2H_6MnN_4O_4$, $2H_2O$
A.Braibanti, A.Tiripicchio, A.M.M.Lanfredi, M.Camellini
Acta Cryst., **23**, 248, 1967
Residue 1 also classified in 84

83.C   **Cobalt(iii) penta - ammine acetate chloride perchlorate**
$C_2H_8CoN_5O_2{}^{2+}$, Cl^- , $ClO_4{}^-$
For complete entry see 81.12

83.C   **bis - Thiosemicarbazidato nickel(ii) (red form)**
$C_2H_8N_6Ni_2S_2$
For complete entry see 85.5

83.8   **Copper(ii) chloride bis(semicarbazide) complex**
$C_2H_{10}CuN_6O_2{}^{2+}$, $2Cl^-$
M.Nardelli, G.F.Gasparri, P.Boldrini, G.G.Battistini
Acta Cryst., **19**, 491, 1965
Residue 1 also classified in 84

83.C   **Nickel(ii) di(thiosemicarbazide) sulfate (β form)**
$C_2H_{10}N_6NiS_2{}^{2+}$, O_4S^{2-}
For complete entry see 85.6

83.C   **Nickel(ii) dithiosemicarbazide sulfate trihydrate (α form)**
$C_2H_{10}N_6NiS_2{}^{2+}$, O_4S^{2-} , $3H_2O$
For complete entry see 85.7

83.9   **Zinc chloride bis(semicarbazide) complex**
$C_2H_{10}N_6Zn^{2+}$, $2Cl^-$
M.Nardelli, G.F.Gasparri, P.Boldrini, G.G.Battistini
Acta Cryst., **19**, 491, 1965
Residue 1 also classified in 84

83.C   **Diaquo di(thiosemicarbazide) nickel(ii) dinitrate**
$C_2H_{14}N_6NiO_2S_2{}^{2+}$, $2NO_3{}^-$
For complete entry see 85.8

83.10 **Zinc hydrazinecarboxylate hydrazine complex**
$C_2H_{14}N_8O_4Zn$
A.Ferrari, A.Braibanti, G.Bigliardi, A.M.Lanfredi
Z. Kristallogr., **122**, 259, 1965
Also classified in 84

83.11 **5 - Methoxy - 1,3,8,10 - tetrathia - 2,4,7,9 - tetra - azadec - 5 - ene - 4,7 - diato - S(1),S(10) - nickel(ii)**
$C_3H_4N_4NiOS_4$
U.Thewalt, J.Weiss *Z. Naturforsch., B*, **23**, 1265, 1968
Also classified in 85

83.12 **5 - Hydroxy - 6 - methoxy - 1,3,8,10 - tetrathia - 2,4,7,9 - tetra - azadecane - 4,7 - diato - S(1),S(10) - nickel(ii)**
$C_3H_6N_4NiO_2S_4$
J.Weiss, U.Thewalt *Z. Anorg. Allg. Chem.*, **355**, 271, 1967
Also classified in 85

83.13 **Hydrazinium tris(hydrazinecarboxylato - N,O) - nickelate(ii) monohydrate**
$C_3H_9N_6NiO_6{}^-$, $H_5N_2{}^+$, H_2O
M.K.Gusejnova *Acta Cryst.*, **21**, A142, 1966
Residue 1 also classified in 84

83.14 **Hydrazinium tris(hydrazinecarboxylato - N,O) - nickelate(ii) monohydrate**
$C_3H_9N_6NiO_6{}^-$, $H_5N_2{}^+$, H_2O
A.Braibanti, A.M.Manotti Lanfredi, A.Tiripicchio
Z. Kristallogr., **124**, 335, 1967
Residue 1 also classified in 84

83.15 **Potassium zinc hydrazinecarboxylate**
$C_3H_9N_6O_6Zn^-$, K^+
A.Braibanti, G.Bigliardi, A.M.M.Lanfredi, A.Tiripicchio
Nature, **211**, 1174, 1966
Residue 1 also classified in 84

83.C **bis(Thiosemicarbazide) silver(i) thiocyanate**
$C_3H_{10}AgN_7S_3$
For complete entry see 85.14

83.C **Nickel trithiosemicarbazide dinitrate**
$C_3H_{15}N_9NiS_3{}^{2+}$, $2NO_3{}^-$
For complete entry see 85.15

83.C   Nickel trithiosemicarbazide dinitrate monohydrate
$C_3H_{15}N_9NiS_3{}^{2+}$, $2NO_3{}^-$, H_2O
For complete entry see 85.16

83.16   Silver nitrate - pyrazine complex
$(C_4H_4AgN_2{}^+)_n$, $nNO_3{}^-$
R.G.Vranka, E.L.Amma   *Inorg. Chem.*, **5**, 1020, 1966

83.17   Succinonitrile - silver nitrate complex
$C_4H_4Ag_2N_2{}^{2+}$, $2NO_3{}^-$
T.Nomura, Y.Saito   *Bull. Chem. Soc. Jap.*, **39**, 1468, 1966

83.18   bis(Acetonitrile) niobium(iv) bromide
$C_4H_6Br_4N_2Nb$
T.A.Dougherty   *Dissert. Abstr. (B)*, **28**, 83, 1967

83.19   Zinc chloride - acetonitrile complex
$C_4H_6Cl_2N_2Zn$
I.V.Isakov, E.V.Evonkova   *Dokl. Akad. Nauk S. S. S. R.*, **145**, 801, 1962

83.20   Di(copper(ii) chloride) di - acetonitrile complex
$C_4H_6Cl_4Cu_2N_2$
R.D.Willett, R.E.Rundle   *J. Chem. Phys.*, **40**, 838, 1964

83.21   Tri(copper(ii) chloride) diacetonitrile complex
$C_4H_6Cl_6Cu_3N_2$
R.D.Willett, R.E.Rundle   *J. Chem. Phys.*, **40**, 838, 1964

83.22   Copper(ii) nitrate bis(acetonitrile) complex
$C_4H_6CuN_4O_6$
B.Duffin   *Acta Cryst. (B)*, **24**, 396, 1968
Also classified in 84

83.23   Potassium bis - biureto cuprate(ii) tetrahydrate
$C_4H_6CuN_6O_4{}^{2-}$, $2K^+$, $4H_2O$
H.C.Freeman, J.E.W.L.Smith, J.C.Taylor   *Acta Cryst.*, **14**, 407, 1961

83.24   Nickel(ii) glyoximate
$C_4H_6N_4NiO_4$
R.K.Murmann, E.O.Schlemper   *Acta Cryst.*, **23**, 667, 1967

83.25   Nickel(ii) glyoximate
$C_4H_6N_4NiO_4$
M.Calleri, G.Ferraris, D.Viterbo   *Acta Cryst.*, **22**, 468, 1967

83.26   **bis(Glyoximato) platinum(ii)**
$C_4H_6N_4O_4Pt$
G.Ferraris, D.Viterbo   *Acta Cryst. (B)*, **25**, 2066, 1969

83.27   **Morpholine - silver iodide complex**
$C_4H_9AgNO^+$, I^-
G.B.Ansell, W.G.Finnegan   *J. Chem. Soc. (D)*, 960, 1969

83.28   **bisAcetamide cadmium(ii) chloride**
$C_4H_{10}CdCl_2N_2O_2$
L.Cavalca, M.Nardelli, L.Coghi   *Nuovo Cimento*, **6**, 278, 1957
Also classified in 84
See also *Int. Distances*, M 115s; *Structure Reports*, **21**, 541, 1957

83.29   **Tetramethylenedinitramine nickel(ii) dihydrate**
$C_4H_{10}N_4NiO_5$
D.M.Liebig, J.H.Robertson, M.R.Truter   *J. Chem. Soc. (A)*, 879, 1966
Also classified in 84

83.30   **Dithiocyanato bis(thiosemicarbazide) nickel(ii)**
$C_4H_{10}N_8NiS_4$
J.Garaj, M.Dunaj-Jurco   *Chem. Communic.*, 518, 1968
Also classified in 85

83.C   **Copper(ii) diammine acetate**
$C_4H_{12}CuN_2O_4$
For complete entry see 81.31

83.C   **Copper(ii) diammine acetate**
$C_4H_{12}CuN_2O_4$
For complete entry see 81.32

83.31   **Diaquo bis(glycinato) nickel(ii)**
$C_4H_{12}N_2NiO_6$
H.C.Freeman, J.M.Guss, R.L.Sinclair   *Chem. Communic.*, 485, 1968
Also classified in 48

83.C   **Trinitro (diethylenetriamine) cobalt(iii)**
$C_4H_{13}CoN_6O_6$
For complete entry see 76.2

83.C   **Trioxo(diethylenetriamine) molybdenum(vi)**
$C_4H_{13}MoN_3O_3$
For complete entry see 76.3

83.C   Zinc dihydrazine diacetate
$C_4H_{14}N_4O_4Zn$
For complete entry see 81.35

83.32   bis(Biguanide) nickel(ii) chloride dihydrate
$C_4H_{14}N_{10}Ni^{2+}$, $2Cl^-$, $2H_2O$
T.C.Creitz, R.Gsell, D.L.Wampler   *J. Chem. Soc. (D)*, 1371, 1969

83.33   Diamine bis(acetamidine) platinum(ii) chloride monohydrate
$C_4H_{18}N_6Pt^{2+}$, $2Cl^-$, H_2O
N.C.Stephenson   *J. Inorg. Nucl. Chem.*, **24**, 801, 1962

83.34   Di - μ - hydroxo bis(dimethylamine copper(ii)) sulfate monohydrate
$C_4H_{22}Cu_2N_4O_2{}^{2+}$, O_4S^{2-}, H_2O
Y.Iitaka, K.Shimizu, T.Kwan   *Acta Cryst.*, **20**, 803, 1966
Residue 1 also classified in 84

83.35   Pyridinium peroxo chromate
$C_5H_5CrNO_5$
R.Stomberg   *Ark. Kemi*, **22**, 29, 1964

83.36   1,4 - Dimethyltetrazadiene iron tricarbonyl
$C_5H_6FeN_4O_3$
R.J.Doedens   *Chem. Communic.*, 1271, 1968

83.37   Piperidine - silver iodide complex
$C_5H_{11}AgIN$
G.B.Ansell, L.A.Burkardt, W.G.Finnegan   *J. Chem. Soc. (D)*, 459, 1969

83.38   Tripotassium molybdenum hexathiocyanate - acetic acid monohydrate
$C_6MoN_6S_6{}^{3-}$, $3K^+$, $C_2H_4O_2$, H_2O
K.Eriks, J.R.Knox   *Acta Cryst.*, **21**, A140, 1966

83.39   Copper(ii) imidazole
$C_6H_6CuN_4$
J.A.J.Jarvis, A.F.Wells   *Acta Cryst.*, **13**, A53, 1960

83.40   Dibromo - 2,5 - dimethylpyrazine nickel(ii)
$(C_6H_8Br_2N_2Ni)_n$
F.D.Ayres, P.Pauling, G.B.Robertson   *Inorg. Chem.*, **3**, 1303, 1964

83.41   Zinc chloride - bis(imidazole) complex
$C_6H_8Cl_2N_4Zn$
B.K.S.Lundberg   *Acta Cryst.*, **21**, 901, 1966

83.42 **Dimethylpiperazine palladous chloride**
$C_6H_{10}Cl_2N_2Pd$
O.Hassel, B.F.Pedersen *Proc. Chem. Soc.*, 394, 1959
See also *Int. Distances*, M 132s; *Structure Reports*, **23**, 709, 1959

83.C **bis(Thioacetamide) nickel(ii) thiocyanate**
$C_6H_{10}N_4NiS_4$
For complete entry see 85.35

83.43 **Dipotassium bis(trimethylenedinitramine) nickelate(ii) tetrahydrate**
$C_6H_{12}N_8NiO_8{}^{2-}$, $2K^+$, $4H_2O$
D.M.Liebig, J.H.Robertson *J. Chem. Soc.*, 5801, 1965

83.44 **Ethylene bidiguanide silver(iii) nitrate**
$C_6H_{14}AgN_{10}{}^+$, $NO_3{}^-$
N.R.Kunchur *Nature*, **217**, 539, 1968

83.45 **Copper(ii) chloride ethylene bis(biguanide) monohydrate**
$C_6H_{16}Cl_2CuN_{10}$, H_2O
N.R.Kunchur, M.Mathew *Chem. Communic.*, 86, 1966

83.46 **1 - (2 - Aminoethyl)biguanide - cyanoguanidine - copper(ii) sulfate monohydrate**
$C_6H_{16}CuN_{10}{}^{2+}$, O_4S^{2-}, H_2O
L.Coghi, A.Mangia, M.Nardelli, G.Pelizzi, L.Sozzi
Chem. Communic., 1475, 1968

83.47 **bis(Trimethylamine) oxo vanadium(iv) dichloride**
$C_6H_{18}Cl_2N_2OV$
J.E.Drake, J.Vekris, J.S.Wood *J. Chem. Soc. (A)*, 1000, 1968
Also classified in 84

83.48 **bis - Trimethylamine chromium trichloride**
$C_6H_{18}Cl_3CrN_2$
G.W.A.Fowles, P.T.Greene, J.S.Wood *Chem. Communic.*, 971, 1967

83.49 **bis - Trimethylamine titanium bromide**
$C_6H_{18}N_2Ti^{3+}$, $3Br^-$
B.J.Russ, J.S.Wood *Chem. Communic.*, 745, 1966

83.C **DL - β(Chloroaquo - triethylenetetramine) cobalt(iii) perchlorate**
$C_6H_{20}ClCoN_4O^{2+}$, $2ClO_4{}^-$
For complete entry see 76.33

83.50 **bis(1,3 - Diaminopropane) copper(ii) nitrate**
$C_6H_{20}CuN_4{}^{2+}$, $2NO_3{}^-$
A.Pajunen *Suomen Kemistil. (B)*, **42**, 15, 1969

83.51 **Diaquo bis(1,3 - diaminopropane) nickel(ii) nitrate**
$C_6H_{24}N_4NiO_2{}^{2+}$, $2NO_3{}^-$
A.Pajunen *Suomen Kemistil. (B)*, **41**, 232, 1968

83.C **Vanadyl(iv) diaquopyridine - 2,6 - dicarboxylate dihydrate**
$C_7H_7NO_7V$, $2H_2O$
For complete entry see 81.49

83.C **Nickel(ii) S - phenylisocyanato aminodithionitride S - aminodithionitride**
$C_7H_7N_5NiOS_4$
For complete entry see 85.38

83.52 **Triaquo - (2,6 - pyridine dicarboxylato) copper**
$C_7H_9CuO_7$
R.V.Chastain *Dissert. Abstr.*, **27**, 124B, 1966
Also classified in 84

83.C **Glycylglycinato - aquo - imidazole - copper(ii) hydrate**
$C_7H_{12}CuN_4O_4$, $1.5H_2O$
For complete entry see 82.23

83.C **cis - Diethylenetriamine chromium tricarbonyl**
$C_7H_{13}CrN_3O_3$
For complete entry see 76.44

83.C **Zinc(ii) 2,2',2'' - triaminotriethylamine thiocyanate isothiocyanate**
$C_7H_{15}N_5SZn^+$, CNS^-
For complete entry see 76.45

83.C **β,β',β'' - Triaminotriethylamine - isothiocyanato copper(ii) thiocyanate**
$C_7H_{18}CuN_5S^+$, CNS^-
For complete entry see 76.46

83.53 **Copper(ii) violurate tetrahydrate**
$C_8H_4CuN_6O_8$, $4H_2O$
M.Hamelin *C. R. Acad. Sci., Fr., C*, **266**, 19, 1968
Residue 1 also classified in 84

83.54 **bis(Succinonitrile) copper(i) perchlorate**
$(C_8H_8CuN_4{}^+)_n$, $nClO_4{}^-$
J.F.Blount, H.C.Freeman, P.Hemmerich, C.Sigwart
Acta Cryst. (B), **25**, 1518, 1969

83.55   **bis(4 - Aminopyrimidine - 2 - one) copper(ii) chloride**
bis - Cytosine dichloro copper(ii)
$C_8H_{10}Cl_2CuN_6O_2$
J.A.Carrabine, M.Sundaralingam   *Chem. Communic.*, 746, 1968
Also classified in 3

83.C   **Disodium glycylglycylglycylglycinato nickelate(ii) octahydrate**
$C_8H_{10}N_4NiO_5{}^{2-}$, $2Na^+$, $8H_2O$
For complete entry see 82.27

83.56   **Copper(ii) dimethylglyoxime (at −140 ° C)**
$C_8H_{14}CuN_4O_4$
E.Frasson, R.Bardi, S.Bezzi   *Acta Cryst.*, **12**, 201, 1959
See also *Int. Distances*, M 147s; *Structure Reports*, **23**, 600, 1959

83.C   **Copper(ii) 2 - keto - 3 - ethoxybutyraldehyde bis(thiosemicarbazonate)**
(anorthic form)
$C_8H_{14}CuN_6OS_2$
For complete entry see 85.46

83.C   **Biacetyl bis(mercaptoethylimine) nickel(ii)**
$C_8H_{14}N_2NiS_2$
For complete entry see 85.47

83.57   **Nickel dimethylglyoxime**
$C_8H_{14}N_4NiO_4$
D.E.Williams, G.Wohlauer, R.E.Rundle
J. Amer. Chem. Soc., **81**, 755, 1959
See also *Int. Distances*, M 147s; *Structure Reports*, **23**, 602, 1959

83.58   **Palladium dimethylglyoxime**
$C_8H_{14}N_4O_4Pd$
D.E.Williams, G.Wohlauer, R.E.Rundle
J. Amer. Chem. Soc., **81**, 755, 1959
See also *Int. Distances*, M 148s; *Structure Reports*, **23**, 602, 1959

83.59   **Palladium dimethylglyoxime**
$C_8H_{14}N_4O_4Pd$
C.Panattoni, E.Frasson, R.Zannetti   *Gazz. Chim. Ital.*, **89**, 2132, 1959
See also *Int. Distances*, M 148s; *Structure Reports*, **23**, 602, 1959

83.60   **Platinum dimethylglyoxime**
$C_8H_{14}N_4O_4Pt$
E.Frasson, C.Panattoni, R.Zannetti   *Acta Cryst.*, **12**, 1027, 1959
See also *Int. Distances*, M 148s; *Structure Reports*, **23**, 602, 1959

83.C **2,2′,2″ - Triaminotriethylamine nickel(ii) dithiocyanate**
$C_8H_{18}N_6NiS_2$
For complete entry see 76.47

83.61 **Cobalt(iii) bis(dimethylglyoximino) diamine nitrate**
$C_8H_{20}CoN_6O_4^+$, NO_3^-
K.S.Viswanathan, N.R.Kunchur   *Acta Cryst.*, **14**, 675, 1961

83.62 **bis - (β - Aminobutyrate) - diaquo copper(ii)**
$C_8H_{20}CuN_2O_6$
R.F.Bryan, R.J.Poljak, K.Tomita   *Acta Cryst.*, **14**, 1125, 1961
Also classified in 84

83.C **bis(N,N - Dimethyl - β - mercaptoethylamine) nickel(ii)**
$C_8H_{20}N_2NiS_2$
For complete entry see 85.56

83.63 **bis(2,2′ - Imino - bis(acetamidoxime)) nickel(ii) chloride dihydrate**
$C_8H_{20}N_{10}NiO_4^{2+}$, $2Cl^-$, $2H_2O$
E.C.Lingafelter, D.L.Cullen
Proc. 10th Internat. Conf. Coord. Chem. Tokyo, 85, 1967

83.C **DαβS(Chloro cobalt(iii) tetraethylene - penta - amine) perchlorate**
$C_8H_{23}ClCoN_5^{2+}$, $2ClO_4^-$
For complete entry see 76.48

83.C **DαβR(Chloro cobalt(iii) tetraethylene - penta - amine) perchlorate**
$C_8H_{23}ClCoN_5^{2+}$, $2ClO_4^-$
For complete entry see 76.49

83.64 **4 - (2 - Aminoethyl) - 1,4,7,10 - tetra - azadecane azido cobalt(iii) nitrate monohydrate**
$C_8H_{23}CoN_8^{2+}$, $2NO_3^-$, H_2O
D.A.Buckingham, P.A.Marzilli, I.E.Maxwell, A.M.Sargeson, H.C.Freeman
J. Chem. Soc. (D), 473, 1969
Residue 1 also classified in 76

83.65 **tetrakis - (1,3 - Dimethyltriazeno copper(i))**
$C_8H_{24}Cu_4N_{12}$
J.E.O′Connor, G.A.Janusonis, E.R.Corey   *Chem. Communic.*, 445, 1968

83.66 **Octamethylphosphonitrilium copper(ii) trichloride**
$C_8H_{25}Cl_3CuN_4P_4$
J.Trotter, S.H.Whitlow, N.L.Paddock   *J. Chem. Soc. (D)*, 695, 1969
Also classified in 64

83.C   **bis(Diethylenetriamine) copper(ii) nitrate**
$C_8H_{26}CuN_6{}^{2+}$, $2NO_3{}^-$
For complete entry see 76.51

83.C   **bis(Di(2 - aminoethyl)amine) nickel(ii) chloride monohydrate**
$C_8H_{26}N_6Ni^{2+}$, $2Cl^-$, H_2O
For complete entry see 76.52

83.C   **Tetra(ethylamine) platinum(ii) dibromo tetra(ethylamine) platinum(iv) tetrabromide**
$C_8H_{28}Br_2N_4Pt^{2+}$, $C_8H_{28}N_4Pt^{2+}$, $4Br^-$
For complete entry see 83.67

83.C   **Wolfram's red salt**
$C_8H_{28}Cl_2N_4Pt^{2+}$, $C_8H_{28}N_4Pt^{2+}$, $4Cl^-$, $4H_2O$
For complete entry see 83.68

83.67   **Tetra(ethylamine) platinum(ii) dibromo tetra(ethylamine) platinum(iv) tetrabromide**
$C_8H_{28}N_4Pt^{2+}$, $C_8H_{28}Br_2N_4Pt^{2+}$, $4Br^-$
B.M.Craven, D.Hall   *Acta Cryst.*, **21**, 177, 1966
Residue 2 classified in 83

83.68   **Wolfram's red salt**
$C_8H_{28}N_4Pt^{2+}$, $C_8H_{28}Cl_2N_4Pt^{2+}$, $4Cl^-$, $4H_2O$
B.M.Craven, D.Hall   *Acta Cryst.*, **14**, 475, 1961
Residue 2 classified in 83

83.69   **μ - Dimethylureylene - bis(tricarbonyl iron)**
$C_9H_6Fe_2N_2O_7$
R.J.Doedens   *Inorg. Chem.*, **7**, 2323, 1968

83.C   **Tetra - n - butylammonium tribromo(quinoline) nickelate (ii)**
$C_9H_7Br_3NNi^-$, $C_{16}H_{36}N^+$
For complete entry see 3.80

83.C   **Diglycylglycinato - aquo - imidazole - copper(ii) monohydrate**
$C_9H_{15}CuN_5O_5$, H_2O
For complete entry see 82.36

83.C   **Cobalt(ii) chloride - bis(2 - dimethylaminoethyl) methylamine complex**
$C_9H_{23}Cl_2CoN_3$
For complete entry see 76.57

83.70   D - tris(Trimethylenediamine) cobalt(iii) bromide monohydrate
$C_9H_{30}CoN_6^{3+}$, $3Br^-$, H_2O
Y.Saito, T.Nomura, F.Marumo   *Bull. Chem. Soc. Jap.*, **41,** 530, 1968

83.71   Oxido di(peroxido) - 2,2′ - dipyridyl chromium(vi) (orthorhombic form)
$C_{10}H_8CrN_2O_5$
R.Stomberg, I.-B.Ainalem   *Acta Chem. Scand.*, **22,** 1439, 1968

83.72   Oxido di(peroxido) - 2,2′ - dipyridyl chromium(vi) (triclinic form)
$C_{10}H_8CrN_2O_5$
R.Stomberg, I.-B.Ainalem   *Acta Chem. Scand.*, **22,** 1439, 1968

83.73   bis - (6 - Aminopurine) copper(ii) tetrahydrate (triclinic form)
$C_{10}H_8CuN_{10}$, $4H_2O$
E.Sletten   *Acta Cryst. (B),* **25,** 1480, 1969

83.74   Cadmium dipyridine dichloride
$C_{10}H_{10}CdCl_2N_2$
R.Zannetti   *Gazz. Chim. Ital.*, **90,** 1428, 1960

83.75 α - Dipyridine cobalt dichloride
$C_{10}H_{10}Cl_2CoN_2$
J.D.Dunitz   *Acta Cryst.*, **10,** 307, 1957
See also *Int. Distances,* M 156s; *Structure Reports,* **21,** 616, 1957

83.76   Dipyridine copper(ii) chloride
$C_{10}H_{10}Cl_2CuN_2$
S.S.Kobalkina   *Dokl. Akad. Nauk S. S. S. R.*, **110,** 1013, 1956
See also *Int. Distances,* M 157s; *Structure Reports,* **20,** 605, 1956

83.77   Dipyridine copper(ii) chloride
$C_{10}H_{10}Cl_2CuN_2$
J.D.Dunitz   *Acta Cryst.*, **10,** 307, 1957
See also *Int. Distances,* M 157s; *Structure Reports,* **21,** 616, 1957

83.78   Zinc dipyridine dichloride
$C_{10}H_{10}Cl_2N_2Zn$
Yu.A.Sokolova, L.O.Atovmyan, M.A.Porai-Koshits
Zh. Strukt. Khim., **7,** 855, 1966

83.79   Glutaronitrile copper(i) chloride
$(C_{10}H_{12}CuN_4^+)_n$, nCl_2Cu^-
J.O.Martin   *Dissert. Abstr. (B),* **28,** 559, 1967

83.80 **Ammonium dithiocyanato - bis - dimethylglyoxime cobalt(iii) trihydrate**
$C_{10}H_{14}CoN_6O_4S_2^-$, H_4N^+ , $3H_2O$
I.D.Samus' *Acta Cryst.*, **21**, A150, 1966

83.81 **Copper diaquo dipyridine sulphate**
$C_{10}H_{14}CuN_2O_2^{2+}$, O_4S^{2-}
E.Cannilo, G.Giuseppetti *Atti Accad. Nazion. Lincei, R. C.,
Cl. Sci. Fis. Mat. Nat.*, **36**, 878, 1964

83.C **Glycylglycinato bis(imidazole) copper(ii) perchlorate**
$C_{10}H_{15}CuN_6O_3^+$, ClO_4^-
For complete entry see 82.38

83.C **Dichloro(tetrahydrogen ethylenediaminetetra - acetato) palladium(ii) pentahydrate**
$C_{10}H_{16}Cl_2N_2O_8Pd$, $5H_2O$
For complete entry see 76.68

83.82 **Copper methylethylglyoxime**
$C_{10}H_{18}CuN_4O_4$
E.Frasson, C.Panattoni *Acta Cryst.*, **13**, A54, 1960

83.83 **Nickel methylethylglyoxime**
$C_{10}H_{18}N_4NiO_4$
E.Frasson, C.Panattoni *Acta Cryst.*, **13**, 893, 1960

83.C **Silver thiocyanate - tri - n - propyl phosphine complex**
$C_{10}H_{21}AgNS$
For complete entry see 85.64

83.84 **bis(2 - Amino - 2 - methyl - 3 - butanone oximato) nickel(ii) chloride monohydrate**
$C_{10}H_{23}N_4NiO_2^+$, Cl^- , H_2O
E.O.Schlemper *Inorg. Chem.*, **7**, 1130, 1968

83.85 **Chloro - N,N' - di(3 - aminopropyl) piperazine nickel(ii) chloride**
$C_{10}H_{24}ClN_2Ni^+$, Cl^-
N.A.Bailey, J.G.Gibson, E.D.McKenzie *J. Chem. Soc. (D)*, 741, 1969

83.86 **1,4,8,11 - Tetra - azacyclotetradecane nickel(ii) chloride**
$C_{10}H_{24}ClN_4Ni^+$, Cl^-
B.Bosnich, R.Mason, P.J.Pauling, G.B.Robertson, M.L.Tobe
Chem. Communic., 97, 1965

83.87   bis - (1,4 - Diazacycloheptane) copper(ii) nitrate hemihydrate
$C_{10}H_{24}CuN_4{}^{2+}$, $2NO_3{}^-$, $0.5H_2O$
M.S.Hussain, H.Hope   *Acta Cryst. (B)*, **25**, 1866, 1969

83.C   (+) - (N,N,N',N' - tetrakis(2' - Aminoethyl) - 1,2 - diaminoethane) cobalt(iii) hexacyanocobaltate(iii) dihydrate
$C_{10}H_{28}CoN_6{}^{3+}$, $C_6CoN_6{}^{3-}$, $2H_2O$
For complete entry see 76.73

83.88   Di - μ(3) - methylimido - tris(tricarbonyl iron)
$C_{11}H_6Fe_3N_2O_9$
R.J.Doedens   *Inorg. Chem.*, **8**, 570, 1969

83.89   Cobalt(ii) chloride - pyridine - 2 - aldehyde 2 - pyridylhydrazone complex (β form)
$C_{11}H_{10}CoN_4{}^{2+}$, $2Cl^-$
M.Gerloch   *J. Chem. Soc. (A)*, 1317, 1966

83.90   Acetylacetone - mono - (o - hydroxyanil) copper(ii)
$C_{11}H_{11}CuNO_2$
G.A.Barclay, B.F.Hoskins   *J. Cheme Soc.*, **1979, 1965**
Also classified in 84

83.C   Phenylethynyl (isopropylamine) gold(i)
$C_{11}H_{14}AuN$
For complete entry see 71.23

83.C   bis(N,N - Dimethyldithiocarbamato) pyridine zinc benzene solvate
$2C_{11}H_{17}N_3S_4Zn$, C_6H_6
For complete entry see 80.22

83.91   cis - Dichloro - bis - phenanthroline cobalt(iii) chloride trihydrate
$C_{12}H_8Cl_2CoN_2{}^+$, Cl^- , $3H_2O$
A.V.Ablov, A.Yu.Kon, T.I.Malinovskii
Dokl. Akad. Nauk S. S. S. R., **167**, 1051, 1966

83.92   Zinc chloride - 1,10 - phenanthroline complex
$C_{12}H_8Cl_2N_2Zn$
C.W.Reimann, S.Block, A.Perloff   *Inorg. Chem.*, **5**, 1185, 1966

83.93   Oxodiperoxo chromium(vi) 1,10 - phenanthroline complex
$C_{12}H_8CrN_2O_5$
R.Stomberg   *Ark. Kemi*, **24**, 111, 1965

83.94 **1,10 - Phenanthroline - mercury(i) nitrate complex**
$C_{12}H_8HgN_2{}^{2+}$, $2NO_3{}^-$
R.C.Elder, J.Halpern, J.S.Pond *J. Amer. Chem. Soc.*, **89**, 6877, 1967

83.95 **Cobalt(ii) thiocyanate - pyridine complex**
$C_{12}H_{10}CoN_4S_2$
M.A.Porai-Koshits, G.N.Tishchenko *Kristallografija*, **4**, 239, 1960

83.96 **bis(Pyridine - 2 - carboxamido) nickel(ii) dihydrate**
$C_{12}H_{10}N_4NiO_2$, $2H_2O$
Y.Nawata, H.Iwasaki, Y.Saito *Bull. Chem. Soc. Jap.*, **40**, 515, 1967
Residue 1 also classified in 84

83.97 **bis(Pyridine - 2 - carboxamido) nickel(ii) dihydrate**
$C_{12}H_{10}N_4NiO_2$, $2H_2O$
S.C.Chang, D.Y.Park, N.C.Li *Inorg. Chem.*, **7**, 2144, 1968

83.C **π - Cyclopentadienyl - dicarbonyl(4 - carboethoxy - 5 - hydroxy - 1,2,3 - molybdeno - diazacyclopenta - 3,5 - diene)**
$C_{12}H_{12}MoN_2O_5$
For complete entry see 71.27

83.98 **Nickel(ii) picolinate tetrahydrate**
$C_{12}H_{12}N_2NiO_6$, $2H_2O$
H.Loiseleur, G.Thomas, B.Chevrier, D.Grandjean
Chem. Communic., 182, 1967
Residue 1 also classified in 84

83.C **Potassium bis(nitrilotriacetato) zirconate(iv) monohydrate**
$C_{12}H_{12}N_2O_{12}Zr^{2-}$, $2K^+$, H_2O
For complete entry see 81.70

83.99 **bis(o - Phenylenediamino) nickel(ii)**
$C_{12}H_{12}N_4Ni$
G.S.Hall, R.H.Soderberg *Inorg. Chem.*, **7**, 2300, 1968

83.100 **Dichloro bis - (2 - methylpyridine) copper(ii)**
$C_{12}H_{14}Cl_2CuN_2$
V.F.Duckworth, N.C.Stephenson *Acta Cryst. (B)*, **25**, 1795, 1969

83.101 **4 - Methylpyridine - copper(ii) chloride complex**
$C_{12}H_{14}Cl_2CuN_2$
V.F.Duckworth, D.P.Graddon, N.C.Stephenson, E.C.Watton
Inorg. Nucl. Chem. Letters, **3**, 557, 1967

83.102   Dibromotetra(pyrazole) nickel(ii)
$C_{12}H_{16}Br_2N_8Ni$
A.D.Mighell, C.W.Reimann, A.Santoro   *Acta Cryst. (B)*, **25**, 595, 1969

83.103   Cobalt(ii) nicotinate tetrahydrate
$C_{12}H_{16}CoN_2O_8$
A.Anagnostopoulos, M.G.B.Drew, R.A.Walton
J. Chem. Soc. (D), 1241, 1969

83.104   Tetra(imidiazole) copper(ii) iodide
$C_{12}H_{16}CuN_8^{2+}$, $2I^-$
F.Akhtar, D.M.L.Goodgame, G.W.Rayner-Caniham, A.C.Skapski
Chem. Communic., 1389, 1968

83.C   bis - π - Cyclopentadienyl (2 - aminoethanethiolato) molybdenum iodide
$C_{12}H_{16}MoNS^+$, I^-
For complete entry see 73.40

83.105   trans - Diaquo - bis(pyridine - 2 - carboxamide) nickel(ii) chloride
$C_{12}H_{16}N_4NiO_4^{2+}$, $2Cl^-$
A.Masuko, T.Nomura, Y.Saito   *Bull. Chem. Soc. Jap.*, **40**, 511, 1967
Residue 1 also classified in 84

83.106   Nickel(ii) chloride - pyrazole complex
$C_{12}H_{16}N_8Ni^{2+}$, $2Cl^-$
C.W.Reimann, A.D.Mighell, F.A.Mauer   *Acta Cryst.*, **23**, 135, 1967

83.107   Diaquo - bis(2,2' - biimidazole) nickel(ii) nitrate (orthorhombic form)
$C_{12}H_{16}N_8NiO_2^{2+}$, $2NO_3^-$
A.D.Mighell, C.W.Reimann, F.A.Mauer   *Acta Cryst. (B)*, **25**, 60, 1969
Residue 1 also classified in 32

83.108   N,N' - Ethylene bis(acetylacetoneiminato) copper(ii) monohydrate
$C_{12}H_{18}CuN_2O_2$, H_2O
D.Hall, H.J.Morgan, T.N.Waters   *J. Chem. Soc. (A)*, 677, 1966
Residue 1 also classified in 84

83.109   N,N' - Ethylene - bis(acetylacetoneiminato) copper(ii) hemihydrate
$C_{12}H_{18}CuN_2O_2$, $0.5H_2O$
G.R.Clark, D.Hall, T.N.Waters   *J. Chem. Soc. (A)*, 223, 1968
Residue 1 also classified in 84

83.110 **Copper(ii) N,N' - ethylene - bis(acetylacetone - iminate) methylammonium perchlorate**
$C_{12}H_{18}CuN_2O_2$, CH_6N^+ , ClO_4^-
N.F.Curtis, E.N.Baker, D.Hall, T.N.Waters   *Chem. Communic.*, 675, 1966
Residue 1 also classified in 84

83.111 **(N,N' - Ethylene bis(1 - acetonylethylideniminato) - O,O',N,N') - oxo vanadium(iv)**
$C_{12}H_{18}N_2O_3V$
L.J.Boucher, D.E.Bruins, T.F.Yen, D.L.Weaver
J. Chem. Soc. (D), 363, 1969
Also classified in 84

83.C **trans - Dichloro cis - 2 - butene α - phenylethylamine platinum(ii)**
$C_{12}H_{19}Cl_2NPt$
For complete entry see 72.38

83.112 **NN, - Ethylene bis(acetylacetoneiminato) copper(ii)**
$C_{12}H_{20}CuN_2O_2$
D.Hall, A.D.Rae, T.N.Waters   *J. Chem. Soc.*, 5897, 1963
Also classified in 84

83.C **Aquo - N,N' - ethylene bis(acetylacetoneiminato) copper(ii)**
$C_{12}H_{20}CuN_2O_3$
For complete entry see 77.17

83.113 **bis - 1 - Aminocyclopentane carboxylato copper(ii)**
$C_{12}H_{20}CuN_2O_4$
G.A.Barclay, F.S.Stephens   *J. Chem. Soc.*, 2027, 1963
Also classified in 84

83.114 **Palladium(ii) chloride bis(cyclohexanone oxime)**
$C_{12}H_{22}Cl_2N_2O_2Pd$
M.Tanimura, T.Mizushima, Y.Kinoshita
Bull. Chem. Soc. Jap., **40**, 2777, 1967

83.115 **Hexamethylylenetetramine manganous chloride dihydrate**
$C_{12}H_{24}Cl_2MnN_4$, $2H_2O$
Y.-C.Tang, J.H.Sturdivant   *Acta Cryst.*, **5**, 74, 1952
See also *Int. Distances*, M 205; *Structure Reports*, **16**, 640, 1952

83.116 **Bromoazido - 1,1,7,7 - tetraethylenetriamine copper(ii)**
$C_{12}H_{28}BrCuN_6$
Z.Dori   *Chem. Communic.*, 714, 1968

83.117 Copper(ii) bis(2 - imino - 4 - amino - 4 - methylpentane) nitrate
$C_{12}H_{28}CuN_4{}^{2+}$, $2NO_3{}^-$
F.Hanic, M.Serator *Chem. Zvesti*, **18**, 572, 1964

83.118 bis(2 - Imino - 4 - amino - 4 - methylpentane) nickel(ii) nitrate
$C_{12}H_{28}N_4Ni^{2+}$, $2NO_3{}^-$
F.Hanic, D.Machajdik *Chem. Zvesti*, **23**, 3, 1969

83.C 1,1,7,7 - Tetraethyldiethylenetriamine cobalt(ii) chloride
$C_{12}H_{29}Cl_2CoN_3$
For complete entry see 76.75

83.119 tris(2 - Dimethylaminoethyl)amine copper(ii) bromide
$C_{12}H_{30}BrCuN_4{}^+$, Br^-
M.di Vaira, P.L.Orioli *Acta Cryst. (B)*, **24**, 595, 1968.
Residue 1 also classified in 76

83.C tris - (2 - Dimethylaminoethyl)amine iron(ii) bromide
$C_{12}H_{30}BrFe^+$, Br^-
For complete entry see 76.76

83.C tris - (2 - Dimethylaminoethyl)amine manganese(ii) bromide
$C_{12}H_{30}BrMn^+$, Br^-
For complete entry see 76.77

83.120 tris(2 - Dimethylaminoethyl)amine nickel(ii) bromide
$C_{12}H_{30}BrN_4Ni^+$, Br^-
M.di Vaira, P.L.Orioli *Acta Cryst. (B)*, **24**, 595, 1968
Residue 1 also classified in 76

83.C Tri(2 - dimethylaminoethyl)amine cobalt(ii) bromide
$C_{12}H_{30}Br_2CoN_4$
For complete entry see 76.79

83.121 bis(μ - (Tri - 1,2,4 - triazolo - N(11),N(2) - triaquo nickel) nickel hexanitrate
: (ihydrate
$C_{12}H_{30}N_{18}N_{13}O_6{}^{6+}$, $6NO_3{}^-$, $2H_2O$
C.W.Reimann, N.Zocchi *Chem. Communic.*, 272, 1968

83.122 bis(Di - (3 - aminopropyl)amine) nickel(ii) perchlorate
$C_{12}H_{34}N_6Ni^{2+}$, $2ClO_4{}^-$
P.Paoletti, S.Biagini, M.Cannas *J. Chem. Soc. (D)*, 513, 1969

83.123 **Tungsten hexadimethylamide**
$C_{12}H_{36}N_6W$
D.C.Bradley, N.H.Chisholm, C.E.Heath, M.B.Hursthouse
J. Chem. Soc. (D), 1261, 1969

83.124 **trans - bis(Cobalt tetracarbonyl) - bis(pyridine) platinum**
$C_{13}H_5Co_2NO_4Pt$
D.Moras, J.Dehand, R.Weiss   *C. R. Acad. Sci., Fr., C*, **267**, 1471, 1968

83.C **Di - iodocarbomethoxycarbonyl (2,2' - bipyridyl) iridium**
$C_{13}H_{11}I_2IrN_2O_3$
For complete entry see 71.33

83.125 **Cyclopropane bis pyridine platinum(iv) dichloride**
$C_{13}H_{16}Cl_2N_2Pt$
N.A.Bailey, R.D.Gillard, M.Keeton, R.Mason, D.R.Russell
Chem. Communic., 396, 1966
Also classified in 71

83.126 **Cyclopropane bispyridine platinum(iv) tetrachloride carbon tetrachloride solvate**
$C_{13}H_{16}Cl_4N_2Pt$, CCl_4
N.A.Bailey, R.D.Gillard, M.Keeton, R.Mason, D.R.Russell
Chem. Communic., 396, 1966
Residue 1 also classified in 71

83.C **Copper(ii) pyridoxylidene ($\pm$) - valine complex**
$C_{13}H_{16}CuN_2O_4$
For complete entry see 82.55

83.C **Manganese(ii) pyridoxylidene valine tetrahydrate**
$C_{13}H_{16}MnN_2O_4$, $4H_2O$
For complete entry see 82.56

83.127 **Dibromo - (1 - (o - methoxyphenyl) - 2,6 - diazaoctane) nickel(ii)**
$C_{13}H_{22}Br_2N_2NiO$
P.L.Orioli, M.di Vaira   *J. Chem. Soc. (A)*, 2078, 1968
Also classified in 84

83.C **bis(O,O - Diethyldithiophosphato) nickel(ii) pyridine**
$C_{13}H_{25}NNiO_4P_2S_4$
For complete entry see 85.70

83.C **μ - Chloro - (2,2' - dipyridyl tricarbonylmolybdenum) - methyldichlorotin**
$C_{14}H_{11}Cl_3MoN_2O_3Sn$
For complete entry see 69.28

83.128 Dichloro - (2,9 - dimethyl - 1,10 - phenanthroline) nickel(ii) chloroform
solvate
$2C_{14}H_{12}Cl_2N_2N$, $2CHCl_3$
H.S.Preston, C.H.L.Kennard   *Chem. Communic.*, 89, 1969

83.129 Zinc(ii) chloride 2,9 - dimethyl - 1,10 - phenanthroline complex
$C_{14}H_{12}Cl_2N_2Zn$
H.S.Preston, C.H.L.Kennard   *Chem. Communic.*, 708, 1967

83.130 Dichloro - bis(4 - vinylpyridine) cobalt(ii)
$C_{14}H_{14}Cl_2CoN_2$
L.J.Admiraal, G.Gafner   *Chem. Communic.*, 1221, 1968

83.131 Dichloro - bis(4 - vinylpyridine) cobalt(ii) (violet β form)
Di(4 - vinylpyridine) cobalt dichloride
$C_{14}H_{14}Cl_2CoN_2$
M.Laing, E.Horsfield   *J. Chem. Soc. (D)*, 902, 1969

83.132 bis(4 - Vinylpyridine) copper(ii) dichloride
$C_{14}H_{14}Cl_2CuN_2$
M.Laing, E.Horsfield   *Chem. Communic.*, 735, 1968

83.133 Dichloro aquo - (2,9 - dimethyl - 1,10 - phenanthroline) copper(ii)
$C_{14}H_{14}Cl_2CuN_2O$
H.S.Preston, C.H.L.Kennard   *Chem. Communic.*, 1167, 1967

83.C Copper(ii) acetate tris - pyridine complex
$C_{14}H_{16}CuN_2O_4$, C_5H_5N
For complete entry see 81.79

83.C Chloro - (2 - methoxycyclo - octa - 1,5 - dienyl) - pyridine platinum
$C_{14}H_{18}ClNOPt$
For complete entry see 71.36

83.134 Dichloro di - p - toluidine cobalt(ii)
$C_{14}H_{18}Cl_2CoN_2$
T.I.Malinovskii   *Kristallografija*, **2**, 734, 1957
See also *Int. Distances*, M 182s; *Structure Reports*, **21**, 598, 1957

83.135 bis - p - Toluidine zinc chloride
$C_{14}H_{18}Cl_2N_2Zn$
A.B.Ablov, T.I.Malinovskii   *Dokl. Akad. Nauk S. S. S. R.*, **123**, 677, 1958
See also *Int. Distances*, M 182s; *Structure Reports*, **22**, 714, 1958

83.136   Di - iodo di - p - toluidine cobalt(ii)
$C_{14}H_{18}CoI_2N_2$
T.I.Malinovskii   *Kristallografija*, **3**, 364, 1958
See also *Int. Distances*, M 182s; *Structure Reports*, **22**, 715, 1958

83.137   Calcium trans - 1,2 - diaminocyclohexane - N,N' - tetra - acetato aquo ferrate(iii) octahydrate
$2C_{14}H_{20}FeN_2O_4^-$, Ca^{2+} , $8H_2O$
G.H.Cohen, J.L.Hoard   *J. Amer. Chem. Soc.*, **88**, 3228, 1966
Residue 1 also classified in 84

83.C   cls - Di - μ - dimethylamido - bis(cyclopentadienyl - nitrosyl chromium)
$C_{14}H_{22}Cr_2N_4O_2$
For complete entry see 73.60

83.C   trans - Di - μ - dimethylamido - bis(cyclopentadienyl - nitrosyl chromium)
$C_{14}H_{22}Cr_2N_4O_2$
For complete entry see 73.61

83.138   (tetrakis - Dimethylformamido) nickel di - isoselenocyanate
$C_{14}H_{28}N_6NiO_4Se_2$
G.V.Tsintsadze, M.A.Porai-Koshits, A.S.Antsyshkina
Zh. Strukt. Khim., **8**, 296, 1967
Also classified in 84

83.139   Zinc tripyridyl chloride (form ii)
$C_{15}H_{11}Cl_2N_3Zn$
F.W.B.Einstein, B.R.Penfold   *Acta Cryst.*, **20**, 924, 1966

83.140   Copper(ii) 1 - (2 - pyridylazo) - 2 - naphtholato perchlorate monohydrate
$C_{15}H_{12}CuN_3O_2^+$, ClO_4^-
S.Ooi, D.Carter, Q.Fernando   *Chem. Communic.*, 1301, 1967
Residue 1 also classified in 84

83.141   2,2',2'' - Terpyridine zinc chloride (form ii)
$C_{15}H_{15}Cl_2N_3Zn$
D.E.C.Corbridge, E.G.Cox   *J. Chem. Soc.*, 594, 1956
See also *Int. Distances*, M 183s; *Structure Reports*, **20**, 611, 1956

83.142   Copper(ii) 1,6 - bis(2' - pyridyl) - 6 - methoxy - 2,5 - diazahex - 1 - ene bromide perchlorate
$C_{15}H_{18}BrCuN_4O^+$, ClO_4^-
B.F.Hoskins, F.D.Williams   *Chem. Communic.*, 798, 1966

83.143 Bromo(2,12 - dimethyl - 3,7,11,17 - tetra - azabicyclo(11.3.1)heptadeca -
1(17),2,11,13,15 - pentaene) nickel(ii) bromide monohydrate
$C_{15}H_{22}BrN_4Ni^+$, Br^-, H_2O
E.B.Fleischer, S.W.Hawkinson *Inorg. Chem.*, **7**, 2312, 1968

83.144 Nickel(ii) dehydroacetic acid monoimide complex
$C_{16}H_{16}N_2NiO_6$
S.Kiryu *Acta Cryst.*, **23**, 392, 1967
Also classified in 84

83.145 Collidine mercury(ii) chloride complex
$C_{16}H_{22}Cl_4Hg_2N_2$
S.Kulpe *Z. Anorg. Allg. Chem.*, **349**, 314, 1966

83.146 O - Methyl - (Co - C) - carboxymethyl - bis(dimethyl -
glyoximato)pyridinato - cobalt
Carbomethoxymethyl - bis(dimethylglyoximato) pyridinato - cobalt
$C_{16}H_{24}CoN_5O_6$
P.G.Lenhert *Chem. Communic.*, 980, 1967

83.147 Nickel(ii) perchlorate - 5,7,7,12,12,14 - hexamethyl - 1,4,8,11 - tetra -
azacyclotetradeca - 4,8,10,14 - tetraene complex (α form)
$C_{16}H_{28}N_4Ni^{2+}$, $2ClO_4^-$
I.E.Maxwell, M.F.Bailey *Chem. Communic.*, 883, 1966

83.148 Nickel(ii) perchlorate - 5,5,7,12,14,14 - hexamethyl - 1,4,8,11 - tetra -
azacyclotetradeca - 7,11 - diene complex (form B,cis form)
$C_{16}H_{32}N_4Ni^{2+}$, $2ClO_4^-$
R.R.Ryan, B.T.Kilbourn, J.D.Dunitz *Chem. Communic.*, 910, 1966

83.149 Nickel(ii) perchlorate - 5,5,7,12,14,14 - hexamethyl - 1,4,8,11 - tetra -
azacyclotetradeca - 7,11 - diene complex (A α form,trans isomer)
$C_{16}H_{32}N_4Ni^{2+}$, $2ClO_4^-$
M.F.Bailey, I.E.Maxwell *Chem. Communic.*, 908, 1966

83.150 Iron(iii) 2,13 - dimethyl - 3,6,9,12,18 - penta - azabicyclo(12.3.1)octadeca -
1(18),2,12,14,16 - pentaene thiocyanate perchlorate
$C_{17}H_{23}FeN_7S_2^+$, ClO_4^-
E.B.Fleischer, S.Hawkinson *J. Amer. Chem. Soc.*, **89**, 720, 1967

83.151 Cobalt(ii) bis - N,N - (2 - diethylaminoethyl) - 2 - methylthioethylamine
isothiocyanate butanol solvate
$C_{17}H_{35}CoN_5S_3$, $0.5C_4H_{10}O$
P.Dapporto, M.di Vaira, L.Sacconi *J. Chem. Soc. (D)*, 153, 1969

83.152   o - Semidine bis(iron tricarbonyl)
$C_{18}H_{10}Fe_2N_2O_6$
P.E.Baikie, O.S.Mills   *Chem. Communic.*, 707, 1966

83.153   trans - bis(Manganese tetracarbonyl) - bis(pyridine) platinum
$C_{18}H_{10}Mn_2N_2O_8Pt$
D.Moras, J.Dehand, R.Weiss   *C. R. Acad. Sci., Fr., C,* **267,** 1471, 1968

83.154   Gold(iii) 2,2' - biquinolyl chloride
$C_{18}H_{12}AuCl_3N_2$
R.J.Charlton, C.M.Harris, H.Patil, N.C.Stephenson
Inorg. Nucl. Chem. Letters, **2,** 409, 1966

83.155   Copper(ii) 8 - hydroxyquinolinate (β form)
$C_{18}H_{12}CuN_2O_2$
G.J.Palenik   *Acta Cryst.*, **17,** 687, 1964
Also classified in 84

83.156   Copper(ii) 8 - hydroxyquinolinate (β form)
$C_{18}H_{12}CuN_2O_2$
F.Kanamaru, K.Ogawa, I.Nitta   *Bull. Chem. Soc. Jap.*, **36,** 422, 1963
Also classified in 84

83.157   Copper(ii) 8 - hydroxyquinolinate (β form)
$C_{18}H_{12}CuN_2O_2$
J.A.Bevan, D.P.Graddon, J.F.McConnell   *Nature,* **199,** 373, 1963
Also classified in 84

83.158   Copper(ii) 8 - hydroxyquinolinate (α form)
$C_{18}H_{12}CuN_2O_2$
R.C.Hoy, R.H.Morriss   *Acta Cryst.*, **22,** 476, 1967
Also classified in 84

83.C   Copper(ii) 8 - hydroxyquinolinate - bis(1,2,4,5 - tetracyanobenzene) complex (photographic data)
$C_{18}H_{12}CuN_2O_2$, $2C_{10}H_2N_4$
For complete entry see 60.146

83.C   Copper(ii) 8 - hydroxyquinolinate - bis(1,2,4,5 - tetracyanobenzene) complex (diffractometer data)
$C_{18}H_{12}CuN_2O_2$, $2C_{10}H_2N_4$
For complete entry see 60.147

83.C **7,7,8,8 - Tetracyanoquinodimethane - copper(ii) 8 - hydroxyquinolate complex**
$C_{18}H_{12}CuN_2O_2$, $C_{12}H_4N_4$
For complete entry see 60.126

83.C **Dioxo bis(8 - hydroxyquinolinato) molybdenum(vi)**
$C_{18}H_{12}MoN_2O_4$
For complete entry see 84.43

83.159 **8 - Hydroxyquinolinato palladium(ii)**
$C_{18}H_{12}N_2O_2Pd$
C.K.Prout, A.G.Wheeler *J. Chem. Soc. (A)*, 1286, 1966
Also classified in 84

83.C **Chloranil - bis(8 - hydroxyquinolinato) palladium(ii)**
$C_{18}H_{12}N_2O_2Pd$, C_6Cl_4O
For complete entry see 60.70

83.C **Palladium(ii) 8 - hydroxyquinolinate - 1,2,4,5 - tetracyanobenzene complex (disordered)**
$C_{18}H_{12}N_2O_2Pd$, $C_{10}H_2N_4$
For complete entry see 60.148

83.160 **Zinc 8 - hydroxyquinolinate dihydrate**
$C_{18}H_{12}N_2O_2Zn$, $2H_2O$
G.J.Palenik *Acta Cryst.*, **17,** 696, 1964
Residue 1 also classified in 84

83.C **Palladium 8 - mercaptoquinolinate**
$C_{18}H_{12}N_2PdS_2$
For complete entry see 85.80

83.161 **bis(8 - Hydroxyquinoline) silver(i) pyridine solvate**
$C_{18}H_{13}AgN_2O_2$, C_5H_5N
J.E.Fleming, H.Lynton *Canad. J. Chem.*, **46,** 471, 1968
Residue 1 also classified in 84

83.162 **Zinc diaquo - 8 - hydroxyquinoline**
$C_{18}H_{16}N_2O_4Zn$
L.L.Merritt Junior, R.T.Cady, B.W.Mundy *Acta Cryst.*, **7,** 473, 1954
Also classified in 84
See also *Int. Distances,* M 247; *Structure Reports,* **18,** 761, 1954

83.C **Monopyridine copper(ii) acetate (monoclinic form)**
$C_{18}H_{22}Cu_2N_2O_8$
For complete entry see 81.85

83.C   **Monopyridine copper (ii) acetate (orthorhombic form)**
$C_{18}H_{22}Cu_2N_2O_8$
For complete entry see 81.86

83.163   **2,9 - Dimethyl - 1,10 - phenanthroline bis(o,o' - dimethyl dithiophosphato) nickel(ii)**
$C_{18}H_{24}N_2NiO_4P_2S_4$
P.S.Shetty, R.E.Ballard, Q.Fernando   *J. Chem. Soc. (D)*, 717, 1969
Also classified in 85

83.164   **Trimethyl(acetylacetonyl) - 2,2' - bipyridyl platinum**
$C_{18}H_{24}N_2O_2Pt$
A.G.Swallow, M.R.Truter   *Proc. R. Soc., A,* **266,** 527, 1962
Also classified in 71

83.165   **hexakis(Imidazole) nickel(ii) nitrate**
$C_{18}H_{24}N_{12}Ni^{2+}$, $2NO_3^-$
A.Santoro, A.D.Mighell, M.Zocchi, C.W.Reimann
Acta Cryst. (B), **25,** 842, 1969

83.166   **Hexaimidazole zinc(ii) dichloride tetrahydrate**
$C_{18}H_{24}N_{12}Zn^{2+}$, $2Cl^-$, H_2O
C.Sandmark, C.Branden   *Acta Chem. Scand.,* **21,** 993, 1967

83.167   **Di(cyclohexane - 1,2 - dioximato(1)) di - imidazole iron(ii)**
$C_{18}H_{30}FeN_8O_6$
C.K.Prout, T.J.Wiseman   *J. Chem. Soc.,* 497, 1964

83.C   **bis(O,O' - Diethyldithiophosphato) nickel(ii) bis(pyridine)**
$C_{18}H_{30}N_2NiO_4P_2S_4$
For complete entry see 85.81

83.C   **tris(Hexamethyldisilylamine) iron(iii)**
$C_{18}H_{54}FeN_3Si_6$
For complete entry see 63.24

83.168 μ **- Diphenylureylene - bis(tricarbonyl iron)**
$C_{19}H_{10}Fe_2N_2O_7$
J.A.J.Jarvis. B.E.Job, B.T.Kilbourn, R.H.B.Mais, P.G.Owston, P.F.Todd
Chem. Communic., 1149, 1967

83.169 μ **- Diphenylureylene - bis(tricarbonyl iron)**
$C_{19}H_{10}Fe_2N_2O_7$
J.Piron, P.Piret, M.van Meerssche   *Bull. Soc. Chim. Belges,* **76,** 505, 1967

83.C   **Tricarbonyl iron N - p - tolylsulfonyl - 1,4 - dimethoxy - 4 - imino - i - but - 2 - enyl iron tricarbonyl**
$C_{19}H_{15}Fe_2No_{10}S$
For complete entry see 72.54

83.C   **Copper(ii) acetylacetonate - quinoline complex**
$C_{19}H_{21}CuNO_4$
For complete entry see 77.38

83.C   **N - t - Butyl - tri(π - cyclopentadienyl nickel) amine**
$C_{19}H_{24}NNi$
For complete entry see 73.82

83.C   **Nickel(ii) - N - β - diethyl aminoethyl - salicylaldimine catecholate**
$C_{19}H_{24}N_2NiO_3$
For complete entry see 78.32

83.C   **bis(π - Cyclopentadienyl cobalt) bis(t - butylimine) carbonyl**
$C_{19}H_{28}Co_2N_2O$
For complete entry see 73.84

83.C   **Tetra - n - propylammonium o - phenanthroline - bis(maleonitriledithiolato) cobaltate**
$C_{20}H_8CoN_6S_4{}^-$, $C_{12}H_{28}N^+$
For complete entry see 3.76

83.170   **Benzaldehydato p - tolueneimine bis(iron carbonyl)**
$C_{20}H_{13}Fe_2NO_6$
P.E.Baikie, O.S.Mills   *Chem. Communic.*, 707, 1966
Also classified in 71

83.171   **Iodo bis - (2,2' - bipyridyl) copper(ii) iodide**
$C_{20}H_{16}CuIN_4{}^+$, I^-
G.A.Barclay, B.F.Hoskins, C.H.L.Kennard   *J. Chem. Soc.*, 5691, 1963

83.172   **bis(2,2' - Bipyridyl) lanthanum(iii) nitrate**
$C_{20}H_{16}LaN_4{}^{3+}$, $3NO_3{}^-$
A.R.Al-Karaghouli, J.S.Wood   *J. Amer. Chem. Soc.*, **90**, 6548, 1968

83.173   **bis(2,2' - Dipyridyliminato) palladium(ii)**
$C_{20}H_{16}N_6Pd$
H.C.Freeman, M.R.Snow   *Acta Cryst.*, **18**, 843, 1965

83.174   **Terbium(iii) bis(2,2' - bipyridyl) tris(nitrate)**
$C_{20}H_{16}N_7O_9Tb$
D.S.Moss, S.P.Sinha   *Z. Phys. Chem. (Frankfurt)*, **63**, 190, 1969

83.175 **Tetrapyridine nickel(ii) dibromide**
$C_{20}H_{20}Br_2N_4Ni$
A.S.Antsyshkina, M.A.Porai-Koshits *Kristallografija*, **3**, 676, 1958
See also *Int. Distances*, M 193s; *Structure Reports*, **22**, 782, 1958

83.176 **Tetrapyridine cobalt(ii) chloride**
$C_{20}H_{20}Cl_2CoN_4$
M.A.Porai-Koshits *Trudy Inst. Kristallogr., S. S. S. R.*, **10**, 269, 1954
See also *Int. Distances*, M 193s; *Structure Reports*, **18**, 750, 1954

83.177 **Tetrapyridine nickel dichloride**
$C_{20}H_{20}Cl_2N_4Ni$
M.A.Porai-koshits *Trudy Inst. Kristallogr., S. S. S. R.*, **10**, 269, 1954
See also *Int. Distances*, M 193s; *Structure Reports*, **18**, 750, 1954

83.178 **trans - Dichlorotetrapyridine rhodium(iii) hydrogen dinitrate**
$C_{20}H_{20}Cl_2N_4Rh^+$, $HN_2O_6^-$
G.C.Dobinson, R.Mason, D.R.Russell *Chem. Communic.*, 62, 1967

83.179 **bis(3 - Amino - 1 - phenyl - 2 - buten - 1 - ono) copper(ii)**
Copper(ii) benzoylacetone - iminate
$C_{20}H_{20}CuN_2O_2$
G.E.Gurr *Acta Cryst. (B)*, **24**, 1511, 1968
Also classified in 84

83.180 **Diazoaminobenzene copper(i)**
$C_{20}H_{20}Cu_2N_6$
I.D.Brown, J.D.Dunitz *Acta Cryst.*, **14**, 480, 1961

83.C **trans - bis(Acetylacetonato) dipyridine cobalt(ii)**
trans - bis(2,4 - Pentanedionato) dipyridine cobalt(ii)
$C_{20}H_{24}CoN_2O_4$
For complete entry see 77.44

83.C **trans - bis(2,4 - Pentanedionato) dipyridine - nickel(ii)**
trans - bis(Acetylacetonato) dipyridine - nickel(ii)
$C_{20}H_{24}N_2NiO_4$
For complete entry see 77.45

83.C **Copper(ii) bis - 1 - ephedrine - benzene**
$C_{20}H_{26}CuN_2O_2$, $0.67C_6H_6$
For complete entry see 61.21

83.181 bis(α - Hydroxy - α - phenylbutyramidine) copper(ii) dihydrate
$C_{20}H_{26}CuN_4O_2$, $2H_2O$
J.Iball, C.H.Morgan *J. Chem. Soc. (A)*, 52, 1967
Residue 1 also classified in 84

83.182 bis - (+) - Pseudoephedrine copper(ii) (+) - pseudoephedrine dihydrate
$C_{20}H_{28}CuN_2O_2$, $C_{10}H_{15}NO$, $2H_2O$
N.A.Bailey, P.M.Harrison, R.Mason *Chem. Communic.*, 559, 1968
Residue 1 also classified in 84; residue 2 classified in 58

83.183 Ethylenediammonium N - hydroxyethyl ethylenediamine - triacetato iron(iii)
- μ - oxo - N - hydroxyethyl - ethylenediamine triacetato ferrate(iii)
hexahydrate
$C_{20}H_{30}Fe_2N_4O_{15}{}^{2-}$, $C_2H_{10}N_2{}^{2+}$, $6H_2O$
S.J.Lippard, H.Schugar, C.Walling *Inorg. Chem.*, 6, 1825, 1967
Residue 1 also classified in 84; residue 2 classified in 3

83.184 Tribenzo(b,f,j,)(1.5.9)triazacycloduodecine nickel(ii) nitrate
$C_{21}H_{19}N_4NiO_5{}^+$, $NO_3{}^-$
E.B.Fleischer, E.Klem *Inorg. Chem.*, 4, 637, 1965
Residue 1 also classified in 84

83.C bis((Salicylidene - γ - iminopropyl)methylamine)) nickel(ii)
$C_{21}H_{25}N_3NiO_2$
For complete entry see 78.37

83.185 Iron(ii) thiocyanate tetra(pyridine)
$C_{22}H_{20}FeN_6S_2$
I.Sotofte, S.E.Rasmussen *Acta Chem. Scand.*, 21, 2028, 1967

83.186 Dithiocyanato tetrapyridine nickel(ii)
$C_{22}H_{20}N_6NiS_2$
A.S.Antsyshkina, M.A.Porai-Koshits *Kristallografija*, 3, 686, 1958
See also *Int. Distances*, M 197s; *Structure Reports*, 22, 784, 1958

83.C 4 - Methylpyridine bis(o - hydroxyacetophenonato) copper(ii)
$C_{22}H_{21}CuNO_4$
For complete entry see 77.54

83.187 N,N' - Ethylene bis(L - ephedrine) copper(ii) dihydrate
$C_{22}H_{30}CuN_2O_2$, $2H_2O$
M.Motohashi, Y.Amano, T.Uno *Bull. Chem. Soc. Jap.*, 41, 2007, 1968
Residue 1 also classified in 84

83.C   Chloro - π - cyclopentadienyl bis(8 - quinolinolato) titanium(iv)
$C_{23}H_{17}ClN_2O_2Ti$
For complete entry see 73.100

83.188   Chromium(iii) pyridinium O,O' - dihydroxy - trans - azobenzene complex
$C_{24}H_{16}CrN_4O_4^-$, $C_5H_6N^+$
R.Grieb, A.Niggli   *Helv. Chim. Acta,* **48,** 317, 1965
Residue 1 also classified in 84; residue 2 classified in 33

83.189   bis - (3 - Hydroxyl - 1,3 - diphenyltriazene) palladium
$C_{24}H_{20}N_6O_2Pd$
E.F.Meyer Junior, S.H.Simonsen   *Acta Cryst.,* **16,** A67, 1963
Also classified in 84

83.C   bis(Diphenylglyoximato) nickel(ii) - iodine complex
$2C_{24}H_{22}N_4NiO_4$, I_2
For complete entry see 60.155

83.190   cis,cis - 1,3,5 - tris(Pyridine - 2 - carboxaldimino) cyclohexane zinc(ii) perchlorate
$C_{24}H_{24}N_6Zn^{2+}$, $2ClO_4^-$
W.O.Gillum, J.C.Huffman, W.E.Streib, R.A.D.Wentworth
J. Chem. Soc. (D), 843, 1969

83.191   Trimethyl - (8 - quinolinolato) platinum(iv)
$C_{24}H_{30}N_2O_2Pt_2$
J.E.Lydon, M.R.Truter   *J. Chem. Soc.,* 6899, 1965
Also classified in 84

83.C   Nickel(ii) N - γ - dimethylaminopropyl salicylaldiminate
$C_{24}H_{34}N_4NiO_2$
For complete entry see 78.49

83.C   Tetra - n - butyl - o - phenanthroline - bis(toluene - 3,4 - dithiolato) cobaltate
$C_{26}H_{20}CoN_2S_4^-$, $C_{16}H_{36}N^+$
For complete entry see 3.81

83.192   2,2' - Diphenyl bis - (2 - iminomethylenephenolato) copper(ii)
$C_{26}H_{22}CuN_2O_2$
T.P.Cheeseman, D.Hall, T.N.Waters   *Proc. Chem. Soc.,* 379, 1963
Also classified in 84

83.C   Mercury dithizone complex
$C_{26}H_{22}HgN_8S_2$, $2C_5H_5N$
For complete entry see 85.90

83.C   **Copper(ii) acetate - quinoline complex**
$C_{26}H_{26}Cu_2N_2O_8$
For complete entry see 81.87

83.C   **Nickel(ii) 5 - chloro - N - (2' - diethylaminoethyl) - salicylaldiminate**
$C_{26}H_{36}Cl_2N_4NiO_2$
For complete entry see 78.53

83.193   **Uranyl 8 - quinolinolate 8 - quinolinol complex**
$C_{27}H_{19}N_3O_5U$
J.E.Fleming, H.Lynton   *Canad. J. Chem.*, **45**, 1637, 1967
Also classified in 84

83.194   **Dioxo di - 8 - quinolinolato - 8 - quinolinol uranium(vi) chloroform solvate**
$C_{27}H_{19}N_3O_5U$, $CHCl_3$
D.Hall, A.D.Rae, T.N.Waters   *Acta Cryst.*, **22**, 258, 1967
Residue 1 also classified in 84

83.195   **Trichloro (p - methoxyphenylimino) bis(diethylphenyl phosphine) rhenium(v)**
$C_{27}H_{37}Cl_3NOP_2Re$
D.Bright, J.A.Ibers   *Inorg. Chem.*, **7**, 1099, 1968
Also classified in 86

83.196   **Dichloro bis(8 - quinolinato) titanium(iv)**
$C_{28}H_{12}Cl_2N_2O_2Ti$
B.F.Studd, A.G.Swallow   *J. Chem. Soc. (A)*, 1961, 1968
Also classified in 84

83.197   **tris(8 - Quinolinolato) chromium methanol complex**
$C_{28}H_{22}CrN_3O_4$
K.Folting, M.M.Cox, J.W.Moore, L.L.Merritt
Chem. Communic., 1170, 1968
Also classified in 84

83.C   **bis - (2 - Diphenylphosphinoethyl)amine nickel(ii) dibromide**
$C_{28}H_{28}Br_2NNiP_2$
For complete entry see 86.49

83.198   **Nickel(ii) perchlorate tetra(3,5 - dimethylpyridine)**
$C_{28}H_{36}N_4Ni^{2+}$, $2ClO_4^-$
F.Madaule-Aubry, G.M.Brown   *Acta Cryst. (B)*, **24**, 745, 1968

83.199 **Nickel(ii) perchlorate tetra(3,4 - dimethylpyridine)**
$C_{28}H_{36}N_4Ni^{2+}$, $2ClO_4^-$
F.Madaule-Aubry, W.R.Busing, G.M.Brown
Acta Cryst. (B), **24,** 754, 1968

83.200 **Trichloro (p - acetylphenylimino) bis(diethyl phosphine) rhenium(v)**
$C_{28}H_{37}Cl_3NOP_2Re$
D.Bright, J.A.Ibers *Inorg. Chem.,* **7,** 1099, 1968
Also classified in 86

83.C **(N - (2 - Hydroxyethyl) - acetylacetoniminato) copper(ii) tetramer**
$C_{28}H_{44}Cu_4N_4O_8$
For complete entry see 84.59

83.201 **bis(Nickel(ii) N,N - bis(2 - diethylaminoethyl) - 2 - hydroxyethylamine) perchlorate**
$C_{28}H_{64}N_6Ni_2O_2^{2+}$, $2ClO_4^-$
P.Dapporto, L.Sacconi *J. Chem. Soc. (D),* 329, 1969
Residue 1 also classified in 84

83.202 **tris - 2,2' - Dipyridyl vanadium(0)**
$C_{30}H_{24}N_6V$
G.Albrecht *Z. Chem.,* **3,** 182, 1963

83.203 **Hexapyridine iron(ii) tetra - iron - tridecacarbonyl**
$C_{30}H_{30}FeN_6^{2+}$, $C_{13}F_5O_{13}^{2-}$
R.J.Doedens, L.F.Dahl *J. Amer. Chem. Soc.,* **88,** 4847, 1966

83.204 **Di - (3 - methyl - 1 - phenyl - 5 - p - tolylformazyl) nickel(ii)**
$C_{30}H_{30}N_8Ni$
D.Dale *J. Chem. Soc. (A),* 278, 1967

83.C **Di - isothiocyanato - (N,N - bis(2 - diethylaminoethyl) - 2 - diphenylarsinoethylamine) nickel(ii)**
$C_{30}H_{42}AsN_5NiS_2$
For complete entry see 65.50

83.C **Nitridodichloro tris(diethylphenylphosphine) rhenium(v)**
$C_{30}H_{45}Cl_2NP_3Re$
For complete entry see 86.58

83.205 **μ - Oxo - bis(iron(iii) aquo - 2,13 - dimethyl - 3,6,9,12,18 - penta - azabicyclo(12.3.1)octadeca - 1,(18),2,12,14,16 - pentaene) perchlorate**
$C_{30}H_{50}Fe_2N_{10}O_3^{4+}$, $4ClO_4^-$
E.B.Fleischer, S.Hawkinson *J. Amer. Chem. Soc.,* **89,** 720, 1967

83.C bis(N - Phenylsalicylaldiminato) copper(ii) pyridine
$C_{31}H_{45}CuN_3O_2$
For complete entry see 78.58

83.206 Pyridinium bis(2 - hydroxy naphthalene - 1 - azo) - (2' - hydroxy - 4' - nitrobenzene) chromium(iii) pyridine solvate
$C_{32}H_{18}CrN_6O_8{}^-$, $C_5H_6N^+$, $0.5C_5H_5N$
R.Grieb, A.Niggli *Helv. Chim. Acta,* **48,** 317, 1965
Residue 1 also classified in 84; residue 2 classified in 33

83.207 Copper(ii) bis - (benzeneazo - β - naphthol)
$C_{32}H_{22}CuN_4O_2$
J.A.J.Jarvis *Acta Cryst.,* **14,** 961, 1961
Also classified in 84

83.C N,N' - Ethylene bis(salicylaldehydeiminato) cobalt(ii) (inactive form)
$C_{32}H_{28}Co_2N_4O_4$
For complete entry see 78.59

83.C μ - Oxo - bis(iron bis(salicylidene - ethylenediamine)) pyridine solvate
$C_{32}H_{28}Fe_2N_4O_5$, $2C_5H_5N$
For complete entry see 78.60

83.208 bis(1 - m - Tolylazo - 2 - naphtholato) nickel(ii)
$C_{34}H_{26}N_4NiO_2$
N.W.Alcock, R.C.Spencer, R.H.Prince, O.Kennard
J. Chem. Soc. (A), 2383, 1968
Also classified in 84

83.209 bis(Diphenyldiazomethane) tris(tricarbonyl iron)
$C_{35}H_{20}Fe_3N_4O_9$
P.E.Baikie, O.S.Mills *Chem. Communic.,* 1228, 1967

83.210 Ferrous phenanthroline antimony d - tartrate octahydrate
$C_{36}H_{24}FeN_6{}^{2+}$, $C_8H_4O_{12}Sb_2{}^{2-}$, $8H_2O$
D.H.Templeton, A.Zalkin, T.Ueki *Acta Cryst.,* **21,** A154, 1966
Residue 1 also classified in 66

83.211 Pyridinium bis(1 - (4 - bromophenyl) - 3 - methyl - 5 - pyrazolone - 4 - diazo - (3 - methyl - anthranilic acid)) chromate(iii) trihydrate
$C_{36}H_{26}Br_2CrN_8O_6{}^-$, $C_5H_6N^+$, $3H_2O$
H.Jaggi *Helv. Chim. Acta,* **51,** 580, 1968
Residue 1 also classified in 84

83.212 Di - μ - (4,4' - dimethylbenzophenoniminato) bis(tricarbonyl iron)
$C_{36}H_{28}Fe_2N_2O_6$
D.Bright, O.S.Mills *Chem. Communic.*, 245, 1967

83.213 tris(1,3 - Diphenyltriazenido) cobalt(iii) toluene solvate
$C_{36}H_{30}CoN_9$, C_7H_8
M.Corbett, B.F.Hoskins *J. Amer. Chem. Soc.*, **89,** 1530, 1967

83.214 bis - (μ - 4,4' - Dimethylbenzophenone hydrazonato tricarbonyl iron)
$C_{36}H_{30}Fe_2N_4O_6$
M.M.Bagga, P.E.Baikie, O.S.Mills, P.L.Pauson
Chem. Communic., 1106, 1967

83.215 Nickel(ii) bis(β - picoline) nitrite trimer benzene solvate
$C_{36}H_{42}N_{12}Ni_3O_{12}$, C_6H_6
D.M.L.Goodgame, M.A.Hitchman, D.F.Marsham, P.Phavanantha,
D.Rogers *J. Chem. Soc. (D)*, 1383, 1969

83.C Nitrosyldicarbonyl bis(triphenylphosphine) manganese
$C_{38}H_{30}MnNO_3P_2$
For complete entry see 86.85

83.C bis(Triphenylphosphine) di(2,2,2 - trifluoro - 1 - iminoethyl)amine platinum
$C_{40}H_{31}F_6N_3P_2Pt$
For complete entry see 86.92

83.216 μ - Oxo - di(bis(2 - methyl - 8 - quinolinolato) aluminium(iii)
$C_{40}H_{32}Al_2N_4O_5$
Y.Kushi, Q.Fernando *J. Chem. Soc. (D)*, 555, 1969
Also classified in 84

83.C Iodo (tris(2 - diphenylphosphinoethyl)amine) cobalt(ii) iodide
$C_{42}H_{42}CoINP_3^+$, I^-
For complete entry see 86.95

83.217 tris - (2,2',2'' - Terpyridyl) europium(iii) perchlorate (polyhedral form)
$C_{45}H_{33}EuN_9^{3+}$, $3ClO_4^-$
G.H.Frost, F.A.Hart, C.Heath, M.B.Hursthouse
J. Chem. Soc. (D), 1421, 1969

83.218 tetrakis(1,3 - Diphenyltriazenido) dinickel(ii)
$C_{48}H_{40}N_{12}Ni_2$
M.Corbett, B.F.Hoskins *Chem. Communic.*, 1602, 1968

83.C   **Cobalt tris(triphenylphosphine) hydride nitrogen complex diethylether solvate**

$C_{54}H_{46}CoN_2P_3$, $C_4H_{10}O$

For complete entry see 86.112

83.219 μ - **Chloro - bis(chlorotetra(benzimidazole) nickel(ii)) chloride tetra(acetone) solvate**

$C_{56}H_{48}Cl_3N_{16}Ni^+$, Cl^- , $4C_3H_6O$

M.G.B.Drew, D.H.Templeton, A.Zalkin   *Inorg. Chem.*, **7**, 2618, 1968

METAL COMPLEXES (OXYGEN LIGAND)

84.1   **Copper(ii) nitrate - nitromethane complex**
$CH_3CuN_3O_8$
B.Duffin, S.C.Wallwork *Acta Cryst.*, **20**, 210, 1966

84.C   **Tetraphenylarsonium oxo - tetrabromo - acetonitrile rhenate(v)**
$C_2H_3Br_4NORe^-$, $C_{24}H_{20}As^+$
For complete entry see 83.1

84.2   **Monomethylurea cadmium chloride**
$(C_2H_6CdCl_2N_2O)_n$
M.Nardelli, L.Coghi, G.Azzoni *Gazz. Chim. Ital.*, **88**, 235, 1958
See also *Int. Distances,* M 86s; *Structure Reports,* **22**, 640, 1958

84.3   **Cadmium chloride - formamide complex**
$C_2H_6CdCl_2N_2O_2$
A.Mitschler, J.Fischer, R.Weiss *Acta Cryst.*, **22**, 236, 1967

84.C   **bis(Hydrazinecarboxylato - N',O) cadmium**
$C_2H_6CdN_4O_4$
For complete entry see 83.4

84.C   **bis(Hydrazinecarboxylato - N',O) cadmium monohydrate**
$C_2H_6CdN_4O_4$, H_2O
For complete entry see 83.5

84.4   **Dimethylnitrosamine copper(ii) chloride**
$(C_2H_6Cl_2CuN_2O)_n$
U.Klement *Acta Cryst. (B)*, **25**, 2460, 1969

84.C   **bis(Hydrazinecarboxylato - N',O) - manganese(ii) dihydrate**
$C_2H_6MnN_4O_4$, $2H_2O$
For complete entry see 83.7

84.5 bisUrea cadmium dichloride
$C_2H_8CdCl_2N_4O_2$
M.Nardelli, L.Cavalca, G.Fava   *Gazz. Chim. Ital.*, **87**, 137, 1957
See also *Int. Distances*, M 89s; *Structure Reports*, **21**, 529, 1957

84.6 Methanol mercury(ii) chloride complex
$C_2H_8Cl_2HgO_2$
H.Brusset, F.Madaule-Aubry   *Bull. Soc. Chim. Fr.*, 3121, 1966

84.C Copper(ii) chloride bis(semicarbazide) complex
$C_2H_{10}CuN_6O_2{}^{2+}$, $2Cl^-$
For complete entry see 83.8

84.C Zinc chloride bis(semicarbazide) complex
$C_2H_{10}N_6Zn^{2+}$, $2Cl^-$
For complete entry see 83.9

84.C Zinc hydrazinecarboxylate hydrazine complex
$C_2H_{14}N_8O_4Zn$
For complete entry see 83.10

84.C Hydrazinium tris(hydrazinecarboxylato - N,O) - nickelate(ii) monohydrate
$C_3H_9N_6NiO_6{}^-$, $H_5N_2{}^+$, H_2O
For complete entry see 83.13

84.C Hydrazinium tris(hydrazinecarboxylato - N,O) - nickelate(ii) monohydrate
$C_3H_9N_6NiO_6{}^-$, $H_5N_2{}^+$, H_2O
For complete entry see 83.14

84.C Potassium zinc hydrazinecarboxylate
$C_3H_9N_6O_6Zn^-$, K^+
For complete entry see 83.15

84.C Ammonium diperoxodioxalato niobate monohydrate
$C_4NbO_{12}{}^{3-}$, $3H_4N^+$, H_2O
For complete entry see 81.16

84.C Copper(ii) nitrate bis(acetonitrile) complex
$C_4H_6CuN_4O_6$
For complete entry see 83.22

84.C cis - bis(Thioacetohydroxamato) nickel(ii)
$C_4H_8N_2NiO_2S_2$
For complete entry see 85.18

84.7   trans - bis(Thioacetohydroxamato) nickel(ii)
$C_4H_8N_2NiO_2S_2$
T.Sato, Y.Tsukuda, M.Shiro, H.Koyama   *J. Chem. Soc. (B)*, 125, 1969
Also classified in 85

84.C   Silver perchlorate - dioxan
$3C_4H_8O_2$, Ag^+ , ClO_4^-
For complete entry see 60.51

84.C   Dibromo bis(1,2 - dimethoxyethane) titanium(iii) tetrabromo(1,2 - dimethoxyethane) - titanate(iii)
$C_4H_{10}Br_4O_2Ti^-$, $C_8H_{20}Br_2O_4Ti^+$
For complete entry see 84.27

84.C   bisAcetamide cadmium(ii) chloride
$C_4H_{10}CdCl_2N_2O_2$
For complete entry see 83.28

84.8   bis - Biuret cadmium chloride
$C_4H_{10}CdCl_2N_6O_4$
L.Cavalca, M.Nardelli, G.Fava   *Acta Cryst.*, **13**, 594, 1960

84.9   bis - Biuret - zinc chloride
$C_4H_{10}Cl_2N_6O_4$
M.Nardelli, G.Fava, G.Giraldi   *Acta Cryst.*, **16,** 343, 1963

84.10   Copper(ii) chloride - biuret complex
$C_4H_{10}CuN_6O_4^{2+}$, $2Cl^-$
H.C.Freeman, J.E.W.L.Smith   *Acta Cryst.*, **20**, 153, 1966

84.C   Tetramethylenedinitramine nickel(ii) dihydrate
$C_4H_{10}N_4NiO_5$
For complete entry see 83.29

84.C   cis - bis(Dimethylsulfoxide) palladium(ii) nitrate
$C_4H_{12}N_2O_8PdS_2$
For complete entry see 85.31

84.11   Titanium tetramethoxide
$C_4H_{12}O_4Ti$
D.A.Wright, D.A.Williams   *Acta Cryst. (B),* **24**, 1107, 1968

84.12   Tetramethanol dichloro cobalt(ii)
$C_4H_{16}Cl_2CoO_4$
H.Brusset, H.Gillier-Pandraud, I.Bkouche-Waksman
Bull. Soc. Fr. Mineral. Cristallogr., **91,** 549, 1968

84.13 Copper(ii) sulfate ethanolamine complex tetrahydrate
$C_4H_{22}CuN_2O_6{}^{2+}$, O_4S^{2-}
G.K.Abdullaev, Kh.S.Mamedov *Zh. Strukt. Khim.*, **6**, 171, 1965

84.C Di - μ - hydroxo bis(dimethylamine copper(ii)) sulfate monohydrate
$C_4H_{22}Cu_2N_4O_2{}^{2+}$, O_4S^{2-}, H_2O
For complete entry see 83.34

84.14 3,5 - Dibromopyridine oxide - mercury(ii) chloride complex
$C_5H_3Br_2NO$, Cl_2Hg
F.Genet, J.-L.Leguen *Acta Cryst. (B)*, **25**, 2029, 1969

84.15 Croconato triaquo copper(ii)
$C_5H_6CuO_8$
M.D.Glick, G.L.Downs, L.F.Dahl *Inorg. Chem.*, **3**, 1712, 1964

84.16 Croconato triaquo manganese(ii)
$C_5H_6MnO_8$
M.D.Glick, L.F.Dahl *Inorg. Chem.*, **5**, 289, 1966

84.17 Croconato triaquo zinc(ii)
$C_5H_6O_8Zn$
M.D.Glick, G.L.Downs, L.F.Dahl *Inorg. Chem.*, **3**, 1712, 1964

84.C Piperidinium gadolinium tetrakis(benzoylacetonate)
$C_5H_{12}N^+$, $C_{40}H_{36}GdO_8{}^-$
For complete entry see 77.66

84.18 Dioxodichloro bis(N,N - dimethylformamido) molybdenum(vi)
$C_6H_{14}Cl_2MoN_2O_4$
L.R.Florian, E.R.Corey *Inorg. Chem.*, **7**, 722, 1968

84.19 Penta(copper(ii) chloride) di - n - propanol complex
$C_6H_{16}Cl_{10}Cu_5O_2$
R.D.Willett, R.E.Rundle *J. Chem. Phys.*, **40**, 838, 1964

84.20 Penta(copper(ii) chloride) di - n - propanol complex
$C_6H_{16}Cl_{10}Cu_5O_2$
J.V.Ugro *Dissert. Abstr. (B)*, **28**, 1891, 1967

84.C bis(Trimethylamine) oxo vanadium(iv) dichloride
$C_6H_{18}Cl_2N_2OV$
For complete entry see 83.47

84.21 bis(Trimethylphosphine oxide) cobalt(ii) dinitrate
$C_6H_{18}CoN_2O_8P_2$
F.A.Cotton, R.H.Soderberg *J. Amer. Chem. Soc.*, **85**, 2402, 1963

84.22 **Methyl vanadate**
$C_6H_{18}O_8V_4$
C.N.Caughlan, H.M.Smith, K.Watenpaugh *Inorg. Chem.*, **5**, 2131, 1966

84.C **Hexafluoro acetylacetonato - bis(carbonyl) rhodium(i)**
$C_7HF_6O_4Rh$
For complete entry see 77.2

84.C **Acetylacetonato - bis(carbonyl) rhodium(i)**
$C_7H_7O_4Rh$
For complete entry see 77.3

84.C **Triaquo - (2,6 - pyridine dicarboxylato) copper**
$C_7H_9CuO_7$
For complete entry see 83.52

84.23 **Silver perfluorobutyrate dimer**
$C_8Ag_2F_2F_{14}O_4$
A.E.Blakeslee, J.L.Hoard *J. Amer. Chem. Soc.*, **78**, 3029, 1956
See also *Int. Distances*, M 143s; *Structure Reports*, **20**, 477, 1956

84.C **Copper(ii) violurate tetrahydrate**
$C_8H_4CuN_6O_8$, $4H_2O$
For complete entry see 83.53

84.24 **Iron chloride - di(dimethyl sulphide) complex**
$C_8H_{12}Cl_2FeO_4S_4{}^+$, Cl_4Fe^-
M.J.Bennett, F.A.Cotton, D.L.Weaver *Acta Cryst.*, **23**, 581, 1967

84.25 **Manganese(ii) dichlorophosphate di - ethylacetate**
$C_8H_{16}Cl_4MnO_8P_2$
J.Danielsen, S.E.Rasmussen *Acta Chem. Scand.*, **17**, 1971, 1963

84.26 **Titanium(iv) chloride - ethyl acetate complex**
$C_8H_{16}Cl_8O_4Ti_2$
L.Brun *Acta Cryst.*, **20**, 739, 1966

84.27 **Dibromo bis(1,2 - dimethoxyethane) titanium(iii) tetrabromo(1,2 - dimethoxyethane) - titanate(iii)**
$C_8H_{20}Br_2O_4Ti^+$, $C_4H_{10}Br_4O_2Ti^-$
G.W.A.Fowles, T.E.Lester, J.S.Wood *J. Inorg. Nucl. Chem.*, **31**, 657, 1969
Residue 2 classified in 84

84.28 **Diethoxy titanium dichloride**
$C_8H_{20}Cl_4O_4Ti_2$
W.Haase, H.Hoppe *Acta Cryst. (B)*, **24**, 281, 1968

84.C bis - (β - Aminobutyrate) - diaquo copper(ii)
$C_8H_{20}CuN_2O_6$
For complete entry see 83.62

84.29 Iron(iii) dichloride tetra(dimethyl sulfoxide) tetrachloro ferrate(iii)
$C_8H_{24}Cl_2FeO_4S_4^+$, Cl_4Fe^-
M.J.Bennett, F.A.Cotton, D.L.Weaver *Nature,* **212,** 286, 1966

84.30 Mercury(ii) chloride - quinoline - N - oxide complex
$C_9H_7Cl_2HgNO$
A.T.McPhail, G.A.Sim *Chem. Communic.,* 21, 1966

84.31 Di - μ - (pyridine oxide) - bis (copper(ii) chloride)
$C_{10}H_{10}Cl_4Cu_2N_2O_2$
H.L.Schafer, J.C.Morrow, H.M.Smith *J. Chem. Phys.,* **42,** 504, 1965

84.32 Di - μ - (pyridine oxide) - bis (copper(ii) chloride)
$C_{10}H_{10}Cl_4Cu_2N_2O_2$
R.S.Sager, R.J.Williams, W.H.Watson *Inorg. Chem.,* **6,** 951, 1967

84.C Potassium copper(ii) ethylenediamine tetra - acetate trihydrate
$C_{10}H_{12}CuN_2O_8^{2-}$, $2K^+$, $3H_2O$
For complete entry see 76.60

84.C Sodium terbium(iii) ethylenediamine - tetra - acetate octahydrate
$C_{10}H_{12}N_2O_8Tb^-$, Na^+
For complete entry see 76.62

84.C Lithium monaquo - ethylenediaminetetra - acetato ferrate(iii) dihydrate
$C_{10}H_{14}FeN_2O_9^-$, Li^+, $2H_2O$
For complete entry see 76.63

84.C Rubidium monoaquo - ethylenediaminetetra - acetato ferrate(iii) monohydrate
$C_{10}H_{14}FeN_2O_9^-$, Rb^+
For complete entry see 76.64

84.C Manganese(ii) monohydrogen ethylenediamine tetra - acetate decahydrate
bisHydrogen (ethylenedinitrilo) tetra - acetato trimanganese decahydrate
$2C_{10}H_{14}MnN_2O_9^-$, $H_8MnO_4^{2+}$, $4H_2O$
For complete entry see 76.65

84.C Hydrogen aquoethylenediaminetetra - acetato chromate(iii)
$C_{10}H_{15}CrN_2O_9$
For complete entry see 76.66

84.C **Hydrogen aquoethylenediaminetetra - acetato ferrate(iii)**
$C_{10}H_{15}FeN_2O_9$
For complete entry see 76.67

84.C **Potassium lanthanum(iii) ethylenediaminetetra - acetate octahydrate**
$C_{10}H_{18}KLaN_2O_{11}$, $5H_2O$
For complete entry see 76.71

84.C **Lanthanum(iii) hydrogen ethylenediaminetetra - acetate septahydrate**
$C_{10}H_{21}LaN_2O_{12}$, $3H_2O$
For complete entry see 76.72

84.C **Acetylacetone - mono - (o - hydroxyanil) copper(ii)**
$C_{11}H_{11}CuNO_2$
For complete entry see 83.90

84.33 **Copper(ii) nitrosophenylhydroxylamine**
$C_{12}H_{10}Cu\dot{N}_4O_4$
L.M.Shkol'nikova, E.A.Shugam *Zh. Strukt. Khim.*, **4,** 380, 1963

84.C **bis(Pyridine - 2 - carboxamido) nickel(ii) dihydrate**
$C_{12}H_{10}N_4NiO_2$, $2H_2O$
For complete entry see 83.96

84.C **Nickel(ii) picolinate tetrahydrate**
$C_{12}H_{12}N_2NiO_6$, $2H_2O$
For complete entry see 83.98

84.C **Potassium bis(nitrilotriacetato) zirconate(iv) monohydrate**
$C_{12}H_{12}N_2O_{12}Zr^{2-}$, $2K^+$, H_2O
For complete entry see 81.70

84.C **trans - Diaquo - bis(pyridine - 2 - carboxamide) nickel(ii) chloride**
$C_{12}H_{16}N_4NiO_4{}^{2+}$, $2Cl^-$
For complete entry see 83.105

84.34 **Polybis(μ(2 - picoline N - oxide) chloro copper(ii) di - μ - chloro) diaquo copper(ii)**
$(C_{12}H_{18}Cl_6Cu_3N_2O_4)_n$
R.S.Sager, W.H.Watson *Inorg. Chem.*, **7,** 2035, 1968

84.C **N,N' - Ethylene bis(acetylacetoneiminato) copper(ii) monohydrate**
$C_{12}H_{18}CuN_2O_2$, H_2O
For complete entry see 83.108

84.C **N,N′ - Ethylene - bis(acetylacetoneiminato) copper(ii) hemihydrate**
$C_{12}H_{18}CuN_2O_2$, $0.5H_2O$
For complete entry see 83.109

84.C **Copper(ii) N,N′ - ethylene - bis(acetylacetone - iminate) methylammonium perchlorate**
$C_{12}H_{18}CuN_2O_2$, CH_6N^+ , $ClO_4{}^-$
For complete entry see 83.110

84.C **Copper(ii) 3 - methylpentane - 2,4 - dionate (at −93 ° C)**
$C_{12}H_{18}CuO_4$
For complete entry see 77.14

84.C **Copper(ii) ethylacetoacetate**
$C_{12}H_{18}CuO_6$
For complete entry see 77.15

84.C **Copper(ii) ethylacetoacetate (discussion)**
$C_{12}H_{18}CuO_6$
For complete entry see 77.16

84.C **(N,N′ - Ethylene bis(1 - acetonylethylideniminato) - O,O′,N,N′) - oxo vanadium(iv)**
$C_{12}H_{18}N_2O_3V$
For complete entry see 83.111

84.C **NN, - Ethylene bis(acetylacetoneiminato) copper(ii)**
$C_{12}H_{20}CuN_2O_2$
For complete entry see 83.112

84.C **Aquo - N,N′ - ethylene bis(acetylacetoneiminato) copper(ii)**
$C_{12}H_{20}CuN_2O_3$
For complete entry see 77.17

84.C **bis - 1 - Aminocyclopentane carboxylato copper(ii)**
$C_{12}H_{20}CuN_2O_4$
For complete entry see 83.113

84.C **Trimethyl(nonane - 4,6 - dionato) platinum(iv) (neutron diffraction)**
$C_{12}H_{24}O_2Pt$
For complete entry see 77.18

84.35 **Uranyl nitrate - triethyl phosphate complex**
$C_{12}H_{30}N_2O_{16}P_2U$
J.E.Fleming, H.Lynton   *Chem. and Industry*, 1415, 1960

84.C Dibromo - (1 - (o - methoxyphenyl) - 2,6 - diazaoctane) nickel(ii)
$C_{13}H_{22}Br_2N_2NiO$
For complete entry see 83.127

84.36 Copper(ii) tropolone
$C_{14}H_{10}CuO_4$
W.M.Macintyre, J.M.Robertson, R.F.Zahrobsky
Proc. R. Soc., A, **289,** 161, 1965

84.37 Mercury(ii) chloride - azoxyanisole complex acetone solvate
$C_{14}H_{14}Cl_2HgN_2O_3$, $0.5H_2O$
A.T.McPhail, G.A.Sim *Chem. Communic.,* 21, 1966

84.38 Dichloro bis(2,6 - lutidine - N - oxide) copper(ii)
$C_{14}H_{18}Cl_2CuN_2O_2$
R.S.Sager, W.H.Watson *Inorg. Chem.,* **8,** 308, 1969

84.39 Dichloro bis(2,6 - lutidine N - oxide) zinc(ii)
$C_{14}H_{18}Cl_2N_4O_2$
R.S.Sager, W.H.Watson *Inorg. Chem.,* **7,** 1358, 1968

84.C Calcium trans - 1,2 - diaminocyclohexane - N,N' - tetra - acetato aquo ferrate(iii) octahydrate
$2C_{14}H_{20}FeN_2O_4{}^-$, Ca^{2+} , $8H_2O$
For complete entry see 83.137

84.40 bis(Toluene - p - sufinato) tetra - aquo copper(ii)
$C_{14}H_{24}CuO_8S_2$
D.A.Langs, C.R.Hare *Chem. Communic.,* 853, 1967

84.C (tetrakis - Dimethylformamido) nickel di - isoselenocyanate
$C_{14}H_{28}N_6NiO_4Se_2$
For complete entry see 83.138

84.C Copper(ii) 1 - (2 - pyridylazo) - 2 - naphtholato perchlorate monohydrate
$C_{15}H_{12}CuN_3O_2{}^+$, $ClO_4{}^-$
For complete entry see 83.140

84.41 pentakis(Trimethylarsine oxide) nickel(ii) perchlorate
$C_{15}H_{45}As_5NiO_5{}^{2+}$, $2ClO_4{}^-$
S.H.Hunter, K.Emerson, G.A.Rodley *J. Chem. Soc. (D),* 1398, 1969

84.42 Copper(ii) ω - nitroacetophenonate
$C_{16}H_{12}CuN_2O_6$
M.Bonamico, I.Collamati, C.Ercolani, G.Dessy
Chem. Communic., 654, 1967

84.C   **Nickel(ii) dehydroacetic acid monoimide complex**
$C_{16}H_{16}N_2NiO_6$
For complete entry see 83.144

84.C   **($\pm$) - Phenylalanine - (pyridoxylidene - 5 - phosphate) aquo copper(ii)**
$C_{17}H_{18}CuN_2O_8P$
For complete entry see 82.57

84.C   **Copper(ii) 8 - hydroxyquinolinate (β form)**
$C_{18}H_{12}CuN_2O_2$
For complete entry see 83.155

84.C   **Copper(ii) 8 - hydroxyquinolinate (β form)**
$C_{18}H_{12}CuN_2O_2$
For complete entry see 83.156

84.C   **Copper(ii) 8 - hydroxyquinolinate (β form)**
$C_{18}H_{12}CuN_2O_2$
For complete entry see 83.157

84.C   **Copper(ii) 8 - hydroxyquinolinate (α form)**
$C_{18}H_{12}CuN_2O_2$
For complete entry see 83.158

84.C   **Copper(ii) 8 - hydroxyquinolinate - bis(1,2,4,5 - tetracyanobenzene) complex (photographic data)**
$C_{18}H_{12}CuN_2O_2 , 2C_{10}H_2N_4$
For complete entry see 60.146

84.C   **Copper(ii) 8 - hydroxyquinolinate - bis(1,2,4,5 - tetracyanobenzene) complex (diffractometer data)**
$C_{18}H_{12}CuN_2O_2 , 2C_{10}H_2N_4$
For complete entry see 60.147

84.C   **7,7,8,8 - Tetracyanoquinodimethane - copper(ii) 8 - hydroxyquinolate complex**
$C_{18}H_{12}CuN_2O_2 , C_{12}H_4N_4$
For complete entry see 60.126

84.43   **Dioxo bis(8 - hydroxyquinolinato) molybdenum(vi)**
$C_{18}H_{12}MoN_2O_4$
L.O.Atovmyan, Yu.A.Sokolova   *J. Chem. Soc. (D)*, 649, 1969
Also classified in 83

84.C 8 - Hydroxyquinolinato palladium(ii)
$C_{18}H_{12}N_2O_2Pd$
For complete entry see 83.159

84.C Chloranil - bis(8 - hydroxyquinolinato) palladium(ii)
$C_{18}H_{12}N_2O_2Pd$, C_6Cl_4O
For complete entry see 60.70

84.C Palladium(ii) 8 - hydroxyquinolinate - 1,2,4,5 - tetracyanobenzene complex (disordered)
$C_{18}H_{12}N_2O_2Pd$, $C_{10}H_2N_4$
For complete entry see 60.148

84.C Zinc 8 - hydroxyquinolinate dihydrate
$C_{18}H_{12}N_2O_2Zn$, $2H_2O$
For complete entry see 83.160

84.C bis(8 - Hydroxyquinoline) silver(i) pyridine solvate
$C_{18}H_{13}AgN_2O_2$, C_5H_5N
For complete entry see 83.161

84.44 Iron cupferron
$C_{18}H_{15}FeN_4O_4$
D.van der Helm, L.L.Merritt, R.Degeilh *Acta Cryst.*, **18**, 355, 1965

84.45 Ammonium uranyl cupferrate
$C_{18}H_{15}N_6O_8U^-$, H_4N^+
W.S.Arrington *Dissert. Abstr.*, **26**, 3411, 1965

84.C Zinc diaquo - 8 - hydroxyquinoline
$C_{18}H_{16}N_2O_4Zn$
For complete entry see 83.162

84.C bis(Methoxyphenyldimethylarsine) trichloro rhodium(iii)
$C_{18}H_{26}As_2Cl_3O_2Rh$
For complete entry see 86.22

84.46 Dichloro tris(tetrahydrofuran) - p - tolyl chromium(iii)
p - Tolylchromium(iii) dichloride tris(tetrahydrofuran)
$C_{19}H_{31}Cl_2CrO_3$
J.J.Daly, R.P.A.Seeden *J. Chem. Soc. (A)*, 736, 1967
Also classified in 71

84.C Copper(ii) benzoylacetonate
$C_{20}H_{16}CuO_4$
For complete entry see 77.40

227

84.C **Palladium(ii) bis(1 - phenyl - 1,3 - butanedionate)**
$C_{20}H_{18}O_4Pd$
For complete entry see 77.41

84.C **Vanadyl 1 - phenyl - 1,3 - butanedionate**
$C_{20}H_{18}O_5V$
For complete entry see 77.42

84.C **Vanadyl 1 - phenyl - 1,3 - butanedionate (discussion)**
$C_{20}H_{18}O_5V$
For complete entry see 77.43

84.C **bis(3 - Amino - 1 - phenyl - 2 - buten - 1 - ono) copper(ii)**
Copper(ii) benzoylacetone - iminate
$C_{20}H_{20}CuN_2O_2$
For complete entry see 83.179

84.47 **Tetra(pyridine oxide) copper(ii) fluoroborate**
$C_{20}H_{20}CuN_4O_4{}^{2+}, 2BF_4{}^-$
J.D.Lee, D.S.Brown, B.G.A.Melsom *Acta Cryst. (B)*, **25**, 1595, 1969

84.48 **Tetra(pyridine oxide) copper(ii) perchlorate**
$C_{20}H_{20}CuN_4O_4{}^{2+}, 2ClO_4{}^-$
J.D.Lee, D.S.Brown, B.G.A.Melsom *Acta Cryst. (B)*, **25**, 1378, 1969

84.49 **bis(Pyridine - N - oxide) copper(ii) nitrate**
$C_{20}H_{20}Cu_2N_8O_{16}$
S.Scavnicar, B.Matkovic *Acta Cryst. (B)*, **25**, 2046, 1969

84.C **bis(Acetylacetonato) bis(pyridine N - oxide) nickel(ii)**
bis(2,4 - Pentanedionato) bis(pyridine N - oxide) nickel(ii)
$C_{20}H_{24}N_2NiO_6$
For complete entry see 77.46

84.C **Copper(ii) bis - 1 - ephedrine - benzene**
$C_{20}H_{26}CuN_2O_2, 0.67C_6H_6$
For complete entry see 61.21

84.C **bis(α - Hydroxy - α - phenylbutyramidine) copper(ii) dihydrate**
$C_{20}H_{26}CuN_4O_2, 2H_2O$
For complete entry see 83.181

84.C **bis - (+) - Pseudoephedrine copper(ii) (+) - pseudoephedrine dipydrate**
$C_{20}H_{28}CuN_2O_2, C_{10}H_{15}NO, 2H_2O$
For complete entry see 83.182

84.50 **Zinc n - butylphenylphosphinate (monoclinic form)**
$(C_{20}H_{28}O_4P_2Zn)_n$
F.Giordano, L.Randaccio, A.Ripamonti
Acta Cryst. (B), **25**, 1057, 1969
Also classified in 64

84.51 **Zinc n - butylphenylphosphinate (orthorhombic form)**
$(C_{20}H_{28}O_4P_2Zn)_n$
F.Giordano, L.Randaccio, A.Ripamonti
Acta Cryst. (B), **25**, 1057, 1969
Also classified in 64

84.C **trans - Oxotrichloro bis(diethylphenylphosphine) rhenium(v)**
$C_{20}H_{30}Cl_3OP_2Re$
For complete entry see 86.27

84.C **Ethylenediammonium N - hydroxyethyl ethylenediamine - triacetato iron(iii) - μ - oxo - N - hydroxyethyl - ethylenediamine triacetato ferrate(iii) hexahydrate**
$C_{20}H_{30}Fe_2N_4O_{15}{}^{2-}, C_2H_{10}N_2{}^{2+}, 6H_2O$
For complete entry see 83.183

84.52 **tris(Tropolonato) iron(iii)**
Ferric tropolone
$C_{21}H_{15}FeO_6$
T.A.Hamor, D.J.Watkin   *J. Chem. Soc. (D)*, 440, 1969

84.53 **Iron(iii) benzhydroxamate trihydrate**
$C_{21}H_{18}FeN_3O_6, 3H_2O$
H.J.Lindner, S.Gottlicher   *Acta Cryst. (B)*, **25**, 832, 1969

84.C **Tribenzo(b,f,j,)(1.5.9)triazacycloduodecine nickel(ii) nitrate**
$C_{21}H_{19}N_4NiO_5{}^+, NO_3{}^-$
For complete entry see 83.184

84.C **4 - Methylpyridine bis(o - hydroxyacetophenonato) copper(ii)**
$C_{22}H_{21}CuNO_4$
For complete entry see 77.54

84.C **Copper(ii) 3 - phenyl - 2,4 - pentanedionate**
$C_{22}H_{22}CuO_4$
For complete entry see 77.55

84.C **N,N' - Ethylene bis(L - ephedrine) copper(ii) dihydrate**
$C_{22}H_{30}CuN_2O_2, 2H_2O$
For complete entry see 83.187

84.C   **bis(Dipivaloylmethanido) nickel(ii)**
$C_{22}H_{38}NiO_4$
For complete entry see 77.56

84.C   **bis(Dipivaloylmethanido) zinc(ii)**
$C_{22}H_{38}O_4Zn$
For complete entry see 77.57

84.C   **Chloro - π - cyclopentadienyl bis(8 - quinolinolato) titanium(iv)**
$C_{23}H_{17}ClN_2O_2Ti$
For complete entry see 73.100

84.C   **Chromium(iii) pyridinium O,O' - dihydroxy - trans - azobenzene complex**
$C_{24}H_{16}CrN_4O_4{}^-$, $C_5H_6N^+$
For complete entry see 83.188

84.54   **Dichlorodiphenoxy titanium(iv)**
$C_{24}H_{20}Cl_4 O_4Ti_2$
K.Watenpauch, C.N.Caughlan   *Inorg. Chem.*, **5**, 1782, 1966

84.C   **bis - (3 - Hydroxyl - 1,3 - diphenyltriazene) palladium**
$C_{24}H_{20}N_6O_2Pd$
For complete entry see 83.189

84.55   **Cobalt(ii) diphenylmonothiophosphinate**
$(C_{24}H_{20}O_2P_2S_2)_n$
S.Bruckner, M.Calligaris, G.Nardin, L.Randaccio, A.Ripamonti
J. Chem. Soc. (D), 474, 1969
Also classified in 85

84.C   **Trimethyl - (8 - quinolinolato) platinum(iv)**
$C_{24}H_{30}N_2O_2Pt_2$
For complete entry see 83.191

84.C   **Trimethyl 4,6 dioxonyl platinum**
$C_{24}H_{48}O_4Pt_2$
For complete entry see 77.58

84.56   **tris(Octamethylpyrophosphoramide) copper(ii) perchlorate**
$C_{24}H_{72}CuN_{12}O_6P_6{}^{2+}$, $2ClO_4{}^-$
M.D.Jeosten, M.S.Hussain, P.G.Lenhert, J.H.Venable
J. Amer. Chem. Soc., **90**, 5623, 1968

84.57   **Palladium(ii) 2 - (o - hydroxyphenyl)benzoxazole**
$C_{26}H_{16}N_2O_4Pd$
C.E.Urdy   *Dissert. Abstr.*, **24**, 4015, 1964

84.C 2,2 - Diphenyl bis - (2 - iminomethylenephenolato) copper(ii)
$C_{26}H_{22}CuN_2O_2$
For complete entry see 83.192

84.C Uranyl 8 - quinolinolate 8 - quinolinol complex
$C_{27}H_{19}N_3O_5U$
For complete entry see 83.193

84.C Dioxo di - 8 - quinolinolato - 8 - quinolinol uranium(vi) chloroform solvate
$C_{27}H_{19}N_3O_5U$, $CHCl_3$
For complete entry see 83.194

84.C Dichloro bis(8 - quinolinato) titanium(iv)
$C_{28}H_{12}Cl_2N_2O_2Ti$
For complete entry see 83.196

84.C tris(8 - Quinolinolato) chromium methanol complex
$C_{28}H_{22}CrN_3O_4$
For complete entry see 83.197

84.C bis(N - o - Methoxyphenyl - salicylaldiminato) zinc
$C_{28}H_{24}N_2O_4Zn$
For complete entry see 78.55

84.58 Methyltriethyl titanate
$C_{28}H_{42}O_{16}Ti_4$
R.D.Witters, C.N.Caughlan *Nature,* **205,** 1312, 1965

84.59 (N - (2 - Hydroxyethyl) - acetylacetoniminato) copper(ii) tetramer
$C_{28}H_{44}Cu_4N_4O_8$
J.A.Bertrand, J.A.Kelly, C.E.Kirkwood *Chem. Communic.,* 1329, 1968
Also classified in 83

84.C bis(Nickel(ii) N,N - bis(2 - diethylaminoethyl) - 2 - hydroxyethylamine)
perchlorate
$C_{28}H_{64}N_6Ni_2O_2^{2+}$, $2ClO_4^-$
For complete entry see 83.201

84.C Copper(ii) 1,3 - diphenyl - 1,3 - propanedionate
$C_{30}H_{22}CuO_4$
For complete entry see 77.59

84.C Palladium bis(1,3 - diphenyl - propane - 1 - one - 3 - thionate)
$C_{30}H_{22}O_2PdS_2$
For complete entry see 85.92

84.C   **Palladium(ii) dibenzoylmethanate**
$C_{30}H_{22}O_4Pd$
For complete entry see 77.60

84.60   **Nickel(ii) monothiobenzoate ethanol complex**
$C_{30}H_{26}Ni_2O_5S_4$
M.Bonamico, G.Dessy, V.Fares   *J. Chem. Soc. (D)*, 697, 1969
Also classified in 85

84.C   **tris(1 - Phenyl - 1,3 - butanedionato) aquo yttrium(iii)**
$C_{30}H_{29}O_7Y$
For complete entry see 77.61

84.C   **Ammonium tetrakis(4,4,4 - trifluoro - 1 - (2 - thienyl) - 1,3 - butanedione) praseodymate(iii) monohydrate**
$C_{32}H_{16}F_{12}O_8PrS_4{}^-$, H_4N^+, H_2O
For complete entry see 77.65

84.C   **Pyridinium bis(2 - hydroxy naphthalene - 1 - azo) - (2' - hydroxy - 4' - nitrobenzene) chromium(iii) pyridine solvate**
$C_{32}H_{18}CrN_6O_8{}^-$, $C_5H_6N^+$, $0.5C_5H_5N$
For complete entry see 83.206

84.C   **Copper(ii) bis - (benzeneazo - β - naphthol)**
$C_{32}H_{22}CuN_4O_2$
For complete entry see 83.207

84.C   **N,N' - Ethylene bis(salicylaldehydeiminato) cobalt(ii) (inactive form)**
$C_{32}H_{28}Co_2N_4O_4$
For complete entry see 78.59

84.61   **Titanium(iv) ethoxide**
$C_{32}H_{80}O_{16}Ti_4$
J.A.Ibers   *Nature*, **197**, 686, 1963

84.C   **bis(1 - m - Tolylazo - 2 - naphtholato) nickel(ii)**
$C_{34}H_{26}N_4NiO_2$
For complete entry see 83.208

84.C   **Pyridinium bis(1 - (4 - bromophenyl) - 3 - methyl - 5 - pyrazolone - 4 - diazo - (3 - methyl - anthranilic acid)) chromate(iii) trihydrate**
$C_{36}H_{26}Br_2CrN_8O_6{}^-$, $C_5H_6N^+$, $3H_2O$
For complete entry see 83.211

84.62 **Di(triphenylarsenic oxide) mercury(ii) chloride**
$C_{36}H_{30}As_2Cl_2HgO_2$
C.-I.Branden *Acta Chem. Scand.,* **17,** 1363, 1963

84.63 **Triphenylarsine oxide mercury(ii) chloride complex**
$C_{36}H_{30}As_2Cl_4Hg_2O_2$
C.-I.Branden *Ark. Kemi,* **22,** 485, 1964

84.C μ - **Dioxygen - di(N,N' - ethylene bis(salicylaldehydeiminato) cobalt(ii) dimethylformamide)**
$C_{38}H_{42}Co_2N_6O_8$
For complete entry see 78.62

84.64 **bis(Testosterone) - mercury(ii) chloride complex (absolute configuration)**
$C_{38}H_{56}Cl_2HgO_4$
A.Cooper, E.M.Gopalakrishna, D.A.Norton
Acta Cryst. (B), **24,** 935, 1968
Also classified in 51

84.C **bis(Triphenylphosphine) platinum(0) oxygen acetone complex**
$C_{39}H_{36}O_3P_2Pt$
For complete entry see 86.89

84.C μ - **Oxo - di(bis(2 - methyl - 8 - quinolinolato) aluminium(iii)**
$C_{40}H_{32}Al_2N_4O_5$
For complete entry see 83.216

84.65 **Niobium oxide ethoxide**
$C_{40}H_{100}Nb_8O_{30}$
D.C.Bradley, M.B.Hursthouse, P.F.Rodesiler
Chem. Communic., 1112, 1968

84.C **Ferrichrome - A tetrahydrate**
$C_{41}H_{58}FeN_9O_{20}$, $4H_2O$
For complete entry see 82.59

84.66 **Cobalt(ii) diethoxyphosphonylacetylmethanide**
bis(tris(Diethoxy phosphonyl acetyl methano) cobalt) cobalt
$C_{42}H_{84}Co_3O_{24}P_6$
F.A.Cotton, R.Hugel, R.Eiss *Inorg. Chem.,* **7,** 18, 1968

84.67 **Mercury(ii) chloride - cis - 4 - p - chlorophenylthian**
$C_{44}H_{48}Cl_4Hg_2O_4S_4$
R.S.McEwen, G.A.Sim, C.R.Johnson *Chem. Communic.,* 885, 1967

84.C   **Di(μ - diphenylphosphinatoacetylacetonato chromium(iii))**
$C_{44}H_{48}Cr_2O_{12}P_2$
For complete entry see 77.69

84.68   **Cobalt(ii) perchlorate - diphenylmethylarsine oxide complex**
$C_{52}H_{52}As_4ClCoO_8{}^+$, $ClO_4{}^-$
P.Pauling, G.B.Robertson, G.A.Rodley   *Nature,* **207,** 73, 1965

84.69   **μ(4) - Oxo - hexa - μ - chloro - tetrakis((triphenylphosphine oxide) copper(ii))**
$C_{72}H_{60}Cl_6Cu_4O_5P_4$
J.A.Bertrand   *Inorg. Chem.*, **6,** 495, 1967

METAL COMPLEXES
(SULPHUR OR SELENIUM LIGAND)

85.1 Nickel(ii) S - methylaminodithionitride S - aminodithionitride
$CH_4N_4NiS_4$
J.Weiss, M.Ziegler   *Z. Naturforsch., B,* **21,** 891, 1966
Also classified in 83

85.2 Mono(thiosemicarbazide) silver(i) chloride
CH_5AgClN_3S
G.F.Gasparri, A.Mangia, A.Musatti, M.Nardelli
Acta Cryst. (B), **24,** 367, 1968

85.3 Thiosemicarbazide zinc chloride
$CH_5Cl_2N_3SZn$
L.Cavalca, M.Nardelli, G.Branchi   *Acta Cryst.,* **13,** 688, 1960
Also classified in 83

85.C Tetraphenylarsonium bis(trithiocarbonato) nickelate(ii)
$C_2NiS_6{}^{2-}$, $2C_{24}H_{20}As^+$
For complete entry see 65.46

85.4 Mercury methylmercaptide
$C_2H_6HgS_2$
D.C.Bradley, N.R.Kunchur   *J. Chem. Phys.,* **40,** 2258, 1964

85.5 bis - Thiosemicarbazidato nickel(ii) (red form)
$C_2H_8N_6Ni_2S_2$
L.Cavalca, M.Nardelli, G.Fava   *Acta Cryst.,* **15,** 1139, 1962
Also classified in 83

85.6 Nickel(ii) di(thiosemicarbazide) sulfate (β form)
$C_2H_{10}N_6NiS_2{}^{2+}$, O_4S^{2-}
R.G.Hazell   *Acta Chem. Scand.,* **22,** 2171, 1968
Residue 1 also classified in 83

85.7 **Nickel(ii) dithiosemicarbazide sulfate trihydrate (α form)**
$C_2H_{10}N_6NiS_2{}^{2+}$, O_4S^{2-}, $3H_2O$
R.Gronbaek, S.E.Rasmussen *Acta Chem. Scand.*, **16**, 2325, 1962
Residue 1 also classified in 83

85.8 **Diaquo di(thiosemicarbazide) nickel(ii) dinitrate**
$C_2H_{14}N_6NiO_2S_2{}^{2+}$, $2NO_3{}^-$
R.G.Hazell *Acta Chem. Scand.*, **22**, 2809, 1968
Residue 1 also classified in 83

85.C **5 - Methoxy - 1,3,8,10 - tetrathia - 2,4,7,9 - tetra - azadec - 5 - ene - 4,7 - diato - S(1),S(10) - nickel(ii)**
$C_3H_4N_4NiOS_4$
For complete entry see 83.11

85.9 **1,3,5 - Trithian - silver(i) perchlorate monohydrate**
$C_3H_6AgS_3{}^+$, $ClO_4{}^-$, H_2O
R.S.Ashworth, C.K.Prout, A.Domenicano, A.Vaciago
J. Chem. Soc. (A), 93, 1968

85.10 **1,3,5 - Trithian - silver(i) nitrate**
$C_3H_6AgS_3{}^+$, $NO_3{}^-$
R.S.Ashworth, C.K.Prout, A.Domenicano, A.Vaciago
J. Chem. Soc. (A), 93, 1968

85.11 **1,3,5 - Trithian - silver(i) nitrate monohydrate**
$C_3H_6AgS_3{}^+$, $NO_3{}^-$, H_2O
R.S.Ashworth, C.K.Prout, A.Domenicano, A.Vaciago
J. Chem. Soc. (A), 93, 1968

85.12 **1,3,5 - Trithian mercury(ii) chloride complex**
$C_3H_6Cl_2Cl_2Hg$
W.R.Costello, A.T.McPhail, G.A.Sim *J. Chem. Soc. (A)*, 1190, 1966

85.C **5 - Hydroxy - 6 - methoxy - 1,3,8,10 - tetrathia - 2,4,7,9 - tetra - azadecane - 4,7 - diato - S(1),S(10) - nickel(ii)**
$C_3H_6N_4NiO_2S_4$
For complete entry see 83.12

85.13 **Trimethyl sulfonium mercury tri - iodide**
$C_3H_9HgI_3S$
R.H.Fenn *Acta Cryst.*, **20**, 20, 1966

85.14 **bis(Thiosemicarbazide) silver(i) thiocyanate**
$C_3H_{10}AgN_7S_3$
L.C.Capacchi, G.F.Gasparri, M.Ferrari, M.Nardelli
Chem. Communic., 910, 1968
Also classified in 83

85.15 **Nickel trithiosemicarbazide dinitrate**
$C_3H_{15}N_9NiS_3{}^{2+}$, $2NO_3{}^-$
R.G.Hazell *Acta Cryst.*, **21**, A142, 1966
Residue 1 also classified in 83

85.16 **Nickel trithiosemicarbazide dinitrate monohydrate**
$C_3H_{15}N_9NiS_3{}^{2+}$, $2NO_3{}^-$, H_2O
R.G.Hazell *Acta Cryst.*, **21**, A_ 1966
Residue 1 also classified in 83

85.C **Tetraphenylarsonium bis(N - cyanodithio) nickelate(ii)**
$C_4N_4NiS_4{}^{2-}$, $2C_{24}H_{20}As^+$
For complete entry see 65.47

85.17 **Tetrahydrothiophene mercury(ii) chloride complex**
$C_4H_8ClHg^+$, Cl^-
C.-I.Branden *Ark. Kemi*, **22**, 495, 1964

85.C **trans - bis(Thioacetohydroxamato) nickel(ii)**
$C_4H_8N_2NiO_2S_2$
For complete entry see 84.7

85.18 **cis - bis(Thioacetohydroxamato) nickel(ii)**
$C_4H_8N_2NiO_2S_2$
T.Sato, M.Shiro, H.Koyama *J. Chem. Soc. (B)*, 989, 1968
Also classified in 84

85.19 **bis(Dithiobiureto) nickel(ii)**
$C_4H_8N_6NiS_4$
H.Luth, E.A.Hall, W.A.Spofford, E.L.Amma *J. Chem. Soc. (D)*, 520, 1969

85.20 **Copper(i) chloride diethylsulfide complex**
$C_4H_{10}ClCuS_2$
C.Branden *Acta Chem. Scand.*, **21**, 1000, 1967

85.21 **Diethylsulfide mercury(ii) chloride complex**
$C_4H_{10}ClHgS^+$, Cl^-, Cl_2Hg
C.-I.Branden *Ark. Kemi*, **22**, 83, 1964

85.22   **Roussin's red ethyl ester**
$C_4H_{10}Fe_2N_4O_4S_2$
J.T.Thomas, J.H.Robertson, E.G.Cox   *Acta Cryst.*, **11**, 599, 1958
See also *Int. Distances*, M 116s; *Structure Reports*, **22**, 579, 1958

85.23   **Mercury ethylmercaptide**
$C_4H_{10}HgS_2$
D.C.Bradley, N.R.Kunchur   *Canad. J. Chem.*, **43**, 2786, 1965

85.C   **Dithiocyanato bis(thiosemicarbazide) nickel(ii)**
$C_4H_{10}N_8NiS_4$
For complete entry see 83.30

85.24   **Nickel(ii) 2 - hydroxyethyl mercaptide**
$C_4H_{10}NiO_2S_2$
R.O.Gould, R.M.Taylor   *Chem. and Industry*, 378, 1966

85.25   **Nickel(ii) ethylmercaptide**
$C_4H_{10}NiS_2$
P.Woodward, L.F.Dahl, E.W.Abel, B.C.Crosse
J. Amer. Chem. Soc., **87**, 5251, 1965

85.26   **Di - μ - bromo - dibromo - trans - bis(dimethylsulfide) - dipalladium(ii)**
$C_4H_{12}Br_4Pd_2S_2$
D.L.Sales, J.Stokes, P.Woodward   *J. Chem. Soc. (A)*, 1852, 1968

85.27   **Di - μ - (diethylsulfide) bis(platinum dibromide)**
$C_4H_{12}Br_4Pt_2S_2$
P.L.Goggin, R.J.Goodfellow, D.L.Sales, J.Stokes, P.Woodward
Chem. Communic., 31, 1968

85.28   **trans - Dichloro bis(dimethyl sulfoxide) palladium(ii)**
$C_4H_{12}Cl_2O_2PdS_2$
R.J.Williams, W.H.Watson   *Acta Cryst.*, **23**, 788, 1967

85.29   **trans - Dichloro bis(dimethyl sulfoxide) palladium(ii)**
$C_4H_{12}Cl_2O_2PdS_2$
M.J.Bennett, F.A.Cotton, D.L.Weaver   *Acta Cryst.*, **23**, 788, 1967

85.30   **Tetramethylammonium bis(maleonitriledithiolato) nickel(ii)**
$2C_4H_{12}N^+, C_8N_4NiS_4^{2-}$
R.Eisenberg, J.A.Ibers   *Inorg. Chem.*, **4**, 605, 1965
Residue 2 classified in 3

85.31 cis - bis(Dimethylsulfoxide) palladium(ii) nitrate
$C_4H_{12}N_2O_8PdS_2$
D.A.Langs, C.R.Hare, R.G.Little *Chem. Communic.*, 1080, 1967
Also classified in 84

85.32 bis(Dimethyldithiophosphato) nickel(ii)
$C_4H_{12}NiO_4P_2S_4$
V.Kastalsky, J.F.McConnell *Acta Cryst. (B)*, **25**, 909, 1969

85.33 bis(Dimethyldithiophosphinato) nickel(ii)
$C_4H_{12}NiP_2S_4$
P.E.Jones, G.B.Ansell, L.Katz *Acta Cryst. (B)*, **25**, 1939, 1969

85.34 Molybdenum dithioglyoxal complex
$C_6H_6MoS_6$
A.E.Smith, G.N.Schrauzer, V.P.Mayweg, W.Heinrich
J. Amer. Chem. Soc., **87**, 5798, 1965

85.C Disodium di - μ - oxo bis(oxocysteinato molybdenum(v)) pentahydrate
$C_6H_{10}Mo_2N_2O_8S_2{}^{2-}$, $2Na^+$, $5H_2O$
For complete entry see 82.14

85.35 bis(Thioacetamide) nickel(ii) thiocyanate
$C_6H_{10}N_4NiS_4$
L.Capacchi, G.F.Gasparri, M.Nardelli, G.Pelizzi
Acta Cryst. (B), **24**, 1199, 1968
Also classified in 83

85.36 Silver(i) nitrate - bis(1,3,5 - trithian) complex
$C_6H_{12}AgS_6{}^+$, $NO_3{}^-$
A.Domenicano, L.Scaramuzza, A.Vaciago, R.S.Ashworth, C.K.Prout
J. Chem. Soc. (A), 866, 1968

85.37 bis(Imidazoline - 2 - thione) cadmium chloride
$C_6H_{12}CdCl_2N_4S_2$
L.Cavalca, P.Domiano, A.Musatti, P.Sgarabotto
Chem. Communic., 1136, 1968

85.38 Nickel(ii) S - phenylisocyanato aminodithionitride S - aminodithionitride
$C_7H_7N_5NiOS_4$
J.Weiss, U.Thewalt *Z. Anorg. Allg. Chem.*, **343**, 274, 1966
Also classified in 83

85.C   **Chlorotriphenylphosphonium bis(cis - 1,2 - bis(trifluoromethyl)ethene - 1,2 - dithiolato) gold**
$C_8AuF_{12}S_4^-$, $C_{18}H_{15}ClP^+$
For complete entry see 64.52

85.39   **Di(tetra - n - butylammonium) cobalt(ii) bis(maleonitrile dithiolate)**
$C_8CoN_4S_4^{2-}$, $2C_{16}H_{36}N^+$
J.D.Forrester, A.Zalkin, D.H.Templeton *Inorg. Chem.*, **3**, 1500, 1964
Residue 2 classified in 3

85.40   **Tetra - n - butylammonium copper(ii) bis(maleonitrile dithiolate)**
$C_8CuN_4S_4^-$, $C_{16}H_{36}N^+$
J.D.Forrester, A.Zalkin, D.H.Templeton *Inorg. Chem.*, **3**, 1507, 1964
Residue 2 classified in 3

85.41   **Tetra - n - butylammonium bis(maleonitrile dithiolato) iron(iii)**
$C_8FeN_4S_4^-$, $C_{16}H_{36}N^+$
W.C.Hamilton, I.Bernal *Inorg. Chem.*, **6**, 2003, 1967
Residue 2 classified in 3

85.42   **bis(Tetraethylammonium) nitrosyl iron bis(dicyanodithiolate)**
$C_8FeN_5OS_4^{2-}$, $2C_8H_{20}N^+$
A.I.M.Rae *Chem. Communic.*, 1245, 1967
Residue 2 classified in 3

85.43   **Triphenylmethylphosphonium bis(1,2 - dicyanoethylene - 1,2 - dithiolate) nickel(iii)**
$C_8N_4NiS_4^-$, $C_{19}H_{18}P^+$
C.J.Fritchie *Acta Cryst.*, **20**, 107, 1966
Residue 2 classified in 64

85.44   **bis(Mercuric(ii) chloride) - 1,6 - dithiacyclodeca - cis,cis - 3,8 - diene complex**
$C_8H_{12}Cl_4Hg_2S_2$
K.K.Cheung, G.A.Sim *J. Chem. Soc.*, 5988, 1965

85.45   **bis - (cis - 1,2 - bis(Trifluoromethyl)ethylene - 1,2 - dithiolate) cobalt**
$C_8H_{12}CoS_4$
J.H.Enemark, W.N.Lipscomb *Inorg. Chem.*, **4**, 1729, 1965

85.C   **bis(Ethylenethiourea) nickel thiocyanate**
$(C_8H_{12}N_6NiS_4)_n$
For complete entry see 79.31

85.46 Copper(ii) 2 - keto - 3 - ethoxybutyraldehyde bis(thiosemicarbazonate) (anorthic form)
$C_8H_{14}CuN_6OS_2$
M.R.Taylor, E.J.Gabe, J.P.Glusker, J.A.Minkin, A.L.Patterson
J. Amer. Chem. Soc., **88,** 1845, 1966
Also classified in 83

85.47 Biacetyl bis(mercaptoethylimine) nickel(ii)
$C_8H_{14}N_2NiS_2$
Q.Fernando, P.J.Wheatley *Inorg. Chem.*, **4,** 1726, 1965
Also classified in 83

85.48 Mercury(ii) chloride - 1,4 - thioxan complex
$C_8H_{16}Cl_2HgO_2$
R.S.McEwen, G.A.Sim *J. Chem. Soc. (A),* 271, 1967

85.49 Di - μ - (bis - 2 - mercaptoethyl sulphide - dinickel(ii)
$C_8H_{16}Ni_2S_6$
D.J.Baker, D.C.Goodall, D.S.Moss *J. Chem. Soc. (D),* 325, 1969

85.50 2,2' - Dimercaptodiethylsulfide - nickel(ii)
$C_8H_{16}Ni_2S_6$
G.A.Barclay, E.M.McPartlin, N.C.Stephenson
Acta Cryst. (B), **25,** 1262, 1969

85.51 Mercury t - butyl mercaptide
$C_8H_{18}HgS_2$
N.R.Kunchur *Nature,* **204,** 468, 1964

85.52 Tetrabromo - μ - bis(diethylsulfide) - diplatinum(ii)
$C_8H_{20}CdCl_2S_2$
D.L.Sales, J.Stokes, P.Woodward *J. Chem. Soc. (A),* 1852, 1968

85.53 tetrakisThioacetamide copper(i) chloride
$C_8H_{20}ClCuNS$
M.R.Truter, K.W.Rutherford *J. Chem. Soc.,* 1748, 1962

85.54 Pentachloro - bis(2,5 - dithiahexane) dirhenium
$C_8H_{20}Cl_5Re_2S_4$
M.J.Bennett, F.A.Cotton, R.A.Walton *Proc. R. Soc., A,* **303,** 175, 1968

85.55 Cobalt(ii) bis(2,5 - dithiahexane)perchlorate
$C_8H_{20}CoS_4^{2+}, 2ClO_4^-$
F.A.Cotton, D.L.Weaver *J. Amer. Chem. Soc.,* **87,** 4189, 1965

85.56 bis(N,N - Dimethyl - β - mercaptoethylamine) nickel(ii)
$C_8H_{20}N_2NiS_2$
R.L.Girling, E.L.Amma *Inorg. Chem.*, **6**, 2009, 1967
Also classified in 83

85.57 Nickel(ii) O,O' - diethyldithiophosphate
bis(Diethyldithiophosphato) nickel(ii)
$C_8H_{20}NiO_4P_2S_4$
J.F.McConnell, V.Kastalsky *Acta Cryst.*, **22**, 853, 1967

85.58 Nickel(ii) O,O' - diethyldithiophosphate
$C_8H_{20}NiO_4P_2S_4$
Q.Fernando, C.D.Green *J. Inorg. Nucl. Chem.*, **29**, 647, 1967

85.59 Nickel(ii) O,O' - diethyldithiophosphate
$C_8H_{20}NiO_4P_2S_4$
E.A.Glinskaya, M.A.Porai-Koshits *Kristallografija*, **4**, 241, 1960

85.60 bis(Diethyldithiophosphinato) nickel(ii)
$C_8H_{20}NiP_2S_4$
P.S.Shetty, Q.Fernando *Acta Cryst. (B)*, **25**, 1294, 1969

85.61 Zinc diethyldithiophosphate
$C_8H_{20}O_4P_2S_4Zn$
T.Ito, T.Igarashi, H.Hagihara *Acta Cryst. (B)*, **25**, 2303, 1969

85.C π - Cyclopentadienyl cis - 1,2 - bis(trifluoromethyl) ethanedithione cobalt
$C_9H_5CoF_6S_2$
For complete entry see 73.9

85.C π - Cyclopentadienyl(1,2 - dicyanoethene - 1,2 - dithiolato) cobalt
$C_9H_5CoN_2S_2$
For complete entry see 73.10

85.62 Ethylthio iron tricarbonyl dimer
$C_{10}H_{10}Fe_2O_6S_2$
L.F.Dahl, C.H.Wei *Inorg. Chem.*, **2**, 328, 1963

85.C bis - Ethylenethiourea cadmium thiocyanate
$C_{10}H_{12}CdN_6S_4$
For complete entry see 79.33

85.63 bis(Dithioacetylacetonato) cobalt(ii)
$C_{10}H_{14}CoS_4$
R.Beckett, B.F.Hoskins *Chem. Communic.*, 909, 1967

85.C **Nickel(ii) diethyldiselenocarbamate**
$C_{10}H_{20}N_2NiSe_4$
For complete entry see 80.18

85.64 **Silver thiocyanate - tri - n - propyl phosphine complex**
$C_{10}H_{21}AgNS$
C.Panattoni, E.Frasson *Gazz. Chim. Ital.*, **93**, 601, 1963
Also classified in 83

85.65 **Di - μ - iodo - tetramethyl - μ - (dimethylsulfide) - bis(dimethylsulfide) dirhodium**
$C_{10}H_{30}I_2Rh_2S_3$
E.F.Paulus, H.P.Fritz, K.E.Schwarzhans
J. Organometal. Chem., **11**, 647, 1968
Also classified in 71

85.C **Di(tetramethylammonium) vanadium tris(maleodinitrile dithiolate)**
$C_{12}N_6S_6V^{2-}$, $2C_4H_{12}N^+$
For complete entry see 3.40

85.C **bis - π - Cyclopentadienyl (2 - aminoethanethiolato) molybdenum iodide**
$C_{12}H_{16}MoNS^+$, I^-
For complete entry see 73.40

85.C **Nickel(ii) chloride - tetra(ethylenethiourea) complex (triclinic form)**
$C_{12}H_{24}Cl_2N_8NiS_4$
For complete entry see 79.34

85.C **Nickel(ii) chloride - tetra(ethylenethiourea) complex (monoclinic form)**
$C_{12}H_{24}Cl_2N_8NiS_4$
For complete entry see 79.35

85.66 **2,2' - Dimercaptodiethylsulfide palladium(ii) trimer**
$C_{12}H_{24}Pd_3S_9$
E.M.McPartlin, N.C.Stephenson *Acta Cryst. (B)*, **25**, 1659, 1969

85.67 **bis(Diethyl - (2 - mercaptoethyl) - phosphine) platinum(ii) - silver(i) nitrate complex**
$(C_{12}H_{28}AgP_2PtS_2^+)_n$, nNO_3^-
P.Strickler *Helv. Chim. Acta*, **52**, 270, 1969
Residue 1 also classified in 86

85.68 **tris(Diethyldithiophosphato) vanadium(iii)**
$C_{12}H_{30}O_6P_3S_6V$
C.Furlani, P.Porta, A.Sgamellotti, A.A.G.Tomlinson
J. Chem. Soc. (D), 1046, 1969

85.69 1,3,5 - Trithian - copper(i) chloride complex
$C_{13}H_{12}Cu_3S_6$
A.Domenicano, R.Spagna, A.Vaciago   *Chem. Communic.*, 1291, 1968

85.70 bis(O,O - Diethyldithiophosphato) nickel(ii) pyridine
$C_{13}H_{25}NNiO_4P_2S_4$
S.Ooi, Q.Fernando
Proc. 10th Internat. Conf. Coord. Chem. Tokyo, 98, 1967
Also classified in 83, 46

85.71 Di(methylthio - di - iron - hexacarbonyl) sulfide
$C_{14}H_6Fe_4O_{12}S_3$
J.M.Coleman, A.Wojcicki, P.J.Pollick, L.F.Dahl
Inorg. Chem., **6,** 1236, 1967

85.72 Palladium(ii) dithiobenzoate
$C_{14}H_{10}PdS_4$
M.Bonamico, G.Dessy   *Chem. Communic.*, 483, 1968

85.73 Triphenylmethylarsonium bis(toluene - 3,4 - dithiolato) cobaltate ethanol solvate
$C_{14}H_{12}CoS_4{}^-$, $C_{19}H_{18}As^+$, $0.5C_2H_6O$
R.Eisenberg, Z.Dori, H.B.Gray, J.A.Ibers   *Inorg. Chem.*, **7,** 741, 1968
Residue 2 classified in 65

85.C Tri(cyclopentadienyl nickel) - di - μ - sulfide
$C_{15}H_{15}N_{13}S_2$
For complete entry see 73.64

85.74 Tricobalt tetra - carbonyl penta(ethylmercaptide)
$C_{15}H_{25}O_7S_5$
C.H.Wei, L.F.Dahl   *J. Amer. Chem. Soc.*, **90,** 3960, 1968

85.75 Mercury dimethoxide - phenyl isothiocyanate complex
$C_{16}H_{16}HgN_2O_2S_2$
R.S.McEwen, G.A.Sim   *J. Chem. Soc. (A)*, 1552, 1967

85.76 Nickel(ii) phenyl - O - ethyldithiophosphonate
$C_{16}H_{20}NiP_2S_4$
H.Hartung   *Z. Chem.*, **7,** 241, 1967

85.77 bis(μ - Ethylmercaptide μ - ethylthioxanthate) di(iron ethylthioxanthate)
$C_{16}H_{30}Fe_2S_{14}$
D.Coucouvanis, S.J.Lippard, J.A.Zubieta
J. Amer. Chem. Soc., **91,** 761, 1969

85.C   Dichloro tetrakis(trimethylenethiourea) nickel(ii)
$C_{16}H_{32}Cl_2N_8NiS_4$
For complete entry see 79.36

85.78 μ - Oxo bis(bis(diethyldithiophosphato) oxomolybdenum(v)) di(1,2 - dichlorobenzene) solvate
$C_{16}H_{40}Mo_2O_{11}P_4S_8$, $2C_6H_4Cl_2$
J.R.Knox, C.K.Prout   *Acta Cryst. (B)*, **25**, 2281, 1969

85.79   Zinc diethyldithiophosphinate dimer
$C_{16}H_{40}P_4S_8Zn_2$
M.Calligaris, G.Nardin, A.Ripamonti   *Chem. Communic.*, 1014, 1968

85.C   bis(π - Cyclopentadienyl) (toluene - 3,4 - dithiolato) molybdenum
$C_{17}H_{16}MoS_2$
For complete entry see 73.75

85.80   Palladium 8 - mercaptoquinolinate
$C_{18}H_{12}N_2PdS_2$
J.Ozols, A.Ozola, A.Ievins   *Acta Cryst.*, **21**, A147, 1966
Also classified in 83

85.C   2,9 - Dimethyl - 1,10 - phenanthroline bis(o,o' - dimethyl dithiophosphato) nickel(ii)
$C_{18}H_{24}N_2NiO_4P_2S_4$
For complete entry see 83.163

85.81   bis(O,O' - Diethyldithiophosphato) nickel(ii) bis(pyridine)
$C_{18}H_{30}N_2NiO_4P_2S_4$
S.Ooi, Q.Fernando   *Inorg. Chem.*, **6**, 1558, 1967
Also classified in 83

85.82   Hexacobalt - tetra(ethylmercaptide) - undecacarbonyl sulfide
$C_{19}H_{20}Co_6O_{11}S_5$
C.H.Wei, L.F.Dahl   *J. Amer. Chem. Soc.*, **90**, 3977, 1968

85.C   Iridium(iii) benzylacetophenone dichloride bis(dimethylsulfoxide)
$C_{19}H_{25}Cl_2IrO_3S_2$
For complete entry see 71.42

85.C   Tetra - n - propylammonium o - phenanthroline - bis(maleonitriledithiolato) cobaltate
$C_{20}H_8CoN_6S_4^-$, $C_{12}H_{28}N^+$
For complete entry see 3.76

85.83 μ,μ'(cis - Stilbene - α,β - dithiolato - bis(tricarbonyl iron)
$C_{20}H_{10}Fe_2O_6S_2$
H.P.Weber, R.F.Bryan   *J. Chem. Soc. (A)*, 182, 1967

85.C   Cyclopentadienyl nickel sulfide
$C_{20}H_{20}Ni_5S_4$
For complete entry see 73.91

85.84   Pentacobalt decacarbonyl penta(ethylmercaptide)
$C_{20}H_{25}Co_5O_{10}S_5$
C.H.Wei, L.F.Dahl   *J. Amer. Chem. Soc.*, **90**, 3969, 1968

85.85   tris - (Dithiobenzoato) chromium(iii)
$C_{21}H_{15}CrS_6$
M.Bonamico, G.Dessy   *Ric. Sci.*, **38**, 1106, 1968

85.C   trans - Di - μ - phenylthio - dinitrosyl bis - (π - cyclopentadienyl) dichromium(i)
$C_{22}H_{20}Cr_2N_2O_2S_2$
For complete entry see 73.99

85.86   Tetra - n - butylammonium bis(bis(1,2,3,4 - tetrachloro - benzene - 5,6 - dithiolato) cobaltate)
$C_{24}Cl_{16}Co_2S_8^{2-}$, $2C_{16}H_{36}N^+$
M.J.Baker-Hawkes, Z.Dori, R.Eisenberg, H.B.Gray
J. Amer. Chem. Soc., **90**, 4253, 1968
Residue 2 classified in 3

85.C   Phenyltrimethylammonium 1,1 - dicyanoethylene - 2,2 - dithiol cuprate
$C_{24}Cu_8N_{12}S_{12}^{4-}$, $4C_9H_{14}N^+$
For complete entry see 3.69

85.87   bis(Phenoxathiin) mercury(ii) chloride
$C_{24}H_{16}Cl_2HgO_2S_2$
K.K.Cheung, R.S.McEwen, G.A.Sim   *Nature,* **205,** 383, 1965

85.C   Di - μ - phenylthio - bis(cyclopentadienyl carbonyl iron) (at $-160\ °C$)
$C_{24}H_{20}Fe_2O_2S_2$
For complete entry see 73.102

85.88   bis(Diphenyldithiophosphinato) nickel(ii)
$C_{24}H_{20}NiP_2S_4$
P.Porta, A.Sgamellotti, N.Vinciguerra   *Inorg. Chem.*, **7**, 2625, 1968

85.C   **Cobalt(ii) diphenylmonothiophosphinate**
$(C_{24}H_{20}O_2P_2S_2)_n$
For complete entry see 84.55

85.89   **Thio - p - toluoyl disulfido bis(dithio - p - toluato) iron(iii)**
$C_{24}H_{21}FeS_7$
D.Coucouvanis, S.J.Lippard   *J. Amer. Chem. Soc.*, **91**, 307, 1969

85.C   **Tetra - n - butyl - o - phenanthroline - bis(toluene - 3,4 - dithiolato) cobaltate**
$C_{26}H_{20}CoN_2S_4{}^-$, $C_{16}H_{36}N^+$
For complete entry see 3.81

85.90   **Mercury dithizone complex**
$C_{26}H_{22}HgN_8S_2$, $2C_5H_5N$
M.M.Harding   *J. Chem. Soc.*, 4136, 1958
Residue 1 also classified in 83
See also *Int. Distances*, M 204s; *Structure Reports*, **22**, 724, 1958

85.C   **tris(N,N - Di - n - butyldithiocarbamato) iron(iii)**
$C_{27}H_{54}FeN_3S_6$
For complete entry see 80.40

85.91   **bis(Dithiobenzil) nickel (at 115 °K)**
$C_{28}H_{20}NiS_4$
D.Sartain, M.R.Truter   *J. Chem. Soc. (A)*, 1264, 1967

85.92   **Palladium bis(1,3 - diphenyl - propane - 1 - one - 3 - thionate)**
$C_{30}H_{22}O_2PdS_2$
E.A.Shugam, L.M.Shkol'nikova, S.E.Livingstone
Zh. Strukt. Khim., **8**, 550, 1967
Also classified in 84

85.C   **Nickel(ii) monothiobenzoate ethanol complex**
$C_{30}H_{26}Ni_2O_5S_4$
For complete entry see 84.60

85.93   **bis(Dithiophenylacetato) nickel(ii)**
$C_{32}H_{28}Ni_2S_8$
M.Bonamico, G.Dessy, V.Fares   *J. Chem. Soc. (D)*, 1106, 1969

85.94   **Cadmium thioglycol sulfate hydrate**
$C_{32}H_{80}Cd_{10}O_{16}S_{16}{}^{4+}$, $2O_4S^{2-}$, $4H_2O$
P.Strickler   *J. Chem. Soc. (D)*, 655, 1969

85.C **Methylzinc isopropyl sulfide octamer**
$C_{32}H_{80}S_8Zn_8$
For complete entry see 71.54

85.95 **Palladium n - propylmercaptide**
$C_{36}H_{84}Pd_6S_{12}$
N.R.Kunchur *Acta Cryst. (B),* **24**, 1623, 1968

85.96 **bis(Dithiobenzoate) nickel(ii) trimer**
$C_{42}H_{30}Ni_3S_{12}$
M.Bonamico, G.Dessy, V.Fares *J. Chem. Soc. (D),* 324, 1969

85.97 **Rhenium tri(cis - 1,2 - diphenylethene - 1,2 - dithiolate)**
$C_{42}H_{30}ReS_6$
R.Eisenberg, J.A.Ibers *Inorg. Chem.,* **5**, 411, 1966

85.98 **tris(cis - 1,2 - Diphenylethene - 1,2 - dithiolato) vanadium**
$C_{42}H_{30}S_6V$
R.Eisenberg, H.B.Gray *Inorg. Chem.,* **6**, 1844, 1967

METAL COMPLEXES
(PHOSPHINE OR ARSINE LIGAND)

86.1   **Tribromo (trimethylphosphine) gold**
$C_3H_9AuBr_3P$
M.F.Perutz, O.Weisz   *J. Chem. Soc.*, 438, 1946
See also *Int. Distances*, M 154; *Structure Reports*, **10,** 210, 1946

86.2   **bis(Trimethylarsine dibromo palladium(ii))**
$C_6H_{18}As_2Br_4Pd_2$
A.F.Wells   *Proc. R. Soc., A,* **167,** 169, 1938
See also *Int. Distances*, M 135s; *Strukturbericht*, **6,** 220, 1938

86.3   **Di - μ - chloro - trans - di(chloro - trimethylarsine platinum(ii))**
$C_6H_{18}As_2Cl_4Pt_2$
S.F.Watkins   *Chem. Communic.*, 504, 1968

86.4   **cis - Dichloro bis(trimethylphosphine) platinum(ii)**
$C_6H_{18}Cl_2P_2Pt$
G.G.Messmer, E.L.Amma, J.A.Ibers   *Inorg. Chem.*, **6,** 725, 1967

86.5   **bis(Iron tricarbonyl) tetramethyltetra - arsine**
$C_{10}H_{12}As_4Fe_2O_6$
B.M.Gatehouse   *J. Chem. Soc. (D)*, 948, 1969

86.6   **Di - μ - dimethylphosphido - bis(iodotricarbonyl iron)**
$C_{10}H_{12}Fe_2I_2O_6P_2$
G.R.Davies, R.H.B.Mais, P.G.Owston, D.T.Thompson
J. Chem. Soc. (A), 1251, 1968

86.C   **Phenylethynyl(trimethylphosphine) silver(i)**
$C_{11}H_{14}AgP$
For complete entry see 72.29

86.C   **Phenylethynyl(trimethylphosphine) copper(i)**
$C_{11}H_{14}CuP$
For complete entry see 72.30

86.7   Methyl bis(3 - propyldimethylarsine)arsine nickel dibromide
        $C_{11}H_{27}As_3Br_2Ni$
        G.A.Mair, H.M.Powell, D.E.Henn   *Proc. Chem. Soc.*, 415, 1960

86.8 μ - Tetramethyldiphosphine - bis(tetracarbonyl iron)
        $C_{12}H_{12}Fe_2O_8P_2$
        J.A.J.Jarvis, R.H.B.Mais, P.G.Owston, D.T.Thompson
        J. Chem. Soc. (A), 622, 1968

86.C   bis(Diethyl - (2 - mercaptoethyl) - phosphine) platinum(ii) - silver(i) nitrate
        complex
        $(C_{12}H_{28}AgP_2PtS_2{}^+)_n$, $nNO_3{}^-$
        For complete entry see 85.67

86.9   Dibromo bistriethylphosphine nickel
        $C_{12}H_{30}Br_2NiP_2$
        V.Scatturin, A.Turco   *J. Inorg. Nucl. Chem.*, **8**, 447, 1958
        See also *Int. Distances*, M 173s; *Structure Reports*, **22**, 642, 1958

86.10   trans - Platinum(ii) bromide bis(triethylphosphine) complex
        $C_{12}H_{30}Br_2P_2Pt$
        G.G.Messmer, E.L.Amma   *Inorg. Chem.*, **5**, 1775, 1966

86.11   trans Platinum(ii) chloride bis(triethylphosphine) complex
        $C_{12}H_{30}Cl_2P_2Pt$
        G.G.Messmer, E.L.Amma   *Inorg. Chem.*, **5**, 1775, 1966

86.12   Hexachloro bis(triethylphosphine) dirhenium(iii)
        $C_{12}H_{30}Cl_6P_2Re_2$
        F.A.Cotton, B.M.Foxman   *Inorg. Chem.*, **7**, 2135, 1968

86.13   bis - Triethylphosphine hydrobromo platinum
        $C_{12}H_{31}BrP_2Pt$
        P.G.Owston, J.M.Partridge, J.M.Rowe   *Acta Cryst.*, **13**, 246, 1960

86.14   Trimethylarsine - cuprous iodide tetramer
        $C_{12}H_{36}As_4Cu_4I_4$
        A.F.Wells, Z.KristallogR.
        See also *Int. Distances*, M 236; *Strukturbericht*, **4**, 284, 1936

86.15   o - Phenylene bisdimethylarsine tricarbonyl iron(0)
        $C_{13}H_{16}As_2FeO_3$
        D.S.Brown, G.W.Bushnell   *Acta Cryst.*, **22**, 296, 1967

86.16 **bis(Triethylphosphine) platinum(ii) carbonyl chloride pentafluorosilicate**
$C_{13}H_{30}ClOP_2Pt^+$, F_5Si^-
H.C.Clark, P.W.R.Corfield, K.R.Dixon, J.A.Ibers
J. Amer. Chem. Soc., **89**, 3360, 1967

86.17 **Pentaethyl cyclopentaphosphine molybdenum carbonyl**
$C_{14}H_{25}MoO_4P_5$
M.A.Bush, P.Woodward   *J. Chem. Soc. (A)*, 1221, 1968

86.C **Ethoxy (phenylamino) carbene dichloro triethylphosphine platinum(ii)**
$C_{15}H_{25}Cl_2OPPt$
For complete entry see 71.38

86.18 **trans - Di - iodo bis(dimethylphenylphosphine) palladium(ii) (monoclinic form)**
$C_{16}H_{22}I_2P_2Pd$
N.A.Bailey, R.Mason   *J. Chem. Soc. (A)*, 2594, 1968

86.19 **trans - Di - iodo bis(dimethylphenylphosphine) palladium(ii) (orthorhombic form)**
$C_{16}H_{22}I_2P_2Pd$
N.A.Bailey, R.Mason   *J. Chem. Soc. (A)*, 2594, 1968

86.20 **Cyano tris(3 - dimethylarsinopropyl)phosphine nickel(ii) perchlorate**
$C_{16}H_{36}As_3NNiP^+$, ClO_4^-
D.L.Stevenson, L.F.Dahl   *J. Amer. Chem. Soc.*, **89**, 3424, 1967

86.C **Tetraphenylarsonium tri - iodo(triphenylphosphine) nickelate(ii)**
$C_{18}H_{15}I_3NiP^-$, $C_{24}H_{20}As^+$
For complete entry see 65.42

86.21 **Dichloro bis(tris - (2 - cyanoethyl)phosphine) nickel(ii) (β form)**
$C_{18}H_{24}Cl_2N_6Ni$
M.G.B.Drew, D.F.Lewis, R.A.Walton   *J. Chem. Soc. (D)*, 326, 1969

86.22 **bis(Methoxyphenyldimethylarsine) trichloro rhodium(iii)**
$C_{18}H_{26}As_2Cl_3O_2Rh$
R.Graziani, G.Bombieri, L.Volponi, C.Panattoni
Chem. Communic., 1284, 1967
Also classified in 84

86.23 **bis(Tricarbonyltriethylphosphine cobalt) mercury**
$C_{18}H_{30}Co_2HgO_6P_2$
R.F.Bryan, A.R.Manning   *Chem. Communic.*, 1316, 1968

86.24   trans - Dichloro bis(tripropylphosphine) - μ,μ' - dichloro diplatinum
$C_{18}H_{42}Cl_4P_2Pt_2$
M.Black, R.H.B.Mais, P.G.Owston   *Acta Cryst. (B)*, **25**, 1760, 1969

86.25 μ - Hydrido - μ - diphenylphosphido - bis(tetracarbonyl - manganese
$C_{20}H_{11}MnO_8$
R.J.Doedens, W.T.Robinson, J.A.Ibers
J. Amer. Chem. Soc., **89**, 4323, 1967

86.26   Dichloro bis(o - phenylene bisdimethylarsine) nickel(iii) chloride
$C_{20}H_{30}As_4Cl_2Ni^+$, Cl^-
P.Kreisman, R.Marsh, J.R.Preer, H.B.Gray
J. Amer. Chem. Soc., **90**, 1067, 1968

86.27   trans - Oxotrichloro bis(diethylphenylphosphine) rhenium(v)
$C_{20}H_{30}Cl_3OP_2Re$
H.W.W.Ehrlich, P.G.Owston   *J. Chem. Soc.*, 4368, 1963
Also classified in 84

86.28   bis(Tetracarbonyl triethylphosphine manganese(0))
$C_{20}H_{30}Mn_2O_8P_2$
M.J.Bennett, R.Mason   *J. Chem. Soc. (A)*, 75, 1968

86.29   bis(o - Phenylenedimethylamine dimethylarsine) trichloro rhodium(iii)
$C_{20}H_{32}As_2Cl_3N_2Rh$
G.Bombieri, R.Graziani, C.Panattoni, L.Volponi
Chem. Communic., 977, 1967

86.30   Di - iodo bis(o - phenylene bis(dimethylarsine)) gold(iii) iodide (form i)
$C_{20}H_{32}As_4AuI_2^+$, I^-
V.F.Duckworth, C.M.Harris, N.C.Stephenson
Inorg. Nucl. Chem. Letters, **4**, 419, 1968

86.31   Dichloro di(o - phenylene bisdimethylarsine) platinum(ii)
$C_{20}H_{32}As_4Cl_2Pt$
N.C.Stephenson   *Acta Cryst.*, **17**, 1517, 1964

86.32   Cobalt(ii) perchlorate - o - phenylene bis(dimethylarsine) complex
(orthorhombic form)
$C_{20}H_{32}As_4Co^{2+}$, $2ClO_4^-$
F.W.B.Einstein, G.A.Rodley   *J. Inorg. Nucl. Chem.*, **29**, 347, 1967

86.33   Di - iodo di(o - phenylene bisdimethylarsine) nickel(ii)
$C_{20}H_{32}As_4I_2Ni$
N.C.Stephenson   *Acta Cryst.*, **17**, 592, 1964

86.34 Di - iodo - di - (o - phenylene bis(dimethylarsine)) palladium(ii)
$C_{20}H_{32}As_4I_2Pd$
N.C.Stephenson *J. Inorg. Nucl. Chem.*, **24**, 797, 1962

86.35 Di - iodo - di - (o - phenylene bis(dimethylarsine)) platinum(ii)
$C_{20}H_{32}As_4I_2Pt$
N.C.Stephenson *J. Inorg. Nucl. Chem.*, **24**, 791, 1962

86.36 tetrakis(4 - Methyl - 2,6,7 - trioxa - 1 - phosphabicyclo(2.2.2)octane - silver(i) perchlorate
$C_{20}H_{36}AgO_{12}P_4{}^+, ClO_4{}^-$
J.V.Ugro *Dissert. Abstr. (B)*, **28**, 1891, 1967

86.37 bis - (Tripropylphosphine) thiocyanate chloro platinum (α form)
$C_{20}H_{42}Cl_2N_2P_2Pt_2S_2$
P.G.Owston, J.M.Rowe *Acta Cryst.*, **13**, 253, 1960

86.38 Palladium(ii) bis(tri - n - propyl phosphine) thiocyanate
$C_{20}H_{42}N_2P_2PdS_2$
E.Frasson, C.Panattoni, A.Turco *Nature*, **199**, 803, 1963

86.39 Nitrosyltricarbonyl (triphenylphosphine) manganese
$C_{21}H_{15}MnNO_4P$
J.H.Enemark, J.A.Ibers *Inorg. Chem.*, **7**, 2339, 1968

86.C π - Allyl(triphenylphosphine) palladium - trichlorotin acetone solvate
$C_{21}H_{20}Cl_3PPdSn$, $0.5C_2H_6O$
For complete entry see 72.58

86.40 Triphenylphosphine gold cobalt tetracarbonyl
$C_{22}H_{15}AuCoO_4P$
B.T.Kilbourn, T.L.Blundell, H.M.Powell *Chem. Communic.*, 444, 1965

86.C Triphenylphosphine methallyl palladium chloride
$C_{22}H_{22}ClPPd$
For complete entry see 72.59

86.41 bis(o - Dimethylarsinophenyl)methylarsine copper manganese pentacarbonyl
$C_{22}H_{23}As_3CuMnO_5$
B.T.Kilbourn, T.L.Blundell, H.M.Powell *Chem. Communic.*, 444, 1965

86.42 Hydrido(2 - naphthyl) ruthenium(ii) bis(1,2 - bis(dimethylphosphino) ethane
$C_{22}H_{40}P_4Ru$
S.D.Ibekwe, B.T.Kilbourn, U.A.Raeburn, D.R.Russell
J. Chem. Soc. (D), 433, 1969

86.43 Di - μ - dimethylphosphido bis(tricarbonyl - triethylphosphine molybdenum
$C_{22}H_{42}Mo_2O_6P_4$
R.H.B.Mais, P.G.Owston, D.T.Thompson *J. Chem. Soc. (A)*, 1735, 1967

86.44 Tricarbonyl(triphenylphosphine) - σ - tetrafluoroethyl cobalt(i)
$C_{23}H_{16}CoF_4O_3P$
J.B.Wilford, H.M.Powell *J. Chem. Soc. (A)*, 2092, 1967
Also classified in 71

86.C π - Allyl triphenyl phosphine dicarbonyl iodo iron (monoclinic form)
$C_{23}H_{20}FeIO_2P$
For complete entry see 72.62

86.C π - Allyl - triphenylphosphine dicarbonyl iodo iron (triclinic form)
$C_{23}H_{20}FeIO_2P$
For complete entry see 72.63

86.C Tetraphenylarsonium tetrakis(trifluoroacetato) cobaltate(ii)
$2C_{24}H_{20}As^+$, $C_8CoF_{12}O_8^{2-}$
For complete entry see 81.50

86.45 Palladium(ii) o - phenylene bis(o - dimethylarsino - phenylmethylarsine)
chloride perchlorate benzene solvate
$C_{24}H_{30}As_4ClPd^+$, ClO_4^-, C_6H_6
T.L.Blundell, H.M.Powell *J. Chem. Soc. (A)*, 1650, 1967

86.46 ((o - Allylphenyl)(dimethylarsine)tribromo) - β - ((ethoxymethyl) - 2 -
(dimethylarsino) phenylethyl) platinum(iv)
$C_{24}H_{35}As_2Br_3OPt$
M.A.Bennett, G.J.Erskine, J.Lewis, R.Mason, R.S.Nyholm, G.B.Robertson,
A.D.C.Towl *Chem. Communic.*, 395, 1966
Also classified in 71

86.C 2 - (trans - bis(Triethylphosphine) - chloro palladium) azobenzene
$C_{24}H_{39}ClN_2P_2Pd$
For complete entry see 71.49

86.C Tetracarbonyl (diphenyl - 2 - (prop - cis - 1 - enyl)phenylphosphine)
molybdenum(0)
$C_{25}H_{17}MoO_4P$
For complete entry see 72.66

86.47 Methylmethoxycarbene triphenylphosphine tetracarbonyl chromium
$C_{25}H_{21}CrO_5P$
O.S.Mills, A.D.Redhouse *Chem. Communic.*, 814, 1966
Also classified in 71

86.C Tricarbonyl - π - cyclopentadienyl tungstio (triphenylphosphine) gold
$C_{26}H_{20}AuO_3PW$
For complete entry see 73.109

86.48 Dichloro bis(diphenylphosphino)ethylamine palladium(ii)
$C_{26}H_{25}Cl_2NP_2Pd$
D.S.Payne, J.A.A.Mokuolu, J.C.Speakman *Chem. Communic.*, 599, 1965

86.C But - 2 - enyl cobalt buta - 1,3 - diene triphenylphosphine benzene solvate
$C_{26}H_{28}CoP$, $0.5C_6H_6$
For complete entry see 72.67

86.C Allene - di - iron hexacarbonyl - triphenylphosphine
$C_{27}H_{19}Fe_2O_6P$
For complete entry see 72.68

86.C trans - Dicarbonyl - π - cyclopentadienyl (triphenylphosphine) molybdenum acetyl
$C_{27}H_{23}MoO_3P$
For complete entry see 71.51

86.C Trichloro (p - methoxyphenylimino) bis(diethylphenyl phosphine) rhenium(v)
$C_{27}H_{37}Cl_3NOP_2Re$
For complete entry see 83.195

86.49 bis - (2 - Diphenylphosphinoethyl)amine nickel(ii) dibromide
$C_{28}H_{28}Br_2NNiP_2$.
P.L.Orioli, L.Sacconi *Chem. Communic.*, 1310, 1968
Also classified in 83

86.50 Platinum(ii) chloride hydride diphenylethyl phosphine complex
$C_{28}H_{31}ClPt$
R.Eisenberg, J.A.Ibers *Inorg. Chem.*, **4,** 773, 1965

86.C Trichloro (p - acetylphenylimino) bis(diethyl phosphine) rhenium(v)
$C_{28}H_{37}Cl_3NOP_2Re$
For complete entry see 83.200

86.51 trans - bis(Phenylethynyl) bis(triethylphosphine) nickel(ii)
$C_{28}H_{40}NiP_2$
W.A.Spofford, P.D.Carfagna, E.L.Amma *Inorg. Chem.*, **6,** 1553, 1967
Also classified in 71

86.52 trans - bis(Phenylethynyl) bis(triethylphosphine) nickel(ii)
$C_{28}H_{40}NiP_2$
G.R.Davies, R.H.B.Mais, P.G.Owston   *J. Chem. Soc. (A)*, 1750, 1967
Also classified in 71

86.53 Tri - iron undecacarbonyl triphenylphosphine
$C_{29}H_{15}Fe_3O_{11}$
D.J.Dahm, R.A.Jacobson   *J. Amer. Chem. Soc.*, **90**, 5106, 1968

86.54 Triphenylphosphine tri - iron undecacarbonyl
$C_{29}H_{15}Fe_3O_{11}P$
D.J.Dahm, R.A.Jacobson   *Chem. Communic.*, 496, 1966

86.C π - Cyclopentadienyl - γ - pentafluorophenyl (triphenylphosphine) nickel
$C_{29}H_{20}F_5NiP$
For complete entry see 73.112

86.C π - Cyclopentadienyl (triphenylphosphine) - phenyl nickel
$C_{29}H_{25}NiP$
For complete entry see 73.113

86.C Chlorobis(pentafluorophenyl) triphenylphosphine gold(iii)
$C_{30}H_{15}AuClF_{10}$
For complete entry see 71.52

86.55 μ - Tetraphenyldiphosphine bis(tricarbonyl nickel)
$C_{30}H_{20}Ni_2O_6P_2$
R.H.B.Mais, P.G.Owston, D.T.Thompson, A.M.Wood
J. Chem. Soc. (A), 1744, 1967

86.C π - Cyclopentadienyl triphenylphosphine - monocarbonyl - σ - phenyl iron
$C_{30}H_{25}FeOP$
For complete entry see 73.114

86.56 bis(Diphenylphosphino)ethylamine molybdenum tetracarbonyl
$C_{30}H_{25}MoNO_4P_2$
D.S.Payne, J.A.A.Mokuolu, J.C.Speakman   *Chem. Communic.*, 599, 1965

86.C Methallyl (bis - 1,2 - (diphenylphosphino)ethane) nickel bromide
$C_{30}H_{31}BrNiP_2$
For complete entry see 72.70

86.57 π - Tetraphenylborato bis(trimethoxy phosphine) rhodium(i)
$C_{30}H_{38}BO_6P_2Rh$
M.J.Nolte, G.Gafner, L.M.Haines   *J. Chem. Soc. (D)*, 1406, 1969
Also classified in 74

86.58   **Nitridodichloro tris(diethylphenylphosphine) rhenium(v)**
$C_{30}H_{45}Cl_2NP_3Re$
P.W.R.Corfield, R.J.Doedens, J.A.Ibers   *Inorg. Chem.*, **6,** 197, 1967
Also classified in 83

86.C   **Tri - μ - chloro - hexakis(diethylphenylphosphine) diruthenium(ii) trichloro - tris(diethylphenylphosphine) ruthenate(ii)**
$C_{30}H_{45}Cl_3P_3Ru^-$, $C_{60}H_{90}Cl_3P_6Ru_2^+$
For complete entry see 86.117

86.59   **tris(Phenyldiethylphosphine) nonachloro trirhenium**
$C_{30}H_{45}Cl_9P_3Re$
F.A.Cotton, J.T.Mague   *Inorg. Chem.*, **3,** 1094, 1964

86.60   **bis(Tricarbonyl tri - n - butylphosphine cobalt) (monoclinic form, disordered)**
$C_{30}H_{54}Co_2O_6P_2$
R.F.Bryan, A.R.Manning   *Chem. Communic.*, 1316, 1968

86.61   **bis(Tricarbonyl tri - n - butylphosphine cobalt) (cubic form)**
$C_{30}H_{54}Co_2O_6P_2$
J.A.Ibers   *J. Organometal. Chem.*, **14,** 423, 1968

86.62   **Di - μ - chloro - bis(μ - diphenylphosphido - tetracarbonyl iron palladium) toluene solvate**
$C_{31}H_{20}Cl_2Fe_2O_8P_2Pd_2$, C_7H_8
B.T.Kilbourn, R.H.B.Mais   *Chem. Communic.*, 1507, 1968

86.C   **Iron π - cyclopentadienyl triphenyl phosphine monocarbonyl σ - benzoyl**
$C_{31}H_{25}FeO_2P$
For complete entry see 71.53

86.63   **bis(meso - syn - Ethylene - 1,2 - bis(methylphenyl phosphine)) dichloro ruthenium(ii)**
$C_{32}H_{40}Cl_2P_4Ru$
I.Kawada   *Tetrahedron Letters*, 793, 1969

86.64   **Nona - carbonyl tris (dimethyl phenyl phosphine) tri - iron**
$C_{33}H_{33}Fe_3O_9P_3$
W.S.McDonald, J.-R.Moss, G.Raper, B.L.Shaw, R.Greatex,
N.N.Greenwood   *J. Chem. Soc. (D),* 1295, 1969

86.C   **Diphenylphosphino - cyclopentadienyl - cobalt dimer**
$C_{34}H_{30}Co_2P_2$
For complete entry see 73.117

86.C Diphenylphosphine - cyclopentadienyl - nickel dimer
$C_{34}H_{30}Ni_2P_2$
For complete entry see 73.118

86.65 Dibromo bis(triphenylphosphine) nickel(ii)
$C_{36}H_{30}Br_2NiP_2$
J.A.J.Jarvis, R.H.B.Mais, P.G.Owston *J. Chem. Soc. (A)*, 1473, 1968

86.66 Nitridodichloro bis(triphenylphosphine) rhenium(v)
$C_{36}H_{30}Cl_2NP_2Re$
R.J.Doedens, J.A.Ibers *Inorg. Chem.*, **6**, 204, 1967

86.67 Dichloro bis(triphenylphosphine) nickel(ii)
$C_{36}H_{30}Cl_2NiP_2$
G.Garton, D.E.Henn, H.M.Powell, L.M.Venanzi
J. Chem. Soc., 3625, 1963

86.68 bis(Triphenylphosphine) trichloro indium(iii)
$C_{36}H_{30}Cl_3InP_2$
M.V.Veidis, G.J.Palenik *J. Chem. Soc. (D)*, 586, 1969

86.69 Borohydrido bis(triphenylphosphine) copper(i)
$C_{36}H_{30}CuP_2{}^+$, H_4B^-
S.J.Lippard, K.M.Melmed *Inorg. Chem.*, **6**, 2223, 1967

86.70 μ - Oxido - bis(triphenylphosphine nitrosyl - iridium(i)) benzene solvate
$C_{36}H_{30}Ir_2N_2O_3P_2$, C_6H_6
P.Carty, A.Walker, M.Mathew, G.J.Palenik *J. Chem. Soc. (D)*, 1374, 1969

86.71 Dioxygen bis(triphenylphosphine) platinum benzene solvate
$C_{36}H_{30}O_2P_2Pt$, $1.5C_6H_6$
T.Kashiwagi, N.Yasuoka, N.Kasai, M.Kakudo, S.Takahashi, N.Hagihara
J. Chem. Soc. (D), 743, 1969

86.72 tris(Triphenylphosphine) platinum(0)
$C_{36}H_{30}P_3Pt$
V.Albano, P.L.Bellon, V.Scatturin *Chem. Communic.*, 507, 1966

86.73 tris(Diphenylphosphine) cobalt(ii) bromide
$C_{36}H_{33}Br_2CoP_3$
J.A.Bertrand, D.L.Plymale *Inorg. Chem.*, **5**, 879, 1966

86.74 bis(Tricyclohexylphosphine) nickel chloride
$C_{36}H_{66}Cl_2NiP_2$
P.L.Bellon, V.Albano, V.D.Bianco, F.Pompa, V.Scatturin
Ric. Sci., **33**, 1213, 1963

86.75 Chlorocarbonyl nitrosyl bis(triphenylphosphine) iridium tetrafluoroborate
$C_{37}H_{30}ClIrNO_2P_2{}^+$, $BF_4{}^-$
D.J.Hodgson, J.A.Ibers *Inorg. Chem.*, **7**, 2345, 1968

86.76 Chlorocarbonyl bis(triphenylphosphine) iridium - oxygen complex
$C_{37}H_{30}ClIrO_3P_2$
S.J.La Placa, J.A.Ibers *J. Amer. Chem. Soc.*, **87**, 2581, 1965

86.77 Chlorocarbonyl(sulfur dioxide) bis(triphenylphosphine) iridium
$C_{37}H_{30}ClIrO_3P_2S$
S.J.La Placa, J.A.Ibers *Inorg. Chem.*, **5**, 405, 1966

86.78 trans - bis(Triphenylphosphine) thiocarbonyl rhodium(i) chloride
$C_{37}H_{30}ClP_2RhS$
J.L.de Boer, D.Rogers, A.C.Skapski, P.G.H.Troughton
Chem. Communic., 756, 1966

86.79 Iridium dioxo - carbonyl - iodide bis(triphenyl - phosphine) methylene chloride
$C_{37}H_{30}IIrO_3P_2$, CH_2Cl_2
J.A.McGinnety, R.J.Doedens, J.A.Ibers *Inorg. Chem.*, **6**, 2243, 1967

86.80 bis(Triphenylphosphine) carbonato platinum(ii)
$C_{37}H_{30}O_3P_2Pt$
F.Cariati, R.Mason, G.B.Robertson, R.Ugo *Chem. Communic.*, 408, 1967

86.81 bis(Triphenylphosphine) palladium - carbon disulfide
$C_{37}H_{30}P_2PdS_2$
T.Kashiwagi, N.Yasuoka, T.Ueki, N.Kasai, M.Kakudo, S.Takahashi,
N.Hagihara *Bull. Chem. Soc. Jap.*, **41**, 296, 1968

86.82 bis(Triphenylphosphine) platinum(0) carbon disulfide complex
$C_{37}H_{30}P_2PtS_2$
M.Baird, G.Hartwell Junior, R.Mason, A.I.M.Rae, G.Wilkinson
Chem. Communic., 92, 1967

86.83 Hydridobromocarbonyl tris(triphenylphosphine) osmium(ii)
$C_{37}H_{31}BrOOsP_3$
P.L.Orioli, L.Vaska *Proc. Chem. Soc.*, 333, 1962

86.C trans - bis(Triphenylphosphine)methyl di - iodo rhodium(iii) benzene solvate
$C_{37}H_{33}I_2P_2Rh$, C_6H_6
For complete entry see 71.55

86.84 **bis(Triphenylphosphine) tetrafluoroethylene - rhodium(i) chloride**
$C_{38}H_{30}ClF_4P_2Rh$
P.B.Hitchcock, M.McPartlin, R.Mason *J. Chem. Soc. (D)*, 1367, 1969
Also classified in 72

86.85 **Nitrosyldicarbonyl bis(triphenylphosphine) manganese**
$C_{38}H_{30}MnNO_3P_2$
J.H.Enemark, J.A.Ibers *Inorg. Chem.*, **6**, 1575, 1967
Also classified in 83

86.C **bis(Triphenylphosphine) ethylene nickel(0)**
$C_{38}H_{34}NiP$
For complete entry see 72.76

86.C **bis(Triphenylphosphine) ethylene nickel(0)**
$C_{38}H_{34}NiP$
For complete entry see 72.77

86.86 **trans - Dimesityl bis(diethylphenylphosphine) cobalt(ii)**
$C_{38}H_{52}CoP_2$
P.G.Owston, J.M.Rowe *J. Chem. Soc.*, 3411, 1963
Also classified in 71

86.87 **bis(Triphenylphosphine) iridium carbonyl iodide tetrafluoroethylene**
$C_{39}H_{30}F_4IIrOP_2$
J.A.Ibers, J.McGinnety, N.Kime
Proc. 10th Internat. Conf. Coord. Chem. Tokyo, 93, 1967
Also classified in 72

86.88 **Tricarbonyl bis(triphenylphosphine) osmium(0)**
$C_{39}H_{30}O_3Os$
J.K.Stalick, J.A.Ibers *Inorg. Chem.*, **8**, 419, 1969

86.C **Iodobis(triphenylphosphine)allene - rhodium**
$C_{39}H_{34}IP_2Rh$
For complete entry see 72.78

86.C **Oxopentachloropropionato bis(triphenyl phosphine) dirhenium(iv)**
$C_{39}H_{35}Cl_5O_3P_2Re_2$
For complete entry see 81.89

86.89 **bis(Triphenylphosphine) platinum(0) oxygen acetone complex**
$C_{39}H_{36}O_3P_2Pt$
R.Ugo, F.Conti, S.Cenini, R.Mason, G.B.Robertson
Chem. Communic., 1498, 1968
Also classified in 84

86.90 **Triphenylphosphine gold - tetracarbonyl triphenoxy phosphine manganese**
$C_{40}H_{30}AuMnO_7P_2$
K.A.I.F.M.Mannan *Acta Cryst.*, **23**, 649, 1967

86.91 **Triphenyl tin tetracarbonyl triphenylphosphine manganese**
$C_{40}H_{30}MnO_4PSn$
R.F.Bryan *J. Chem. Soc. (A)*, 172, 1967

86.92 **bis(Triphenylphosphine) di(2,2,2 - trifluoro - 1 - iminoethyl)amine platinum**
$C_{40}H_{31}F_6N_3P_2Pt$
W.J.Bland, R.D.W.Kemmitt, I.W.Nowell, D.R.Russell
Chem. Communic., 1065, 1968
Also classified in 83

86.C **Dichloro - π - methylallyl bis(triphenylarsine) rhodium**
$C_{40}H_{37}As_2Cl_2Rh$
For complete entry see 72.80

86.C **Triphenylphosphite diphenyl(o - π - cyclopentadienyl - phenyl)phosphite iron iodide**
$C_{41}H_{33}FeIO_6P_2$
For complete entry see 73.119

86.93 **Dicyano tris - (9 - methyl - 9 - phosphfluorene) nickel(ii)**
$C_{41}H_{33}N_2NiP_3$
D.W.Allen, F.G.Mann, I.T.Millar, H.M.Powell, D.Watkin
J. Chem. Soc. (D), 1004, 1969

86.C **bis(Triphenylphosphine) acetylacetonato rhenium chloride**
$C_{41}H_{37}Cl_2O_2P_2Re$
For complete entry see 77.68

86.94 **bis(Cobalt tricarbonyl triphenoxyphosphine)**
$C_{42}H_{30}Co_2O_{12}P_2$
J.A.M.Case *Dissert. Abstr. (B)*, **28**, 2786, 1968

86.C **Tetracyanoethylene - platinum bis(triphenylphosphine)**
$C_{42}H_{30}N_4P_2Pt$
For complete entry see 72.81

86.95 **Iodo (tris(2 - diphenylphosphinoethyl)amine) cobalt(ii) iodide**
$C_{42}H_{42}CoINP_3{}^+ , I^-$
P.L.Orioli, L.Sacconi *J. Chem. Soc. (D)*, 1012, 1969
Residue 1 also classified in 83

86.96 Iodo - tris(2 - diphenylphosphinoethyl)amine nickel(ii) iodide
$C_{42}H_{42}INNiP_3{}^+$, I^-
P.Dapporto, L.Sacconi   *J. Chem. Soc. (D)*, 1091, 1969

86.97 Tetranickel hexacarbonyl tetra(tripropionitrile - phosphine)
$C_{42}H_{48}N_{12}NiO_6$
M.J.Bennett, F.A.Cotton, B.H.C.Winquist
J. Amer. Chem. Soc., **89,** 5366, 1967

86.98 Iridium carbonyl bromide bis(triphenylphosphine) tetracyanoethylene
$C_{43}H_{30}BrIrN_4OP_2$
J.A.McGinnety, J.A.Ibers   *Chem. Communic.*, 235, 1968
Also classified in 72

86.99 Dicyano tris(9 - ethyl - 9 - phosphafluorene) nickel(ii)
$C_{44}H_{39}N_2NiP_3$
D.W.Allen, F.G.Mann, I.T.Millar, H.M.Powell, D.Watkin
J. Chem. Soc. (D), 1004, 1969

86.100 Tetra - iridium decacarbonyl di(triphenylphosphine)
$C_{46}H_{30}Ir_4O_{10}P_2$
V.Albano, P.L.Bellon, V.Scatturin   *Chem. Communic.*, 730, 1967

86.101 Tri - μ - chloro dichloro tetrakis(tri - n - butyl - phosphine) diruthenium
$C_{48}H_{144}Cl_5P_4Ru_2$
G.Chioccola, J.J.Daly   *J. Chem. Soc. (A)*, 1981, 1968

86.C bis(Triphenylphosphine)diphenylacetylene platinum
$C_{50}H_{40}P_2Pt$
For complete entry see 72.83

86.102 Tri - μ - chloro - chloro pentakis(diethylphenyl - phosphine) diruthenium(ii)
n - propyl propionate solvate (monoclinic form)
$C_{50}H_{75}Cl_4P_5Ru_2$, $xC_6H_{12}O_2$
N.W.Alcock, K.A.Raspin   *J. Chem. Soc. (A)*, 2108, 1968

86.C bis(Diphenylphosphino) acetylene bis(π - cyclopentadienyl iron - di - μ -
carbonyl (π - cyclopentadienyl - iron carbonyl))
$C_{52}H_{40}Fe_4O_6P_2$
For complete entry see 73.122

86.103 Dioxo - iridium - bis(tetraphenylethylene diphosphine) hexafluorophosphate
$C_{52}H_{48}IrO_2P_4{}^+$, F_6P^-
J.A.McGinnety, J.A.Ibers   *Chem. Communic.*, 235, 1968

86.104 **Iridium carbonyl di(ethylene bis(diphenyl - phosphine))chloride**
$C_{53}H_{48}IrOP_4{}^+$, Cl^-
J.A.J.Jarvis, R.H.B.Mais, P.G.Owston, K.A.Taylor
Chem. Communic., 906, 1966

86.105 **tris(o - Diphenylarsinophenyl)arsine mercury(ii) bromide dichloromethane chelate**
$C_{54}H_{42}As_4Br_2Hg$, CH_2Cl_2
G.Dyer, D.C.Goodall, R.H.B.Mais, H.M.Powell, L.M.Venanzi
J. Chem. Soc. (A), 1110, 1966

86.106 **tris - (o - Diphenylarsinophenyl)arsine ruthenium dibromide**
$C_{54}H_{42}As_4Br_2Ru$
R.H.B.Mais, H.M.Powell *J. Chem. Soc.*, 7471, 1965

86.107 **Cobalt(ii) chloride tetraphenylborate tris(o - diphenylphosphinophenyl) phosphine**
$C_{54}H_{42}ClCoP_4{}^+$, $C_{24}H_{20}B^-$
T.L.Blundell, H.M.Powell, L.M.Venanzi *Chem. Communic.*, 763, 1967
Residue 2 classified in 62

86.108 **tris(Triphenylphosphine) rhodium(i) chloride**
$C_{54}H_{45}ClP_3Rh$
P.B.Hitchcock, M.McPartlin, R.Mason *J. Chem. Soc. (D)*, 1367, 1969

86.109 **Ruthenium(ii) chloride - triphenylphosphine complex**
$C_{54}H_{45}Cl_2P_3Ru$
S.J.La Placa, J.A.Ibers *Inorg. Chem.*, **4**, 778, 1965

86.110 **Hydridochloro tris(triphenylphosphine) ruthenium(ii) benzene solvate**
$C_{54}H_{46}ClP_3Ru$, C_6H_6
A.C.Skapski, P.G.H.Troughton *Chem. Communic.*, 1230, 1968

86.111 **Cobalt tris(triphenylphosphine) nitrogen hydride**
$C_{54}H_{46}CoN_2P_3$
B.R.Davis, N.C.Payne, J.A.Ibers *J. Amer. Chem. Soc.*, **91**, 1240, 1969

86.112 **Cobalt tris(triphenylphosphine) hydride nitrogen complex diethylether solvate**
$C_{54}H_{46}CoN_2P_3$, $C_4H_{10}O$
J.H.Enemark, B.R.Davis, J.A.McGinnety, J.A.Ibers
Chem. Communic., 96, 1968
Residue 1 also classified in 83

86.113 tris(Triphenylphosphine) carbonyl platinum (absolute configuration)
$C_{55}H_{45}OP_3Pt$
V.G.Albano, P.L.Bellon, M.Sansoni *J. Chem. Soc. (D)*, 899, 1969

86.114 Rhodium carbonyl hydride tris - triphenylphosphine
$C_{55}H_{46}OP_3Rh$
S.J.La Placa, J.A.Ibers *Acta Cryst.*, **18,** 511, 1965

86.115 μ - Sulfido - cyclo(bis(triphenylphosphine) platinum(ii) carbonyl
triphenylphosphine platinum(ii))
$C_{56}H_{45}O_2P_3Pt_2S$
A.C.Skapski, P.G.H.Troughton *J. Chem. Soc. (D)*, 170, 1969

86.116 Hydrido acetato tris(triphenylphosphine) ruthenium(ii)
$C_{57}H_{49}O_2P_3Ru$
A.C.Skapski, F.A.Stephens *J. Chem. Soc. (D)*, 1008, 1969
Also classified in 81

86.117 Tri - μ - chloro - hexakis(diethylphenylphosphine) diruthenium(ii) trichloro -
tris(diethylphenylphosphine) ruthenate(ii)
$C_{60}H_{90}Cl_3P_6Ru_2{}^+$, $C_{30}H_{45}Cl_3P_3Ru^-$
K.A.Raspin *J. Chem. Soc. (A)*, 461, 1969
Residue 2 classified in 86

86.118 Rhenium(iii) hydride bis(triphenylphosphine) 1,2 - bis(diphenylphosphino)
ethane
$C_{62}H_{57}P_4Re$
V.Albano, P.Bellon, V.Scatturin
R. C. Ist. Lombardo Sci. A, **100,** 989, 1966

86.119 Tetra - iridium nonacarbonyl tri(triphenylphosphine)
$C_{63}H_{45}Ir_4O_9P_3$
V.Albano, P.L.Bellon, V.Scatturin *Chem. Communic.*, 730, 1967

86.120 tetrakis(Triphenylphosphine) rhodium(i) hydride benzene solvate
$C_{72}H_{61}P_4Rh$, $0.5C_6H_6$
R.W.Baker, P.Pauling *J. Chem. Soc. (D)*, 1495, 1969

FORMULA INDEX

C₁

F 1

C_4

C_6

C_7

$C_{10}H_{25}BrMgO_2$	67.10	2
$C_{10}H_{26}Al_2O_2$	68.18	2
$C_{10}H_{28}CoN_6^{3+}$, $C_6CoN_6^{3-}$, $2H_2O$	76.73	2
$C_{10}H_{30}I_2Rh_2S_3$	85.65	2
$C_{10}H_{30}N_2Si_4$	63.14	2
$C_{10}H_{30}N_4^{4+}$, $4Cl^-$	59.20	1
$C_{10}H_{30}N_4^{4+}$, $2HO_4P^{2-}$, $6H_2O$	59.21	1
$C_{10}H_{32}N_{10}Ni_2S_2^{2+}$, $2I^-$	76.74	2
$C_{10}H_{33}N_6^{5+}$, $5Cl^-$	3.73	1
$C_{10}H_{45}N_5Si_5$	63.15	2

C_{11}

$C_{11}H_5Co_3O_9$	71.22	2
$C_{11}H_5CoFeHgO_6$	73.31	2
$C_{11}H_5F_7MoO_3$	73.32	2
$C_{11}H_6Fe_3N_2O_9$	83.88	2
$C_{11}H_6NO$	36.5	1
$C_{11}H_7ClO_2$	25.15	1
$C_{11}H_7CrO_5^-$, $C_8H_{12}N^+$	3.58	1
$C_{11}H_7CrO_5^-$, $C_8H_{12}N^+$	3.59	1
$C_{11}H_8Br_2O_3$	1.128	1
$C_{11}H_8CrO_5$	74.7	2
$C_{11}H_8FeO_3$	75.19	2
$C_{11}H_8MoO_3$	75.20	2
$C_{11}H_8O$	27.10	1
$C_{11}H_8O_2$	24.22	1
$C_{11}H_8O_2$	24.23	1
$C_{11}H_8O_2$	28.4	1
$C_{11}H_8O_3$	25.16	1
$C_{11}H_9ClN_2O_3$	40.25	1
$C_{11}H_9FeO_3^+$, BF_4^-	75.21	2
$C_{11}H_9NO_2$	25.17	1
$C_{11}H_9NO_2$, $0.25H_2O$	25.18	1
$C_{11}H_9N_4O_2^+$, Br^-, $2H_2O$	36.6	1
$C_{11}H_9O_3Re$	73.33	2
$C_{11}H_9O_3S^-$, Na^+	11.44	1
$C_{11}H_{10}CoN_4^{2+}$, $2Cl^-$	83.89	2
$C_{11}H_{10}CrO_3$	75.22	2
$C_{11}H_{10}N_2$	27.11	1
$C_{11}H_{10}N_2O_2S$	41.32	1
$C_{11}H_{10}N_2O_3$	40.26	1
$C_{11}H_{10}N_2O_3$	43.26	1
$C_{11}H_{10}O_2$	24.24	1
$C_{11}H_{11}BrN_2O$	32.31	1
$C_{11}H_{11}CuNO_2$	83.90	2
$C_{11}H_{12}NO^+$, Cl^-	24.25	1
$C_{11}H_{13}N_2O_2^+$, Br^-	48.84+	1
$C_{11}H_{13}N_2O_2^+$, Cl_2^-	48.86	1
$C_{11}H_{14}AgP$	72.29	2
$C_{11}H_{14}AuN$	71.23	2
$C_{11}H_{14}CrNO_5$	71.24	2
$C_{11}H_{14}CuP$	72.30	2
$C_{11}H_{15}BrN_2O_4$	45.57+	1
$C_{11}H_{15}BrN_2O_4$	45.59+	1
$C_{11}H_{15}BrN_3O_7^+$, Br^-	58.5	1

$C_{11}H_{15}Cl$	19.33	1
$C_{11}H_{15}NO$	52.19	1
$C_{11}H_{15}N_2O_4^+$, Cl^-, H_2O	48.87	1
$C_{11}H_{15}N_3$	37.15	1
$C_{11}H_{15}O_2Tl$	68.19	2
$C_{11}H_{16}BrN_3O_{10}S$, $.H_2O$	50.4	1
$C_{11}H_{16}N_2O_3$	43.27	1
$C_{11}H_{16}N_5O_3^+$, Br^-	50.5	1
$C_{11}H_{16}P^+$, I^-	64.31	2
$C_{11}H_{17}ClN_5^+$, Br^-	8.47	1
$C_{11}H_{17}ClN_5^+$, Cl^-	8.48	1
$C_{11}H_{17}N_2O_2^+$, Cl_3Ge^-, $0.5H_2O$	58.6	1
$2C_{11}H_{17}N_3S_4Zn$, C_6H_6	80.22	2
$C_{11}H_{18}N_2O_3$	43.28+	1
$C_{11}H_{18}N_3O_8^+$, Br^-	58.7	1
$C_{11}H_{20}O_4$	1.129	1
$C_{11}H_{21}BrO_2$	1.130	1
$C_{11}H_{22}O_2S$	1.131	1
$C_{11}H_{24}I_3Zn^-$, $C_{11}H_{25}S^+$	11.45	1
$C_{11}H_{24}N^+$, Cl^-	34.7	1
$C_{11}H_{24}NO_2^+$, Br^-, $0.5H_2O$	1.132	1
$C_{11}H_{25}S^+$, $C_{11}H_{24}I_3Zn^-$	11.45	1
$C_{11}H_{26}CoN_5O_2^{2+}$, Cl_4Zn^{2-}	82.43	2
$C_{11}H_{26}CoN_5O_2^{2+}$, $2I^-$, $2H_2O$	82.44	2
$C_{11}H_{27}As_3Br_2Ni$	86.7	2
$C_{11}H_{28}N_2^{2+}$, $2I^-$, $0.25H_2O$	3.74	1

C_{12}

$C_{12}Cl_{12}$	28.5	1
$C_{12}Cl_{12}$	31.13	1
$C_{12}Cl_{14}$	22.13	1
$C_{12}Co_2F_5O_7$	73.34	2
$C_{12}Co_2F_6O_6$	75.23	2
$C_{12}F_{10}$	19.34	1
$C_{12}F_{10}Hg$	71.25	2
$C_{12}F_{12}FeO_4$	73.35	2
$C_{12}N_6S_6V^{2-}$, $2C_4H_{12}N^+$	3.40	1
$C_{12}H_2C_{l6}O_2$	59.22	1
$C_{12}H_4Cl_4N_2$	36.7	1
$C_{12}H_4Cl_4N_2$	36.8	1
$C_{12}H_4N_4$	7.29	1
$C_{12}H_4N_4$, $C_{10}H_{16}N_2$	60.124	2
$C_{12}H_4N_4$, $2C_{12}H_4N_4^-$, $2Cs^+$	7.30	1
$C_{12}H_4N_4$, $C_{14}H_{10}$	60.125	2
$C_{12}H_4N_4$, $C_{18}H_{12}CuN_2O_2$	60.126	2
$C_{12}H_4N_4^-$, $C_{13}H_{11}N_2^+$	36.23	1
$C_{12}H_4N_4^-$, $C_{14}H_{16}Cr^+$	74.16	2
$C_{12}H_4N_4^-$, $C_{14}H_{16}Cr^+$, $C_{12}H_4N_4$	60.140	2
$2C_{12}H_4N_4$, $C_{10}H_{16}N_2$	60.127	2
$C_{12}H_5Cl_7O_4$	31.14	1
$C_{12}H_5FeMnO_7$	73.36	2
$C_{12}H_5MnMoO_7$	73.37	2
$C_{12}H_6FeO_8$	72.31	2
$C_{12}H_6Fe_2O_6$	72.32	2
$C_{12}H_6Fe_2O_6$	73.38	2

$C_{12}H_{12}O_{15}Yb^{3-}$, $2ClO_4^-$, $5Na^+$, $6H_2O$		
	81.73	2
$C_{12}H_{12}S_2$, $C_6N_6O_6$	60.74	2
$C_{12}H_{12}V$	75.24	2
$C_{12}H_{13}NO_4$	35.32	1
$C_{12}H_{14}Cl_2CuN_2$	83.100	2
$C_{12}H_{14}Cl_2CuN_2$	83.101	2
$C_{12}H_{14}INO_2S$	35.33	1
$C_{12}H_{14}N^+$, I^-	35.34	1
$C_{12}H_{14}N_2^{2+}$, Cl_4Cu^{2-}	33.66	1
$C_{12}H_{14}N_2^{2+}$, $C_4N_4Ni^{2-}$	33.67	1
$C_{12}H_{14}N_2O_2$	37.16	1
$C_{12}H_{14}N_4O_2$	32.32	1
$C_{12}H_{14}O_4$	13.48	1
$C_{12}H_{15}NO_3$	37.17	1
$C_{12}H_{16}BrFO_7$	45.62	1
$C_{12}H_{16}Br_2N_8Ni$	83.102	2
$C_{12}H_{16}CdN_6O_4$, H_2O	82.45	2
$C_{12}H_{16}Cl_2N_2O_2Sn$	69.24	2
$C_{12}H_{16}CoN_2O_8$	83.103	2
$C_{12}H_{16}CoN_6O_4$, H_2O	82.46	2
$C_{12}H_{16}CuN_6^{2+}$, $2I^-$	83.104	2
$C_{12}H_{16}Cu_2N_6O_8^{2-}$, $2Na^+$, $2H_2O$	82.47	2
$C_{12}H_{16}MoNS^+$, I^-	73.40	2
$C_{12}H_{16}NO_2^+$, Br^-	38.44	1
$C_{12}H_{16}NO_2^+$, Cl^-	38.45	1
$C_{12}H_{16}NO_4^+$, Br^-, $2H_2O$	48.88	1
$C_{12}H_{16}N_4NiO_4^{2+}$, $2Cl^-$	83.105	2
$C_{12}H_{16}N_4O_4$	44.72	1
$C_{12}H_{16}N_6NiO_4$, H_2O	82.48	2
$C_{12}H_{16}N_6O_4Zn$, $2H_2O$	82.49	2
$C_{12}H_{16}N_6O_4Zn$, $5H_2O$	82.50	2
$C_{12}H_{16}N_8Ni^{2+}$, $2Cl^-$	83.106	2
$C_{12}H_{16}N_8NiO_2^{2+}$, $2NO_3^-$	83.107	2
$C_{12}H_{17}BrN_2O_5$	45.63+	1
$C_{12}H_{17}NO_3Si$	63.16	2
$C_{12}H_{18}$	19.40	1
$C_{12}H_{18}$	20.27	1
$C_{12}H_{18}$	23.9	1
$C_{12}H_{18}$, $C_6Cl_4O_2$	60.71	2
$C_{12}H_{18}$, $C_8H_2N_4$	60.101	2
$C_{12}H_{18}Ag_2^{2+}$, $2NO_3^-$	75.25	2
$C_{12}H_{18}Be_4O_{13}$	67.11	2
$C_{12}H_{18}Cl_2O_3Pd_2$	72.36	2
$C_{12}H_{18}Cl_2Ru$	72.37	2
$(C_{12}H_{18}Cl_6Cu_3N_2O_4)_n$	84.34	2
$C_{12}H_{18}CuN_2O_2$, H_2O	83.108	2
$C_{12}H_{18}CuN_2O_2$, $0.5H_2O$	83.109	2
$C_{12}H_{18}CuN_2O_2$, CH_6N^+, ClO_4^-	83.110	2
$C_{12}H_{18}CuO_4$	77.14	2
$C_{12}H_{18}CuO_6$	77.15+	2
$C_{12}H_{18}FeN_6^{2+}$, $2Cl^-$, $3H_2O$	71.28	2
$C_{12}H_{18}Fe_2Ge_3O_6$	69.25	2
$C_{12}H_{18}N_2O_3V$	83.111	2
$C_{12}H_{18}N_4OS^{2+}$, $2Cl^-$, H_2O	41.34	1
$C_{12}H_{18}N_4O_7P_2S$, $4H_2O$	46.18	1
$C_{12}H_{18}N_6O_6$, $0.5H_2O$	48.89	1
$C_{12}H_{18}O_{13}Zn_4$	81.74	2
$C_{12}H_{19}BrO$	27.12	1

$C_{12}H_{19}Cl_2NPt$	72.38	2
$C_{12}H_{19}Co$	72.39	2
$C_{12}H_{19}N_4O_7P_2S^+$, Cl^-, $0.5H_2O$	46.19	1
$C_{12}H_{20}$	21.28	1
$C_{12}H_{20}Br_2O$	20.28	1
$C_{12}H_{20}Cl_2Rh_2$	72.40	2
$C_{12}H_{20}CuN_2O_2$	83.112	2
$C_{12}H_{20}CuN_2O_3$	77.17	2
$C_{12}H_{20}CuN_2O_4$	83.113	2
$C_{12}H_{20}Mo_2O_7S_8$	80.23	2
$C_{12}H_{20}O_2S$	39.73	1
$C_{12}H_{20}O_4$	38.46	1
$C_{12}H_{20}O_6$	38.47	1
$C_{12}H_{22}Br_2N_2O_2$	1.133	1
$C_{12}H_{22}Cl_2N_2O_2Pd$	83.114	2
$C_{12}H_{22}CuN_6O_6^{2+}$, $2NO_3^-$	82.51	2
$C_{12}H_{22}N_2O_2$	34.9	1
$C_{12}H_{22}N_2O_4$	33.68	1
$C_{12}H_{22}O_4$	1.134	1
$C_{12}H_{22}O_{11}$	45.69	1
$C_{12}H_{22}O_{11}$	45.65+	1
$C_{12}H_{22}O_{11}$, Na^+, Br^-, $2H_2O$	45.70	1
$C_{12}H_{23}N_3O_5^{2+}$, $2Cl^-$	58.8	1
$C_{12}H_{24}$	23.10	1
$C_{12}H_{24}Cl_2MnN_4$, $2H_2O$	83.115	2
$C_{12}H_{24}Cl_2N_8NiS_4$	79.34+	2
$C_{12}H_{24}Cr_3O_{16}^+$, Cl^-, $2H_2O$	81.75	2
$C_{12}H_{24}Cu_2O_8$	81.76	2
$C_{12}H_{24}N_2O_2$	1.135+	1
$C_{12}H_{24}N_4S_8Zn_2$	80.24	2
$C_{12}H_{24}O_2$	1.137	1
$C_{12}H_{24}O_2Pt$	77.18	2
$C_{12}H_{24}Pd_3S_9$	85.66	2
$C_{12}H_{25}N_3Nd_2O_{16}^{2+}$, $2Cl^-$, $3H_2O$	82.52	2
$C_{12}H_{26}CuN_2O_5$	82.53	2
$C_{12}H_{26}N_2O_2$	1.138	1
$C_{12}H_{26}N_2O_2$	34.10	1
$C_{12}H_{27}S^+$, F^-, $20H_2O$	61.11	2
$C_{12}H_{27}S^+$, F^-, $23H_2O$	61.12	2
$(C_{12}H_{28}AgP_2PtS_2^+)_n$, nNO_3^-	85.67	2
$C_{12}H_{28}BrCuN_6$	83.116	2
$C_{12}H_{28}CuN_4^{2+}$, $2NO_3^-$	83.117	2
$C_{12}H_{28}N^+$, Br^-	3.75	1
$C_{12}H_{28}N^+$, $C_{20}H_8CoN_6S_4^-$	3.76	1
$C_{12}H_{28}N_2^{2+}$, $2Br^-$	34.11	1
$C_{12}H_{28}N_4Ni^{2+}$, $2NO_3^-$	83.118	2
$C_{12}H_{28}N_4O_2$	64.34	2
$C_{12}H_{28}O_4P_2PbS_4$	69.26	2
$C_{12}H_{28}Si_4$	63.17	2
$C_{12}H_{29}Cl_2CoN_3$	76.75	2
$C_{12}H_{29}N_4O_4PS^+$, HO_4P^-, $3H_2O$	46.20	1
$C_{12}H_{30}AlF^-$, K^+	68.20	2
$C_{12}H_{30}B_3N_3$	62.47	2
$C_{12}H_{30}BrCuN_4^+$, Br^-	83.119	2
$C_{12}H_{30}BrFe^+$, Br^-	76.76	2
$C_{12}H_{30}BrMn^+$, Br^-	76.77	2
$C_{12}H_{30}BrN_4Ni^+$, Br^-	83.120	2
$C_{12}H_{30}BrN_4Zn^+$, Br^-	76.78	2
$C_{12}H_{30}Br_2CoN_4$	76.79	2

C_{13}

C₁₄

C_{15}

C_{16}

C_{17}

$C_{18}H_{16}AsClO$	65.32	2
$C_{18}H_{16}BrNO_2$	31.31	1
$C_{18}H_{16}Cl_2N_2O_2$	31.32	1
$C_{18}H_{16}Mn_2O_4$	72.51	2
$C_{18}H_{16}N_2O_4Zn$	83.162	2
$C_{18}H_{17}BrO_3$	59.38	1
$C_{18}H_{17}NO_3$	24.29	1
$C_{18}H_{18}$	23.16	1
$C_{18}H_{18}$	28.23+	1
$C_{18}H_{18}Br_2$	31.33	1
$C_{18}H_{18}Cl_2$	5.34	1
$C_{18}H_{18}Co_2$	75.61	2
$C_{18}H_{18}N_2PS_2$, ClO_4	64.54	2
$C_{18}H_{18}N_2S_2$	41.43	1
$C_{18}H_{18}N_6$	7.32	1
$C_{18}H_{18}N_6S_2$	41.44	1
$C_{18}H_{18}O$	21.30	1
$C_{18}H_{18}O_2$	29.15	1
$C_{18}H_{19}N_2{}^+$, I^-	35.44	1
$C_{18}H_{20}$	29.16	1
$C_{18}H_{20}$	31.34	1
$C_{18}H_{20}$	31.35	1
$C_{18}H_{20}ClNO_2$	31.36	1
$C_{18}H_{20}ClN_3O_6S_2$, $0.65CH_2Br_2$	58.27	1
$C_{18}H_{20}CuN_2O_2$	78.27+	2
$C_{18}H_{20}CuN_2O_4$	78.29	2
$C_{18}H_{20}NO_3{}^+$, Br^-	58.28	1
$C_{18}H_{20}N_2NiO_2$	78.30	2
$C_{18}H_{20}N_2O_2Pd$	78.31	2
$3C_{18}H_{20}OS$, C_2H_6O	61.16	2
$C_{18}H_{20}O_2$	17.47	1
$C_{18}H_{21}BrO_2$	51.2	1
$C_{18}H_{21}N_2O_2$	40.29	1
$C_{18}H_{22}Co_2O_4$	72.52	2
$C_{18}H_{22}Cu_2N_2O_8$	81.85+	2
$C_{18}H_{22}HgN_2$	71.41	2
$C_{18}H_{22}NO_3{}^+$, Br^-	53.26	1
$C_{18}H_{22}NO_3{}^+$, Br^-, $2H_2O$	58.29	1
$C_{18}H_{22}N_4O_{10}S_2$	47.30	1
$C_{18}H_{22}Ni$	75.62	2
$C_{18}H_{23}BrO_2$, CH_4O	51.3	1
$C_{18}H_{23}N_2S^+$, Cl^-	41.45	1
$C_{18}H_{24}$	24.30	1
$C_{18}H_{24}Cl_2N_6Ni$	86.21	2
$C_{18}H_{24}NO_3{}^+$, I^-	58.30	1
$C_{18}H_{24}N_2NiO_4P_2S_4$	83.163	2
$C_{18}H_{24}N_2O_2Pt$	83.164	2
$C_{18}H_{24}N_{12}Ni^{2+}$, $2NO_3^-$	83.165	2
$C_{18}H_{24}N_{12}Zn^{2+}$, $2Cl^-$, H_2O	83.166	2
$C_{18}H_{24}NiO_2$	75.63	2
$C_{18}H_{24}O_2$, H_2O	51.4	1
$C_{18}H_{24}O_3$	51.5	1
$C_{18}H_{25}NO$	58.31	1
$C_{18}H_{26}As_2Cl_3O_2Rh$	86.22	2
$C_{18}H_{26}BrNO_6$, $0.5C_2H_6O$	58.32	1
$C_{18}H_{26}NO^+$, Br^-, H_2O	58.33	1
$2C_{18}H_{26}NO_6{}^+$, Cl_6Pt^{2-}, $2H_2O$	58.34	1
$C_{18}H_{30}$	29.19	1
$C_{18}H_{30}$	29.20	1

$C_{18}H_{30}$	29.17+	1
$C_{18}H_{30}$	52.30	1
$C_{18}H_{30}$, $0.5CHCl_3$	61.17	2
$C_{18}H_{30}$, $0.39C_6H_{12}$	61.18	2
$C_{18}H_{30}$, $0.225C_7H_{16}$	61.19+	2
$C_{18}H_{30}Ag_2{}^{2+}$, $2ClO_4^-$	72.53	2
$C_{18}H_{30}BrNO$	21.31	1
$C_{18}H_{30}Br_2O$	54.1	1
$C_{18}H_{30}Ca_2O_{12}Sr$	67.18+	2
$C_{18}H_{30}Cl_2$	5.35	1
$C_{18}H_{30}Cl_4O_2Sn_2$	69.34	2
$C_{18}H_{30}Co_2HgO_6P_2$	86.23	2
$C_{18}H_{30}FeN_8O_6$	83.167	2
$C_{18}H_{30}N^+$, I^-	58.35	1
$C_{18}H_{30}N_2NiO_4P_2S_4$	85.81	2
$C_{18}H_{34}O_2$	1.145	1
$C_{18}H_{34}O_2$	1.146	1
$C_{18}H_{35}BrO_2$	1.147	1
$C_{18}H_{35}N_2O_6S^+$, Cl^-, H_2O	50.19	1
$C_{18}H_{36}$	21.32	1
$C_{18}H_{36}AuN_2S_4{}^+$, $AuBr_2^-$	80.34	2
$C_{18}H_{36}O_2$	1.148	1
$C_{18}H_{36}O_6Pt_2$	77.36	2
$C_{18}H_{38}N_4O_{11}{}^{2+}$, O_4S^{2-}, H_2O	50.20	1
$C_{18}H_{38}N_4O_{11}{}^{2+}$, O_4Se^{2-}, H_2O	50.21	1
$C_{18}H_{40}N_2O_4Pt_2$	77.37	2
$C_{18}H_{42}Cl_4P_2Pt_2$	86.24	2
$C_{18}H_{54}B_2N_4Si_6$	63.23	2
$C_{18}H_{54}FeN_3Si_6$	63.24	2

C_{19}

$C_{19}H_{10}Fe_2N_2O_7$	83.168+	2
$C_{19}H_{12}Br_2N_2$	35.45	1
$C_{19}H_{12}N_3O_6$	12.13	1
$C_{19}H_{12}O_2$	29.21	1
$C_{19}H_{12}O_2$	29.22	1
$C_{19}H_{12}O_6$	59.39	1
$C_{19}H_{13}BrOS_4$	39.87	1
$C_{19}H_{14}BrNO_9$, C_3H_7NO	50.22	1
$C_{19}H_{14}Br_2O$	20.31	1
$C_{19}H_{14}Br_2O_2$	17.48+	1
$C_{19}H_{14}ClNO_9$, C_3H_7NO	50.23	1
$C_{19}H_{14}I_2O$	20.32	1
$C_{19}H_{15}{}^+$, ClO_4^-	12.14	1
$C_{19}H_{15}Br$	19.60	1
$C_{19}H_{15}Fe_2No_{10}S$	72.54	2
$C_{19}H_{16}FeO_3$	75.64	2
$C_{19}H_{16}O_3S$	39.88	1
$C_{19}H_{16}P^+$, Cl_4Fe^-	12.15	1
$C_{19}H_{17}P$	64.55	2
$C_{19}H_{18}As^+$, $C_{14}H_{12}CoS_4^-$, $0.5C_2H_6O$	85.73	2
$2C_{19}H_{18}As^+$, Cl_4Ni^{2-}	65.33	2
$C_{19}H_{18}N_3{}^+$, ClO_4^-	12.16	1
$C_{19}H_{18}N_3O_4S_2{}^+$, Cl^-, H_2O	50.24	1
$C_{19}H_{18}O_6$	59.40	1

C_{20}

C_{24}

C_{25}

C_{26}

C_{27}

C_{28}

TRANSITION METAL INDEX

Ag

Au

Cd

Ce

Co

Cr

Cu

77.17, 77.38, 77.40, 77.54, 77.55, 77.59, 78.1, 78.2, 78.3, 78.4, 78.5, 78.6, 78.7, 78.8, 78.16, 78.17, 78.18, 78.19, 78.20, 78.21, 78.26, 78.27, 78.28, 78.29, 78.34, 78.35, 78.40, 78.41, 78.46, 78.50, 78.51, 78.56, 78.57, 78.58, 79.1, 79.7, 79.13, 79.32, 80.9, 80.13, 80.14, 80.15, 80.25, 80.26, 80.37, 81.3, 81.4, 81.6, 81.7, 81.8, 81.15, 81.22, 81.23, 81.27, 81.31, 81.32, 81.33, 81.45, 81.47, 81.48, 81.62, 81.64, 81.65, 81.66, 81.76, 81.77, 81.78, 81.79, 81.80, 81.81, 81.82, 81.84, 81.85, 81.86, 81.87, 82.3, 82.4, 82.5, 82.10, 82.13, 82.15, 82.16, 82.17, 82.18, 82.21, 82.23, 82.25, 82.26, 82.29, 82.35, 82.36, 82.38, 82.39, 82.41, 82.42, 82.47, 82.51, 82.53, 82.55, 82.57, 82.58, 83.2, 83.3, 83.6, 83.8, 83.20, 83.21, 83.22, 83.23, 83.34, 83.39, 83.45, 83.46, 83.50, 83.52, 83.53, 83.54, 83.55, 83.56, 83.62, 83.65, 83.66, 83.73, 83.76, 83.77, 83.79, 83.79, 83.81, 83.82, 83.87, 83.90, 83.100, 83.101, 83.104, 83.108, 83.109, 83.110, 83.112, 83.113, 83.116, 83.117, 83.119, 83.132, 83.133, 83.140, 83.142, 83.155, 83.156, 83.157, 83.158, 83.171, 83.179, 83.180, 83.181, 83.182, 83.187, 83.192, 83.207, 84.1, 84.4, 84.10, 84.13, 84.15, 84.19, 84.20, 84.31, 84.32, 84.33, 84.34, 84.36, 84.38, 84.40, 84.42, 84.47, 84.48, 84.49, 84.56, 84.59, 84.69, 85.20, 85.40, 85.46, 85.53, 85.69, 86.14, 86.41, 86.69

Eu

Vol. 2 83.217

Fe

Vol. 2 60.119, 62.39, 63.24, 69.35, 71.13, 71.20, 71.21, 71.28, 71.34, 71.43, 71.46, 71.53, 72.7, 72.13, 72.14, 72.16, 72.17, 72.18, 72.31, 72.32, 72.33, 72.41, 72.50, 72.54, 72.57, 72.60, 72.61, 72.62, 72.63, 72.65, 72.68, 72.69, 72.71, 72.74, 72.75, 73.2, 73.4, 73.5, 73.13, 73.14, 73.21, 73.22, 73.23, 73.24, 73.25, 73.31, 73.35, 73.36, 73.38, 73.39, 73.43, 73.44, 73.47, 73.48, 73.49, 73.50, 73.51, 73.52, 73.55, 73.56, 73.59, 73.68, 73.69, 73.71, 73.72, 73.76, 73.77, 73.85, 73.86, 73.87, 73.89, 73.90, 73.93, 73.94, 73.95, 73.96, 73.97, 73.98, 73.101, 73.102, 73.103, 73.104, 73.105, 73.106, 73.107, 73.110, 73.114, 73.115, 73.116, 73.119, 73.120, 73.122, 75.8, 75.10, 75.14, 75.19, 75.21, 75.26, 75.29, 75.32, 75.33, 75.38, 75.48, 75.56, 75.57, 75.58, 75.60, 75.64, 75.65, 75.75, 75.79, 76.63, 76.64, 76.67, 76.76, 77.23, 78.13, 78.14, 78.60, 80.6, 80.12, 80.16, 80.40, 82.9, 82.9, 82.59, 83.36, 83.69, 83.88, 83.137, 83.150, 83.152, 83.167, 83.168, 83.169, 83.170, 83.183, 83.185, 83.203, 83.205, 83.209,

83.210, 83.212, 83.214, 84.24, 84.24, 84.29, 84.29, 84.44, 84.52, 84.53, 85.22, 85.41, 85.42, 85.62, 85.71, 85.77, 85.83, 85.89, 86.5, 86.6, 86.8, 86.15, 86.53, 86.54, 86.62, 86.64

Gd

Vol. 2 77.66, 81.13, 81.69

Hg

Vol. 2 71.1, 71.2, 71.3, 71.4, 71.5, 71.8, 71.11, 71.12, 71.15, 71.19, 71.25, 71.26, 71.35, 71.39, 71.41, 73.3, 73.31, 76.19, 79.8, 79.11, 79.12, 79.17, 79.24, 81.53, 83.94, 83.145, 84.6, 84.14, 84.30, 84.37, 84.62, 84.63, 84.64, 84.67, 85.4, 85.12, 85.13, 85.17, 85.21, 85.21, 85.23, 85.44, 85.48, 85.51, 85.75, 85.87, 85.90, 86.23, 86.105

Ho

Vol. 2 77.31

Ir

Vol. 2 71.9, 71.33, 71.42, 75.52, 86.70, 86.75, 86.76, 86.77, 86.79, 86.87, 86.98, 86.100, 86.103, 86.104, 86.119

La

Vol. 2 76.71, 76.72, 77.32, 83.172

Mn

Vol. 1 3.57 **Vol. 2** 69.14, 69.40, 71.10, 72.51, 73.8, 73.36, 73.37, 73.41, 73.63, 73.121, 75.34, 75.42, 76.65, 76.65, 76.77, 77.12, 77.24, 81.9, 81.10, 82.8, 82.56, 83.7, 83.115, 83.153, 84.16, 84.25, 86.25, 86.28, 86.39, 86.41, 86.85, 86.90, 86.91

Mo

Vol. 2 69.28, 71.27, 71.37, 71.51, 72.66, 73.6,

73.15, 73.26, 73.27, 73.29, 73.30, 73.32, 73.37, 73.40, 73.41, 73.66, 73.70, 73.75, 73.96, 74.17, 75.11, 75.12, 75.20, 75.43, 75.44, 75.45, 75.69, 75.76, 76.3, 76.61, 77.53, 80.23, 81.1, 81.24, 81.25, 81.59, 82.14, 83.38, 84.18, 84.43, 85.34, 85.78, 86.17, 86.43, 86.56

Nb

Vol. 2 72.72, 72.79, 72.82, 72.84, 73.11, 81.16, 81.38, 83.18, 84.65

Nd

Vol. 2 77.33, 81.41, 81.43, 81.72, 82.52

Ni

Vol. 1 3.80 **Vol. 2** 60.155, 65.42, 65.46, 65.47, 65.50, 72.6, 72.15, 72.22, 72.27, 72.48, 72.64, 72.70, 72.76, 72.77, 73.74, 73.82, 73.91, 73.112, 73.113, 73.118, 75.28, 75.53, 75.54, 75.62, 75.63, 76.6, 76.20, 76.21, 76.22, 76.23, 76.27, 76.29, 76.31, 76.32, 76.39, 76.41, 76.42, 76.43, 76.47, 76.50, 76.52, 76.69, 76.70, 76.74, 77.13, 77.19, 77.45, 77.46, 77.56, 77.62, 78.9, 78.10, 78.11, 78.22, 78.23, 78.30, 78.32, 78.36, 78.37, 78.42, 78.43, 78.47, 78.49, 78.52, 78.53, 79.18, 79.23, 79.26, 79.28, 79.31, 79.34, 79.35, 79.36, 80.1, 80.3, 80.17, 80.18, 80.27, 80.28, 81.11, 81.36, 82.6, 82.19, 82.22, 82.27, 82.30, 82.31, 82.33, 82.40, 82.48, 83.11, 83.12, 83.13, 83.14, 83.24, 83.25, 83.29, 83.30, 83.31, 83.32, 83.40, 83.43, 83.51, 83.57, 83.63, 83.83, 83.84, 83.85, 83.86, 83.96, 83.97, 83.98, 83.99, 83.102, 83.105, 83.106, 83.107, 83.118, 83.120, 83.122, 83.127, 83.138, 83.143, 83.144, 83.147, 83.148, 83.149, 83.163, 83.165, 83.175, 83.177, 83.184, 83.186, 83.198, 83.199, 83.201, 83.204, 83.208, 83.215, 83.218, 83.219, 84.7, 84.41, 84.60, 85.1, 85.5, 85.6, 85.7, 85.8, 85.15, 85.16, 85.18, 85.19, 85.24, 85.25, 85.30, 85.32, 85.33, 85.35, 85.38, 85.43, 85.47, 85.49, 85.50, 85.56, 85.57, 85.58, 85.59, 85.60, 85.70, 85.76, 85.81, 85.88, 85.91, 85.93, 85.96, 86.7, 86.9, 86.20, 86.21, 86.26, 86.33, 86.49, 86.51, 86.52, 86.55, 86.65, 86.67, 86.74, 86.93, 86.96, 86.97, 86.99

Np

Vol. 2 80.38

Os

Vol. 2 72.34, 86.83, 86.88

Pd

Vol. 2 77.65

Pr

Vol. 2 60.70, 60.70, 60.148, 60.148, 71.49, 72.4, 72.8, 72.9, 72.10, 72.11, 72.19, 72.21, 72.28, 72.36, 72.42, 72.47, 72.58, 72.59, 74.8, 75.3, 75.4, 75.7, 75.27, 75.36, 75.59, 75.80, 75.81, 76.14, 76.68, 77.41, 77.60, 78.31, 78.38, 78.39, 78.44, 78.45, 78.48, 79.19, 79.25, 81.17, 82.12, 83.42, 83.58, 83.59, 83.114, 83.159, 83.173, 83.189, 84.57, 85.26, 85.28, 85.29, 85.31, 85.66, 85.72, 85.80, 85.92, 85.95, 86.2, 86.18, 86.19, 86.34, 86.38, 86.45, 86.48, 86.62, 86.81

Pt

Vol. 2 69.43, 71.29, 71.30, 71.31, 71.36, 71.38, 72.1, 72.2, 72.3, 72.5, 72.38, 72.49, 72.81, 72.83, 73.17, 75.17, 75.18, 75.71, 75.73, 76.80, 77.4, 77.18, 77.35, 77.36, 77.37, 77.58, 78.33, 80.19, 81.18, 81.19, 82.7, 83.26, 83.33, 83.60, 83.67, 83.67, 83.68, 83.68, 83.124, 83.125, 83.126, 83.153, 83.164, 83.191, 85.27, 85.67, 86.3, 86.4, 86.10, 86.11, 86.13, 86.16, 86.24, 86.31, 86.35, 86.37, 86.46, 86.50, 86.71, 86.72, 86.80, 86.82, 86.89, 86.92, 86.113, 86.115

Re

Vol. 2 62.41, 71.47, 73.33, 73.45, 75.40, 76.25, 77.5, 77.68, 81.88, 81.89, 83.1, 83.195, 83.200, 85.54, 85.97, 86.12, 86.27, 86.58, 86.59, 86.66, 86.118

AUTHOR INDEX

Abdullaev, G.K. **2** 84.13
Abe, J. **1** 50.31
Abel, E.W. **2** 85.25
Abel, H. **1** 54.24
Ablov, A.B. **2** 83.135
Ablov, A.V. **2** 66.11, 81.32, 81.76, 81.82, 81.87, 83.91
Ablov, A.W. **2** 60.87
Abowitz, G. **1** 32.5
Abraham, D.J. **1** 32.22
Abrahams, S.C. **1** 11.29, 11.50, 11.55, 11.65, 11.67, 13.54, 24.13 **2** 70.25, 73.30
Abrahamsson, S. **1** 1.131, 1.146, 1.147, 5.32, 16.34, 41.28, 46.13, 50.13, 53.32, 59.62, 59.69
Abramson, H.N. **1** 51.1
Abushanab, E. **1** 58.97
Acton, N. **2** 75.35
Adam, G. **1** 58.104, 58.105
Adamson, G.W. **2** 64.37, 67.5, 71.50, 71.54
Adamson, M.G. **2** 65.39
Addamiano, A. **1** 9.28
Adeoye, S.A. **1** 59.66
Adman, E. **1** 16.30, 20.9, 20.12, 44.63 **2** 60.119
Admiraal, L.J. **1** 58.11 **2** 83.130
Agre, V.M. **2** 69.33, 80.11, 80.36
Agtarap, A. **1** 50.41
Ahmad, M. **2** 73.1, 73.60, 73.61
Ahmed, F.R. **1** 1.14, 16.61, 24.14, 29.14, 33.76, 33.77, 33.78, 35.46, 36.22, 38.57, 58.19, 58.29, 58.41, 58.53, 58.60 **2** 64.32, 64.64, 64.81
Ahmed, N.A.K. **1** 2.29
Aihara, A. **1** 7.31
Aime, S. **1** 25.14
Ainalem, I.-B. **2** 83.71, 83.72
Akamatu, H. **2** 60.141
Akhtar, F. **2** 83.104
Akhtar, M.N. **2** 81.51
Akimoto, H. **1** 24.33
Akiyama, T. **1** 56.5
Akopyan, Z.A. **1** 15.1, 24.11, 24.12 **2** 83.172
Albano, V. **1** 40.25, 40.26 **2** 60.149, 86.72, 86.74, 86.100, 86.118, 86.119
Albano, V.G. **1** 3.41 **2** 71.33, 86.113

Albertsson, J. **2** 81.41, 81.69, 81.71, 81.72, 81.73, 82.52
Albrecht, G. **2** 72.12, 83.202
Alcock, N.W. **1** 39.65 **2** 69.41, 69.47, 73.45, 81.40, 83.208, 86.102
Alden, R.A. **1** 31.39, 52.24, 52.30, 59.65
Alderman, P.R.H. **2** 72.5, 80.5
Aleby, S. **1** 1.150, 1.152, 1.153
Aleksandrov, G.G. **2** 72.73, 72.84, 73.83
Alexander, E. **1** 31.31
Alexander, L.E. **1** 1.97, 7.25, 34.9, 43.14 **2** 63.11, 63.13, 78.12
Alexander, R. **2** 75.2
Ali, S.M. **1** 2.24
Alleaume, M. **1** 11.60, 13.26, 13.31, 16.14, 16.18, 16.19, 16.22
Allegra, G. **1** 23.9, 29.17, 29.18, 29.19, 38.27, 38.28, 38.41 **2** 61.17, 61.18, 61.19, 61.20, 68.1, 68.20, 71.45, 72.20, 72.26, 72.39, 72.43, 72.45, 72.67, 73.18, 74.5, 74.8, 74.24, 74.25, 75.13, 75.48
Allen, C.W. **2** 64.33
Allen, D.W. **2** 86.93, 86.99
Allen, F.H. **1** 52.9, 53.13 **2** 75.41
Allentoff, N. **1** 9.7
Alleyne, A.B. **1** 38.17
Allmann, R. **1** 17.51, 33.82, 34.12, 41.44 **2** 60.105, 60.106, 64.54, 64.59
Almodovar, I. **2** 81.3, 81.4, 81.8, 81.10
Altona, C. **1** 38.8, 38.9, 38.10, 38.11, 38.12, 38.13, 42.3, 42.5
Alver, E. **2** 64.31
Alvino, G. **1** 1.133
Amakasu, O. **1** 58.7, 58.12
Amano, Y. **2** 61.21, 83.187
Amanov, A.Z. **2** 80.19
Ambady, G.K. **1** 2.60, 2.66, 2.68
Ambats, I. **1** 40.32
Amendola, A. **1** 39.18
Amirthalingam, V. **1** 2.35, 41.9 **2** 67.8
Amisimov, K.N. **2** 72.72
Amit, A. **1** 31.6
Amit, A.G. **1** 9.27
Amma, E.L. **2** 68.3, 68.10, 74.2, 74.3, 74.26, 79.1, 79.3, 79.4, 79.7, 79.19, 79.22, 79.28, 83.16, 85.19, 85.56, 86.4, 86.10, 86.11, 86.51

A 1

AUTHOR INDEX

Housty, J. 1 1.75, 1.82, 1.87, 1.94, 1.98, 1.105, 1.106, 1.111, 1.112, 1.114, 1.124, 1.126, 1.129, 1.134, 1.140, 19.16, 41.45, 43.22, 43.26, 43.30
Hovenkamp, S.G. 2 64.4
Howell, P.A. 1 39.11
Hoy, R.C. 2 83.158
Hoy, T.G. 1 13.49, 19.46, 19.47
Hubel, W. 2 72.61, 73.116
Huber-Buser, E. 1 23.6, 23.7
Huber, C.P. 1 53.5, 58.78
Huber, C.S. 1 37.14
Huber, M. 1 47.18 2 69.29
Huber, R. 1 9.1, 28.24, 31.6, 51.28 2 75.61
Hudson, P. 1 38.15, 39.22, 39.24
Huffman, J.C. 2 83.190
Hugel, R. 2 84.66
Hughes, D.O. 1 1.24
Hughes, E.W. 1 3.3, 8.17, 8.31, 33.8, 48.30, 48.83 2 60.8, 65.11
Hughes, R.E. 1 7.2, 7.32, 37.2, 37.3
Hullen, A. 2 76.9
Hulme, R. 2 60.89, 60.114, 60.136, 60.137, 60.138, 69.16, 69.17, 71.20
Hulscher, J.B. 1 38.47
Huml, K. 1 12.10, 16.42, 31.21, 36.14, 36.15, 36.16, 36.17
Hunt, D.J. 1 47.26, 50.12
Hunter, F.D. 1 45.14
Hunter, S.H. 2 84.41
Hursthouse, M.B. 1 2.92, 21.33, 21.34, 50.27, 59.48 2 60.136, 60.137, 60.138, 63.24, 78.61, 83.123, 83.217, 84.65
Huse, G. 1 15.4 2 60.86
Huse, Y. 1 59.9, 59.21
Husebye, S. 2 64.27, 70.6, 70.7, 70.10, 70.14, 70.15, 70.17, 70.18
Hussain, M.S. 2 83.87, 84.56
Huttner, G. 2 75.49
Hvoslef, J. 1 45.12, 45.17, 45.19, 45.20, 45.21 2 60.44, 60.45
Hybl, A. 1 45.76
Hyde, A.J. 1 33.58

Iball, J. 1 26.41, 28.16, 29.7, 29.11, 29.16, 29.21, 29.22, 30.2, 30.8, 30.11, 30.17, 30.20, 36.31, 36.32, 40.29, 40.33, 44.36, 47.2, 47.3, 47.4 2 77.23, 83.181
Ibekwe, S.D. 2 86.42
Ibers, J.A. 1 1.10, 1.13, 1.15, 1.20, 28.15 2 64.52, 74.10, 75.51, 76.39, 76.40, 83.195, 83.200, 84.61, 85.30, 85.73, 85.97, 86.4, 86.16, 86.25, 86.39, 86.50, 86.58, 86.61, 86.66, 86.75, 86.76, 86.77, 86.79, 86.85, 86.87, 86.88, 86.98, 86.103, 86.109, 86.111, 86.112, 86.114
Ichikawa, T. 1 1.66, 1.67, 48.80

Ievins, A. 2 62.51, 85.80
Igarashi, T. 2 85.61
Ignatowicz, A.K. 2 71.37
Ihara, M. 1 58.45, 58.58
Iitaka, Y. 1 1.66, 1.67, 21.22, 21.23, 24.33, 31.16, 41.31, 46.6, 47.13, 47.14, 48.3, 48.4, 48.42, 48.80, 50.9, 50.20, 50.21, 50.34, 50.35, 54.12, 55.1, 55.2, 55.3, 56.5, 56.7, 58.90, 59.9, 59.21, 59.60, 59.63 2 83.34
Ikeda, T. 2 80.4
Ikekawa, T. 1 50.9
Ikemoto, I. 2 60.141, 60.142, 60.153
Imado, S. 1 58.9
Immirzi, A. 1 21.32, 23.9, 29.20 2 61.17, 61.18, 61.19, 68.1, 72.20, 72.43, 72.45, 74.8, 75.48
Inglis, F. 2 65.51
Innes, M. 1 16.34, 59.62
Inouye, H. 1 52.25
Inubushi, Y. 1 58.93
Irie, H. 1 1.108, 58.86
Irving, H.M.N.H. 1 41.35
Isaacs, N.W. 2 69.13
Isakov, I.V. 1 1.33, 16.25 2 83.19
Isaksson, G. 1 41.28
Ishaq, M. 2 73.13, 73.15
Ishii, H. 1 58.93
Islam, K.M.S. 1 19.28, 29.8, 51.16, 51.19, 54.15, 59.54 2 73.102 ·
Isono, Y. 1 50.4
Itai, A. 1 55.2
Itazaki, H. 1 51.15, 51.23
Ito, S. 1 31.16, 58.111
Ito, T. 2 85.61
Ivanov, V.I. 2 77.33
Ivanova, N.S. 2 69.38, 80.21
Iwasaki, F. 1 7.31 2 60.117
Iwasaki, F.F. 1 1.16, 1.17
Iwasaki, H. 1 1.17 2 60.123, 76.35, 76.58, 83.96
Iwashita, Y. 1 7.18
Iwata, M. 2 76.38

Jackobs, J. 1 39.75
Jackson, R.B. 2 75.9
Jacobi, P. 1 54.24
Jacobson, R.A. 1 33.23, 33.46, 38.14, 38.54, 45.26, 45.65 2 62.45, 64.3, 73.36, 75.34, 77.69, 86.53, 86.54
Jaggi, H. 1 1.92, 13.33 2 83.211
Jain, P.C. 2 76.45, 76.46, 78.10, 78.38, 78.39
Jain, S.C. 1 2.72, 2.85
Jakovenko, V.I. 1 11.30 2 62.29, 69.20
James, M.N.G. 1 44.12, 50.14, 50.40
James, V.H.T. 1 51.38, 51.39, 51.40
Jameson, M.B. 1 24.26
Jamet-Delcroix, S. 2 60.27

A 15

A 26

Errata

60.103   Compound name should be:—
Ethyl picrate—cesium ethylate complex

60.131   Last line of entry should read:—
Residue 1 also classified in 43; residue 2 classified in 60, 44

71.37   Reference should be:—
J. Amer. Chem. Soc., **90**, 3242, 1968

73.23   In reference, for A114 read 1088

73.37   This entry should be deleted. The correct entry is **73.41**

75.73   In compound name, for **palladium** read **platinum**

76.47   In reference, for A51 read 1025

79.9   This entry should be classified in 69 not 79

80.2   In reference, for A51 read 1025

80.41   In reference, for A51 read 1025

81.75   Compound name should be:—
Triaquo-hexa-acetato-μ (3)-oxo-trichromium (iii) chloride hexahydrate

83.39   In reference, for A53 read 1027

83.82   In reference, for A54 read 1028

83.128   In reference, for 89 read 819

83.216   This entry should be classified in 68 not 83

85.19   This entry should be classified in 79 not 85

86.68   This entry should be classified in 68 not 86

Errata in formula index

For $(C_2H_2CuO_2)_n$, $4nH_2O$ read $(C_2H_2CuO_4)_n$, $4nH_2O$

For $2C_2H_5O$, Na^+, Br^- read $2C_2H_5NO$, Na^+, Br^-

For $C_3H_9BI_2S$ read $C_3H_9I_2Sb$

For $C_4H_{10}Cl_2N_6O_4$ read $C_4H_{10}Cl_2N_6O_4Zn$

For $C_4H_{16}N_{10}{}^+$, NO_3^- read $C_4H_{16}CoN_{10}{}^+$, NO_3^-

For $C_8H_{16}CuF_4O_8P_2$ read $C_8H_{16}CaF_4O_8P_2$

For $C_{12}H_{10}Br_2Fe$ read $C_{12}H_{10}Br_2Te$

For $C_{12}H_{12}NaO_{15}{}^{3-}$, $3Na^+$, $6H_2O$ read $C_{12}H_{12}NdO_{15}{}^{3-}$, $3Na^+$, $6H_2O$

For $C_{12}H_{24}Cr_3O_{16}{}^+$, Cl^-, $2H_2O$ read $C_{12}H_{24}Cr_3O_{16}{}^+$, Cl^-, $6H_2O$

For $C_{12}H_{28}N_4O_2$ read $C_{12}H_{28}N_4O_2P_2$

For $C_{12}H_{30}BrFe^+$, Br^- read $C_{12}H_{30}BrFeN_4{}^+$, Br^-

For $C_{12}H_{30}BrMn^+$, Br^- read $C_{12}H_{30}BrMnN_4{}^+$, Br^-

For $C_{12}H_{30}N_{18}N_{13}O_6{}^{6+}$, $6NO_3^-$, $2H_2O$ read $C_{12}H_{30}N_{18}Ni_3O_6{}^{6+}$, $6NO_3^-$, $2H_2O$

For $2C_{14}H_{12}Cl_2N_2N$, $2CHCl_3$ read $2C_{14}H_{12}Cl_2N_2Ni$, $2CHCl_3$

For $C_{14}H_{18}Cl_2N_4O_2$ read $C_{14}H_{18}Cl_2N_2O_2Zn$

For $C_{15}H_{15}N_{13}S_2$ read $C_{15}H_{15}Ni_3S_2$

For $C_{15}H_{25}O_7S_5$ read $C_{14}H_{25}Co_3O_4S_5$

For $C_{19}H_{15}Fe_2No_{10}S$ read $C_{19}H_{15}Fe_2NO_{10}S$

For $(C_{24}H_{20}O_2P_2S_2)_n$ read $(C_{24}H_{20}CoO_2P_2S_2)_n$

The following entry should be added to the bibliography:—

77   **Cobalt (iii) acetylacetonate**
        $C_{15}H_{21}CoO_6$
        V.M. Padmanabhan *Proc. Indian Acad. Sci.*, *A*, **47**, 329, 1958
        See also *Int. Distances*, M184s; *Structure Reports*, **22**, 594, 1958